Anechoic Range Design for Electromagnetic Measurements

For a listing of recent titles in the
Artech House *Electromagnetics Library*,
turn to the back of this book.

Anechoic Range Design for Electromagnetic Measurements

Vince Rodriguez

ARTECH HOUSE
BOSTON | LONDON
artechhouse.com

Library of Congress Cataloging-in-Publication Data
A catalog record for this book is available from the U.S. Library of Congress

British Library Cataloguing in Publication Data
A catalog record for this book is available from the British Library.

ISBN 13: 978-1-63081-537-0

Cover design by John Gomes

10 9 8 7 6 5 4 3 2 1

To Rhonda, Halyn, and Eleanna

Contents

Foreword

This book is primarily about the use of radio frequency (RF) absorber in indoor chambers for the measurement of radiated performance of antennas. This book also provides guidance for electromagnetic compatibility (EMC) emission and immunity, radar cross section (RCS), multiple-input multiple-output (MIMO), commercial radome, and target and scene simulator measurement. Thorough discussions of measurement chamber size, chamber layout, absorber size, absorber shape, and absorber placement for indoor far-field, near-field, and compact antenna measurements are presented, backed with measurements of performance. The author, Dr. Vince Rodriguez, is a leading expert in the use and analysis of absorbers and has contributed greatly to the art and science of RF absorber use for antenna measurement.

Electromagnetic boundary conditions for, reflection from, transmission into, and the propagation within lossy materials. In this case, RF absorbers, are presented to give a firm foundation for modeling absorber materials. Dr. Rodriguez's models have resulted in widely used performance graphs for foam absorber versus absorber size (in wavelengths), polarization, and angle of incidence. The basic principles of antenna measurement are presented for each of the major indoor antenna measurement range types. Guidance is given for the selection of the range type best suited to the size and frequency of the antenna. Important variations of each major type of antenna measurement technique are presented. The rectangular and tapered forms of the indoor far-field range, the planar, cylindrical, and spherical surface near-field forms of the near-field range and the serrated and rolled edge forms of the single reflector compact range are functionally explained.

Deep understanding of the three forms of absorber (electric field, magnetic field, and hybrid) is achieved by understanding the basic nature of lossy dielectrics and lossy ferrites, including the variables of the materials such as carbon loading and the effect of material backings. The proper use and placement of pyramidal and wedge foam absorbers, the primary absorbers used for antenna measurement, as well as new insights concerning the power handling ability of absorber are presented. Chamber shielding choices, reasons for shielding, and often-required shielding specifications are presented including the types of shielding that are commercially available. Special consideration is given to chamber doors; RF filters; and air, fluid, signal, and control cable penetrations of the chamber shielding.

Rectangular and tapered indoor far-field measurement anechoic chambers are discussed separately. The ray-tracing method of reflectivity analysis for each type of chamber is presented, which results in an incident angle estimate for the absorber in each part of the chamber, chamber required size, and separation of the test (quiet) zone from the absorber. Recommendations for absorber placement on each surface of the chamber, including corners and edges, are given with particular attention given to the end walls, specular zone, and power density incident on the absorber. The two rival models for the operation of the tapered chamber are discussed and measurements presented to show the low- and high-frequency accuracy limitations of the two models and thus the tapered chamber concept. Polarization purity of the test zone is of special concern.

Planar, cylindrical, and spherical near-field antenna measurement ranges; their theory of operation; and chamber requirements, including size and spacing to absorber, are presented. A reflection and associated absorber incident angle analysis lead to the choice of absorber thickness and placement for the chamber surfaces and equipment. The pattern, gain, and main beam direction of the antennas are considered in the choice of the near-field range type best suited for the antenna. These antenna parameters together with antenna size and frequency are part of the analysis for the optimal size and placement of the absorber for each near-field range type. There are at least three different positioning systems for the spherical near-field measurement technique. Each positioning system is treated separately, as the absorber requirements are unique to each of the three systems. Near-field measurement error and uncertainty analysis focus on the effect of absorber performance in the mitigation of room reflections.

The compact range measurement system theory of operation, required reflector, and chamber size are presented for both the serrated edge and rolled-edge versions of the compact range reflector. The back wall and ceiling absorbers are the most critical as they are shown to have the highest incident field strength. Other special areas for absorber include an absorber fence between the reflector feed antenna and the test zone, and the wall absorber near the reflector edge. The rolled-edge reflector is shown to require greater thickness absorber than the serrated edge on the surfaces near the reflector edge and possibly greater chamber size to accommodate the greater thickness absorber. Severe high-power concentration is shown possible at the reflector feed location. A 13-term error analysis is presented for the compact range including the effect of room reflections.

Indoor anechoic chamber and compact RCS measurement range types are presented. The reflectivity (RCS) of the empty range is shown to be the dominate performance level for RCS measurement ranges. The RCS level is primarily due to the monostatic RCS of the back-wall absorber, whereas in an antenna range the monostatic RCS of the back wall and the bistatic RCS of the specular zone absorber are shown to be important. Other differences between RCS measurement and antenna measurement are presented and include the noticeable backscattering of pyramidal absorber tips for RCS measurement the use of twisted absorber for superior polarization purity. The compact range is shown to be superior to the anechoic chamber for RCS measurement. The size of the compact range chamber must be much larger for RCS measurement than for antenna measurement, as most RCS measurement techniques employ fast pulses and time gating. Back-wall reflections can be time-gated out if the range is long enough. Bigger RCS chambers are better.

Anechoic ranges for EMC measurement are presented. It is shown that most facets of EMC measurement are specified in various EMC standards. The dimensions of the chamber, the absorber requirement, and the location of the equipment under test (EUT) are specified. EMC measurements are often performed at frequencies as low as 50 MHz, deep into the reactive field of the EUT, where absorber is not effective.

Absorber is shown to be more effective at higher frequencies such as 200 MHz to 1 GHz and metal-backed ferrite tile is often used as an absorber. Automotive EMC testing, including emission, immunity, and self-immunity are commonly performed and these tests and test conditions commonly follow the Society of Automotive Engineers (SAE) recommendations. The primary absorber reflectivity requirement is shown to be less than -6 dB in the test zone with absorber incident angles less than 30 degrees. ISO 11451 test setup now requires less than or equal to -10 dB reflectivity in the test zone. It is noted that the requirements for electric vehicles are still evolving. Defense, automotive, and commercial radiated immunity and emission standards are shown to include required range size and absorber specifications.

Several important specialty ranges are given special coverage due to their increasing popularity. Cellphone ranges are a prime example. These ranges are described for the measurement of total radiated power (TRP) and total isotropic sensitivity (TIS), the two performance factors for cell phone antenna performance. These chambers are shown to be versions of spherical near-field chambers with criteria specifically for the cell phone test chambers given by CTIA standards, including chamber size, minimum distances between the cellphone under test and the range antennas, the prescribed use of phantom hands and heads and required maximum range reflectivity levels. MIMO testing chambers for cellphone and other mobile devices is similar to cellphone chambers with the addition of multipath and interference generators capable of multiple angle emitters to simulate real-world conditions.

Commercial radome testing, specifically a commercial aircraft nose radome enclosing a weather radar antenna, is described including the chambers and test zone field requirements. The most common chamber for such radome testing is a compact range with a radome/antenna positioning system, which allows independent antenna and radome positioning. Two primary measurements are required for radome acceptance and possibly after radome repair. These two tests are described: the radome transparency efficiency (TE) test and the pattern test. Both tests are essentially a comparison test of antenna gain and pattern with and without the radome in place. Target and scene simulators are used to test aircraft and missile radar antennas and associated radar systems. The description of the several types of simulators and the required speed of real targets and the time to measure the radar system response.

Chebyshev absorber arrangement is an interesting advanced absorber configuration that can lower the effective reflectivity of a collection of absorber panels. The theory and application of the Chebyshev arrangement to lower reflectivity over a band of frequencies at a specific angle of incidence is presented in a clear and understandable manner. An example, including measured verification, is presented where the reflectivity of such an absorber arrangement lowered the reflectivity by 10 dB over the band of 8 to 11 GHz for 12" absorber at 60-degree incident angle.

Dr. Rodriguez has chosen to include only well-established, time-tested measurement techniques and absorber information based on his own extensive experience.

I know of no other book that describes in understandable detail the proper use of RF absorbing material for antenna measurement. I highly recommend this book to all students and practitioners of antenna measurement.

Edward B. Joy
Professor Emeritus
Georgia Institute of Technology
August 2019

Introduction

The Beginning

In 1992 I saw my first anechoic range for antenna measurements. This was a homemade range that resided in the now-demolished Engineering Science Building at the University of Mississippi (see Figure I.1). The building was renamed 10 years later for the person in charge of the range, one of my mentors, Dr. Charles E. Smith, Sr.

The range was a set of L-shaped frames mounted on casters and balanced with institutional-sized cans of beans filled with concrete. On the long upright absorber was "velcroed." These L-shaped frames were placed together, set in position, and aligned with wooden pegs that extended from a hole in one frame to a hole in the adjacent frame. The floor absorber laid directly on the floor. Across the top, wooden frames extended from one set of uprights to the opposite side. I never used the range, even during my tenure at Ole Miss as a graduate student; other graduate students and professors did use it. My research was more associated with the area of computational electromagnetics, and while I had an office in the Engineering Science Building from 1994 to 1996, I did not perform many measurements other than those associated with coursework.

The Early Years

Upon graduation I started my first job as a visiting professor at a regional university in South Texas. There I continued my work on numerical methods for electromagnetics. Eventually, my professional career moved to companies in the area of electromagnetic measurements, first at ETS-Lindgren and after that at NSI-MI Technologies.

At ETS-Lindgren, I was exposed to anechoic ranges and the basic theory of their design and operation. Specifically, I was introduced to them in the fall of 2000 by Kefeng Liu, who, at the time, was in charge of the RF design of anechoic chambers, a position that I held from the end of 2000 to the end of 2004. In 2004, I moved into the antenna design area. During that four-year period, I learned the basics of anechoic chamber design and RF absorber design. I started to approach anechoic chamber design from my own understanding. Based on my experience. After 2004, even with a shift in my area of work, I continued to design anechoic chambers, including some interesting new models such as the conical tapered anechoic chanber

Figure I.1 The front entrance of the now-demolished Charles E. Smith Engineering Science Building. I was introduced to RF absorbers and RF measurements in this building; in fact, my first office as a graduate student was to the left as you entered the door shown in the picture. *(Photo credit:* Rhonda Rodriguez.)

at the National University of Singapore's Temasek Laboratories and other tapered anechoic chambers in Europe and India.

Developing My Own Approach

It was, however, in when I joined MI-Technologies in the fall of 2014—now known as NSI-MI Technologies—that anechoic chamber design became again an important part of my career. At NSI-MI Technologies, I developed new tools to help with the design of anechoic chambers, including what I called Rodriguez's equations to estimate the anechoic performance of ranges. Steve Barnes of PPG was a great help in those years, supplying me with measured data for ranges I could use to check my tools' accuracy of predicting the performance. NSI-MI's main market is antenna measurements, and it uses ranges of all types including outdoor elevated ranges, tapered chambers, compact antenna test ranges, and near-field and far-field ranges in all their possible configurations. I learned what levels of reflected energy were required to achieve a given level of error and meet a certain level of uncertainty. It was, and continues to be, a great learning experience. This book owes a lot to my colleagues at NSI-MI, including Marion Baggett and Stephen Blalock for their help with RCS; Jeffrey Fordham, Steve Nichols, Pat Pelland, Anil Tellakula, and Dr. Brett Walkenhorst, all of whom were more than helpful in answering my questions; John

McKenna for helping with comments; and Dr. Daniel Janse van Rensburg, who also helped with comments while working on his own book.

About This Book

While at NSI-MI, I was told that I should write a book or monograph on anechoic chambers and RF absorbers. Among the people trying to convince me were Dr. Daniël Janse van Rensburg and Jeff Fordham. However, the seed idea of this book goes back to probably more than 10 years ago when my wife, Rhonda, suggested that I write it. She is the true force and driver that pushed me to put this together. Last but not least, I also must thank NSI-MI's CEO John Breyer, who made sure that NSI-MI gave me the support I needed to complete this work.

What is the goal of this book? Above all I want this book to be a reference, a useful source of information—a place where engineers can find answers and guidance to solve their problems related to the size of range they should build and the best approach for measuring their antenna. In addition, the book aims to help users of antenna ranges get through the marketing hype and really understand the potential limitations of a given anechoic range. Although antenna measurements are my main focus here, the book also includes information on EMC measurements and the ranges recommended for these measurements (Chapter 9) as well as information on ranges for radar cross-section (RCS) measurements (Chapter 8).

The book is written with the assumption that the reader has some basic knowledge of electromagnetic theory; however, Chapter 1 provides a brief refresher on Electromagnetic theory for those readers that feel they need a refresher.

Chapter 2 offers a more proper introduction to the topic covered in the book, providing a survey of different anechoic ranges for different applications. Furthermore, Chapter 2 guides readers through the issues that need to be addressed in the selection of the range for antenna measurements. If an indoor range is possible, then an RF absorber is required; accordingly, absorber technology is the subject of Chapter 3. Does the range need to be shielded? Chapter 4 addresses this topic. Depending on a series of decisions, different potential ranges are discussed: far-field ranges (Chapter 5), near-field ranges (Chapter 6), or compact ranges (Chapter 7).

Chapters 8–10 deal with other special indoor anechoic ranges: RCS ranges in Chapter 8, ranges for electromagnetic compatibility (EMC) in Chapter 9, and hardware in loop ranges in Chapter 10.

In summary, the book aims to be a good starting point for users, designers, and students interested in anechoic rooms for RF measurements.

CHAPTER 1

Basic Electromagnetics

1.1 Introduction

This book is intended for a broad audience of users of indoor anechoic ranges. The audience is expected to have some basic electromagnetic (EM) theory background, as the book does not intend to teach engineering electromagnetics or numerical methods based on EM theory. This chapter serves as a basic refresher, but readers are encouraged to refer to more complete sources, such as the critical and in-depth book by Balanis [1] (or one of the later editions) or the more introductory text by Chen [2].

The chapter starts by defining Maxwell's equations, which describe all EM phenomena, and then discusses wave propagation in lossy media and the rules of reflection at interfaces. These are based on the boundary conditions derived from Maxwell's equations.

In addition, the chapter briefly introduces antennas and antenna terminology and provides a basic overview of numerical methods to guide readers in the common approaches and shows how it becomes very difficult to solve the problem of the anechoic chamber using any specific method.

1.2 Maxwell's Equations

In his treatise on electromagnetics, James Clerk Maxwell did not render his equations in their current form. In fact, the type of formulation that he used was not even the vector formulation that we were taught in school. Most of his treatise followed the nomenclature of the time and used the names of analogous mechanical properties to describe the EM phenomena. Indeed the idea of the ether as a medium for the propagations of waves was still accepted, and a lot of Maxwell's writings talk about the properties of the ether, such as its elasticity, as a mechanical engineer today will talk about the properties of waves in a mechanical media. It was after Maxwell died prematurely in 1879 due to cancer while working on the second edition to the treatise that other physicists and engineers (including George Fitzgerald, Oliver Lodge, Oliver Heaviside, and Heindrich Hertz) interpreted Maxwell's writings [3]. Oliver Heaviside, a self-taught engineer who worked for some time in the telegraph industry and gave us the telegrapher's equations for propagation on a transmission line, produced the current four-equation form of Maxwell's equations. (Maxwell used eight equations to describe the same phenomena so the author owes a debt of gratitude to Heaviside for condensing them into four [3].)

Most people are familiar with Maxwell's equations in the point or derivative form:

$$\nabla \cdot \vec{D} = \rho_v \quad (1.1)$$

$$\nabla \cdot \vec{B} = 0 \quad (1.2)$$

$$\nabla \times \vec{E} = -\frac{\partial \vec{B}}{\partial t} \quad (1.3)$$

$$\nabla \times \vec{H} = \frac{\partial \vec{D}}{\partial t} + \vec{J} \quad (1.4)$$

We know (1.1) as Gauss' law, (1.2) as Gauss' law for magnetism (or the "no magnetic charges law"), (1.3) as Faraday's law, and (1.4) as the Ampère-Maxwell law. The vector fields in (1.1) to (1.4) are the electric field intensity $\vec{D}$ having units of volts per meter; the electric flux density or electric displacement $\vec{D}$ with units of coulombs per square meter; the magnetic field intensity $\vec{H}$ measured in amperes per meter; and the magnetic flux density $\vec{B}$ with units that honor Nikola Tesla. Accordingly, 1 tesla is defined as a newton per ampere per meter (N/Am). The vector $\vec{J}$ is the electric current density, and ρ_v is the charge density. The different vector fields are related by the material properties of permittivity ε and permeability μ, hence $\vec{D} = \varepsilon\vec{E}$ and $\vec{B} = \mu\vec{H}$.

Equations (1.1) to (1.4) can also be written in integral form. To do so, we apply Stokes's theorem: $\int_s (\nabla \times \vec{A}) \cdot d\vec{s} = \oint_c \vec{A} \cdot d\vec{l}$, to (1.3) and (1.4) and the divergence theorem, $\int_v \nabla \cdot \vec{A} dv = \oint_s \vec{A} \cdot d\vec{s}$, to (1.1) and (1.2) to obtain

$$\oint_s \vec{D} \cdot d\vec{s} = Q_{enc} \quad (1.5)$$

$$\oint_s \vec{B} \cdot d\vec{s} = 0 \quad (1.6)$$

$$\oint_c \vec{E} \cdot d\vec{l} = -\frac{\partial}{\partial t} \int_s \vec{B} \cdot d\vec{s} \quad (1.7)$$

$$\oint_c \vec{H} \cdot d\vec{l} = -\frac{\partial}{\partial t} \int_s \vec{D} \cdot d\vec{s} + \vec{I} \quad (1.8)$$

Equations (1.5) to (1.8) are the integral form of Maxwell's equations. In (1.5) integrating the charge density ρ_v we obtain the total enclosed charge in the volume (Q_{enc}); similarly in (1.7) integrating the current density vector we obtain the total current $\vec{I}$ flowing through the surface. Either form of the equations (which are equivalent) describes the EM phenomena as we understand it in a classical physics (nonrelativistic) frame.

1.3 Boundary Conditions and Wave Propagation

1.3.1 Boundary Conditions

From (1.7) and (1.8) we can obtain the EM boundary conditions by applying those equations to a surface at the boundary between two media and making that surface approach the boundary. Medium 1 has the EM material properties ε_1, and μ_1, and medium 2 has the material properties ε_2, and μ_2. On medium 1 we have $\vec{E}_1, \vec{D}_1, \vec{H}_1$, and $\vec{B}_1$ while on medium 2 we have $\vec{E}_2, \vec{D}_2, \vec{H}_2$, and $\vec{B}_2$. Consider the surface shown in Figure 1.1. If we apply (1.7) on the surface with the perimeter made up of segments $\overline{ab}$, $\overline{bc}$, $\overline{cd}$, and $\overline{dc}$ and we let $\overline{ab} \rightarrow 0$ and $\overline{cd} \rightarrow 0$, then the surfaces limited by the segments approach 0, and (1.7) becomes $\vec{E}_{1t} - \vec{E}_{2t} = 0$, where the subscript t indicates tangential to the boundary, and the negative sign is used because as we integrate along the closed perimeter, the direction of $\overline{bc}$ is opposite to $\overline{dc}$. A similar approach can be used with (1.8), and we obtain the boundary condition for the tangential magnetic field across a boundary. However, in this case the right side of (1.8) yields the surface electric current flowing on the boundary, or $\vec{H}_{1t} - \vec{H}_{2t} = \vec{J}_s$.

A similar approach is taken with (1.5) and (1.6), but in this case we choose a small volume across the interface between medium 1 and medium 2. For the small cylinder with height h illustrated in Figure 1.2, we will use this volume to apply the volume integrals in (1.5) and (1.6), and as with the small surface illustrated in Figure 1.1, we will reduce the volume in height and approach the boundary; hence we shall make $h \rightarrow 0$. Forcing that limit, (1.6) becomes $\vec{B}_{1n} - \vec{B}_{2n} = 0$ where the subscript n indicates normal to the boundary. The reason for the negative sign is since the normal vector to a close surface always points outward, it is in an opposite direction in medium 1 than the normal vector to the close surface has in medium 2. Doing the same with (1.5) we end up with $\vec{D}_{1n} - \vec{D}_{2n} = q_s$, where q_s is the sheet of charge that resides at the interface.

Thus, we can write the boundary conditions in terms of the EM vector fields and a vector $\hat{n}_{12}$ that is normal to the boundary. Table 1.1 lists and explains the EM boundary conditions. These are general boundary conditions. Let's examine some specific cases. For example, what happens at the boundary between two lossless media? For the interface between two lossless media where the conductivity $\sigma = 0$, there are no free charges and no flowing currents; hence in that case we

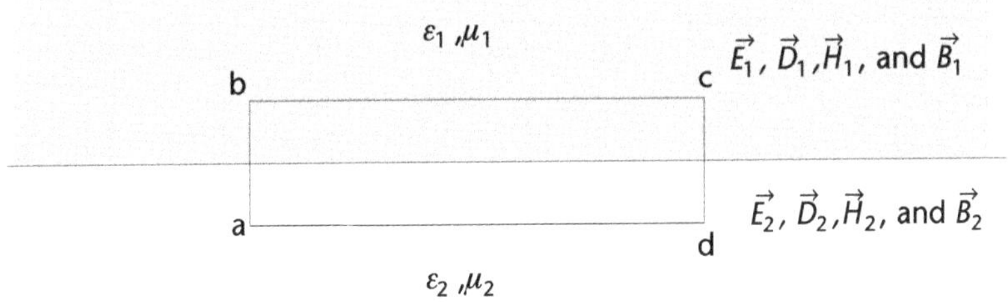

Figure 1.1 Graphical representation of the surface at an interface between two media.

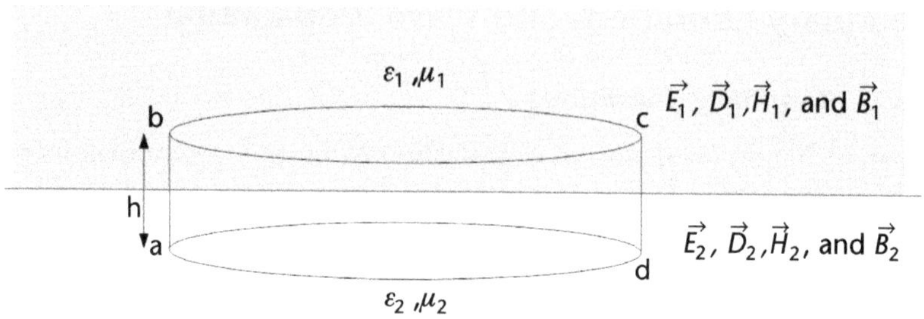

Figure 1.2 The small volume used to derive the boundary conditions for the electric and magnetic flux densities.

have that both the tangential electric field and the tangential magnetic field are continuous across the boundary or $\vec{E}_{1t} = \vec{E}_{2t}$ and $\vec{H}_{1t} = \vec{H}_{2t}$. In addition, the electric field and magnetic field normal to the boundary are such that $\varepsilon_1\vec{E}_{1n} = \varepsilon_2\vec{E}_{2n}$ and $\mu_1\vec{H}_{1n} = \mu_2\vec{H}_{2n}$. Now, let us assume that medium two is an excellent conductor, a perfect electric conductor (PEC), where $\sigma = \infty$, then, on medium 2, $\vec{E}_2, \vec{D}_2, \vec{H}_2$, and $\vec{B}_2$ are all equal to 0. So, the boundary conditions are that $\vec{E}_{1t} = 0$, and the tangential magnetic field is equal to the current flowing on the surface of the PEC, or $\vec{H}_{1t} = \vec{J}_s$. That is, the tangential electric field on the surface of a PEC goes to zero and the tangential magnetic field is equal to the current flowing on the surface of the PEC. The normal magnetic flux density also is zero so $\vec{B}_{1n} = 0$, and the normal electric flux density is equal to the surface charge $\vec{D}_{1n} = q_s$.

1.3.2 Wave Propagation

As EM fields propagate in waves, this section provides a refresher by quickly going over the wave equation derivation. Readers requiring a more in-depth coverage of the derivation might refer to Balanis [1].

Let us start with a mathematical vector identity that holds for all vector fields and that is given by

$$\nabla \times \nabla \times \vec{F} = \nabla(\nabla \cdot \vec{F}) - \nabla^2\vec{F} \tag{1.9}$$

Table 1.1 General Boundary Conditions in Electromagnetics

Boundary Condition	*Explanation*
$\hat{n}_{12} \times (\vec{E}_1 - \vec{E}_2) = 0$	The tangential electric field across an interface is continuous
$\hat{n}_{12} \times (\vec{H}_1 - \vec{H}_2) = \vec{J}_s$	The difference on the tangential magnetic field across an interface is equal to the surface current at the interface
$(\vec{D}_1 - \vec{D}_2) \cdot \hat{n}_{12} = q_s$	The difference on the normal electric flux density across an interface is equal to the surface charge
$(\vec{B}_1 - \vec{B}_2) \cdot \hat{n}_{12} = 0$	The normal magnetic flux density across an interface is continuous

Now consider a source-free medium, with no charges or currents. Starting with Faraday's Law (1.3) let us take the curl on both sides, yielding

$$\nabla \times \nabla \times \vec{E} = -\mu \frac{\partial \left(\nabla \times \vec{H}\right)}{\partial t} \tag{1.10}$$

Using (1.9) we can rewrite the left side of (1.10) and using (1.4) we can rewrite the right side of (1.10) leaving

$$\nabla(\nabla \cdot \vec{E}) - \nabla^2 \vec{E} = -\mu \frac{\partial}{\partial t}\left(\frac{\partial \vec{D}}{\partial t} + \vec{J}\right) \tag{1.11}$$

Being a source-free region (no charges or currents) and using equation (1.1) reduces (1.11) to

$$\nabla^2 \vec{E} - \mu\varepsilon\left(\frac{\partial^2 \vec{E}}{\partial t^2}\right) = 0 \tag{1.12}$$

Equation (1.12) is the wave equation for the electric field. A similar equation can be obtained for the magnetic field following a similar derivation that yields

$$\nabla^2 \vec{H} - \mu\varepsilon\left(\frac{\partial^2 \vec{H}}{\partial t^2}\right) = 0 \tag{1.13}$$

where (1.12) and (1.13) are the homogeneous vector wave equations. For time-harmonic fields, phasor representation can be used, and the wave equations can be written as

$$\nabla^2 \vec{E} - \omega^2 \mu\varepsilon \vec{E} = 0 \tag{1.14}$$

$$\nabla^2 \vec{H} - \omega^2 \mu\varepsilon \vec{H} = 0 \tag{1.15}$$

where ω is the angular frequency associated with the time-varying EM field. $\omega^2\mu\varepsilon$ is often represented as k^2 where $k = \omega\sqrt{\varepsilon\mu}$, is known as the propagation constant. Since the phase velocity is given by $u_p = 1/\sqrt{\varepsilon\mu}$, the propagation constant can be written as $k = 2\pi/\lambda$, where λ is the wavelength. Maxwell's equations can also be written in phasor form as

$$\nabla \cdot \vec{E} = \frac{\rho_v}{\varepsilon} \tag{1.16}$$

$$\nabla \cdot \vec{H} = 0 \tag{1.17}$$

$$\nabla \times \vec{E} = -j\omega\mu\vec{H} \tag{1.18}$$

$$\nabla \times \vec{H} = j\omega\varepsilon\vec{E} + \vec{J} \tag{1.19}$$

It should be noted that in engineering electromagnetics, the phasors are peak phasors and not RMS as is the case in other areas of electrical engineering. Let us assume now an electric field wave propagating. The wave has an E field directed in the x direction (also called an x-direction polarized field). Hence $\vec{E} = E_x\hat{a}_x$, and that has constant magnitude and phase over planes perpendicular to the z direction. This allows us to write (1.14) as

$$\frac{\partial^2 E_x}{\partial z^2} + k^2 E_x = 0 \tag{1.20}$$

We can identify (1.20) as an ordinary differential equation with a solution of the form

$$E_x(z) = E_x^+(z) + E_x^-(z) = E_o^+ e^{-jkz} + E_o^- e^{jkz} \tag{1.21}$$

Equation (1.21) for the solution of (1.20) represents two time-harmonic waves, traveling in opposite directions, one along the z direction and the other along the $-z$ direction. The values of the constants E_o^+ and E_o^- can be determined by the boundary conditions. Let us consider that there is only one wave propagating in the positive z direction, hence $E_o^- = 0$; then we can use (1.18) to find the magnetic field vector propagating with the electric field vector. Doing so we obtain

$$\vec{H} = \hat{a}_y H_y^+(z) = \hat{a}_y \frac{k}{\omega\mu} E_x^+(z) = \hat{a}_y \frac{E_x^+(z)}{\eta} \tag{1.22}$$

where $\eta = \sqrt{\mu/\varepsilon}$, also known as the intrinsic impedance of the medium. For free space $\eta_0 = \sqrt{\mu_o/\varepsilon_o} = 120\pi$, the free-space wave impedance is approximately 377Ω. For free space, we define the phase velocity of the wave as $u_p = 1/\sqrt{\varepsilon_o\mu_o} = c$; where c is the speed of light in vacuo, measured at 299,792,458.00 m/s.

1.3.3 Wave Propagation in Lossy Media

In source-free region with media that has losses (lossy media), $\sigma \neq 0$, we can rewrite (1.19) as

$$\nabla \times \vec{H} = j\omega\varepsilon\vec{E} + \sigma\vec{E} = j\omega\left(\varepsilon + \frac{\sigma}{j\omega}\right)\vec{E} \tag{1.23}$$

From (1.23) we can define a complex permittivity to represent lossy media;

$$\varepsilon_c = \left(\varepsilon + \frac{\sigma}{j\omega}\right) = (\varepsilon' + \varepsilon'') \tag{1.24}$$

The ratio of the imaginary part to the real part of the complex permittivity of a lossy media is known as the loss tangent:

$$\tan\delta = \frac{\varepsilon''}{\varepsilon'} \tag{1.25}$$

The propagation constant of the wave can be rewritten to account for the losses in the medium as

$$\gamma = \alpha + j\beta = j\omega\sqrt{\mu\varepsilon}\sqrt{1 + \frac{\sigma}{j\omega\varepsilon}} \tag{1.26}$$

Equation (1.26) is a general equation, and we can make approximations for good conductors where $\sigma/\omega\varepsilon \gg 1$ and (1.26) becomes

$$\gamma = \alpha + j\beta \cong \sqrt{\pi f \mu \sigma} \tag{1.27}$$

Using (1.27) the skin depth is derived as demonstrated in Chapter 3. For lower loss dielectric where the imaginary part is such that $\sigma/\omega\varepsilon \ll 1$, or to put it another way $\varepsilon'' \ll \varepsilon'$, as is the case with some absorber materials at microwave frequencies, (1.26) can be approximated as

$$\gamma = j\omega\sqrt{\mu\varepsilon'}\sqrt{1 - j\frac{\varepsilon''}{\varepsilon'}} \cong j\omega\sqrt{\mu\varepsilon'}\left[1 - j\frac{\varepsilon''}{2\varepsilon'} + \frac{1}{8}\left(\frac{\varepsilon''}{\varepsilon'}\right)^2\right] \tag{1.28}$$

This is a low-loss dielectric approximation that is done using the binomial expansion. Chapter 3 uses the general equation given in [1]. From (1.28) the attenuation constant becomes

$$\alpha \cong \frac{\omega\varepsilon''}{2}\sqrt{\frac{\mu}{\varepsilon'}} \tag{1.29}$$

And the propagation constant becomes

$$\beta \cong \omega\sqrt{\mu\varepsilon'}\,\frac{\omega\varepsilon''}{2}\left[1 + \frac{1}{8}\left(\frac{\varepsilon''}{\varepsilon'}\right)^2\right] \tag{1.30}$$

The attenuation of the waves caused by the losses on the material (conductivity or loss tangent) is used in absorbing materials as well as in shielding materials for indoor ranges.

1.3.4 Incident and Reflected Waves

This section provides the basic equations for reflection at an interface for plane waves. These are important equations, as the key to indoor ranges is based on the reflectivity of the treatment of the walls. As noted previously, a detailed derivation can be found in [1] or [2]. The reflection coefficient at an interface between two media where a plane wave propagates on a direction normal to the interface from medium 1 to medium 2 is given by

$$\Gamma = \frac{\eta_2 - \eta_1}{\eta_2 + \eta_1} \tag{1.31}$$

The transmission coefficient is given by

$$\tau = \frac{2\eta_2}{\eta_2 + \eta_1} \tag{1.32}$$

From (1.31) we can get the standing wave ratio, which relates the peaks and valleys of a standing wave caused by the summation of two waves propagating in opposite directions:

$$S = \frac{1 - |\Gamma|}{1 + |\Gamma|} \tag{1.33}$$

When the wave traveling to the interface between the two media propagates at an angle θ_i off the normal incidence, the reflected wave will travel at an angle θ_r (as shown in Figure 1.3). The relation between the two angles is

$$\theta_i = \theta_r \tag{1.34}$$

Equation (1.34) is known as Snell's law of reflection. The wave that propagates into the material travels in a different direction that is related to the incident wave by Snell's law of refraction:

$$\frac{\sin\theta_t}{\sin\theta_r} = \frac{\eta_1}{\eta_2} \tag{1.35}$$

What are the values of the reflection coefficient and the transmission coefficients for oblique incidence—that is, when $\theta_i \neq 0$? Looking at Figure 1.4, it is clear that we have two cases that will cause different results related to the boundary

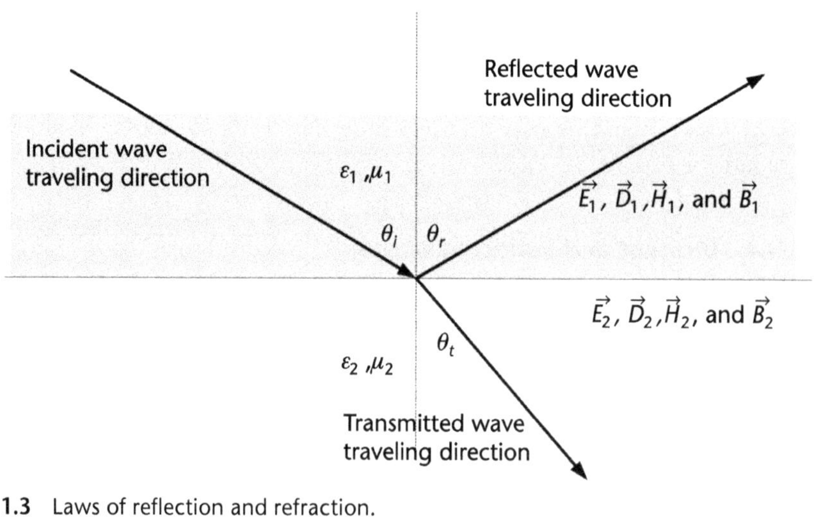

Figure 1.3 Laws of reflection and refraction.

conditions derived above. In one case the electric field is traveling perpendicular to the plane that contains the direction of travel of the incident and reflected waves. This is called perpendicular polarization

In the other case, represented on the right of Figure 1.4 the electric field is polarized so that it is parallel to the plane that contains the direction of travel of the incident and reflected fields. In the first case, the perpendicular polarization, the electric field is tangential to the surface, so it must be continuous across the interface, as we saw in the derivation of the boundary conditions. In the second case, the parallel polarization, a component of the E field will be tangential to the surface while another component of the E field will be normal to the surface. Hence, we have different boundary conditions, and that will yield different equations for the reflection and transmission coefficients.

For perpendicular polarization it can be shown that

$$\Gamma_{\perp} = \frac{\eta_2 \cos\theta_i - \eta_1 \cos\theta_t}{\eta_2 \cos\theta_i + \eta_1 \cos\theta_t} \tag{1.36}$$

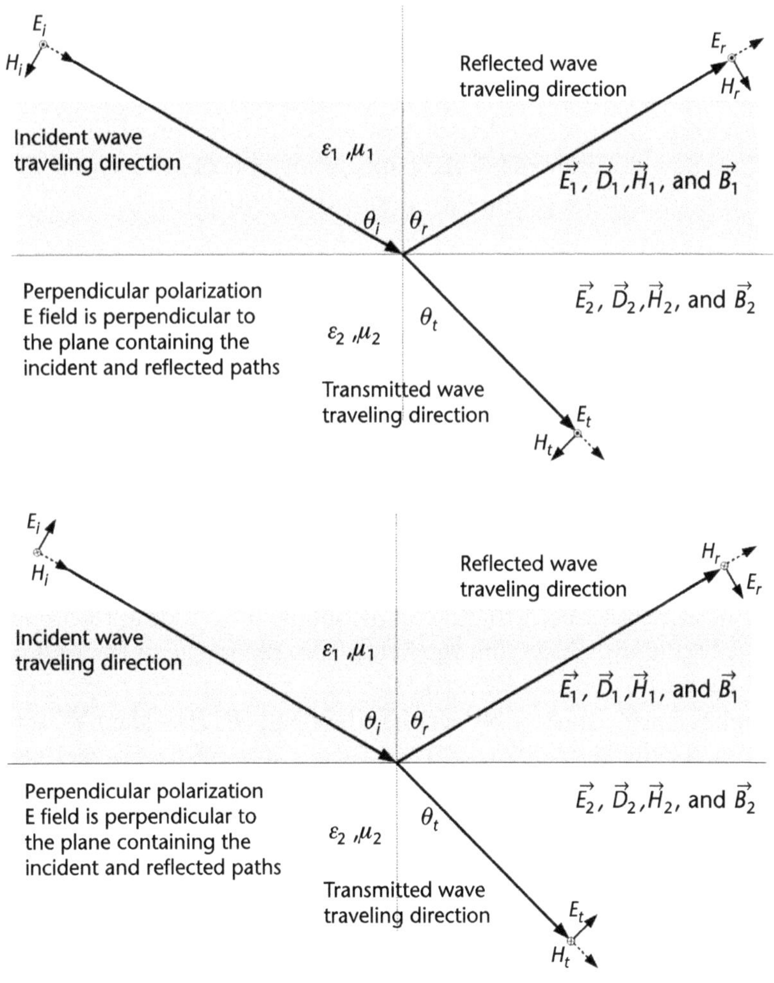

Figure 1.4 (a) Perpendicular and (b) parallel polarizations on a scattering problem.

and

$$\tau_{\perp} = \frac{2\eta_2 \cos\theta_i}{\eta_2 \cos\theta_i + \eta_1 \cos\theta_t} \tag{1.37}$$

while for the parallel polarization case the equations are

$$\Gamma_{\parallel} = \frac{\eta_2 \cos\theta_t - \eta_1 \cos\theta_i}{\eta_2 \cos\theta_t + \eta_1 \cos\theta_i} \tag{1.38}$$

and

$$\tau_{\parallel} = \frac{2\eta_2 \cos\theta_i}{\eta_2 \cos\theta_t + \eta_1 \cos\theta_i} \tag{1.39}$$

These equations show that the reflectivity on a surface when the incoming wave is not normal to the surface depends on the polarization of the field.

1.4 Theorems of Electromagnetics

Without going into the derivations, which can be found in detail in the literature [1], we now review some of the critical theorems of electromagnetic theory, specifically image theory, reciprocity, and surface equivalence.

1.4.1 Image Theory

Image theory is used to analyze the performance of radiating objects, such as antennas near an infinite conductive plane. Typically, the ground is not a perfect conductor. But in many cases, such as electromagnetic compatibility measurement ranges and ground reflection ranges, the ground is treated as a PEC to simplify the problem. Deviating from the PEC case implies that the magnitude of the images is lower, and so may be the phase difference compared to the actual radiating object.

Image theory states that we can remove the infinite obstacle and replace it with an image of the radiating object that exists over the infinite ground. The fields at any location over the ground can be obtained by adding the direct fields from the actual radiating object and the fields from the image of the radiated object.

Figure 1.5 explains the image theory approach for the two principal polarizations of the radiating source. Notice that the image has the same direction as the actual source for the vertically polarized case to account for the reflection coefficient of $\Gamma = 1$, while for the horizontally polarized case the image is flipped to account for the reflection coefficient of $\Gamma = -1$.

1.4.2 Reciprocity Theorem

The reciprocity theorem is extremely important in electromagnetics, especially for anechoic range design. The theorem states that for a linear system—and we have to

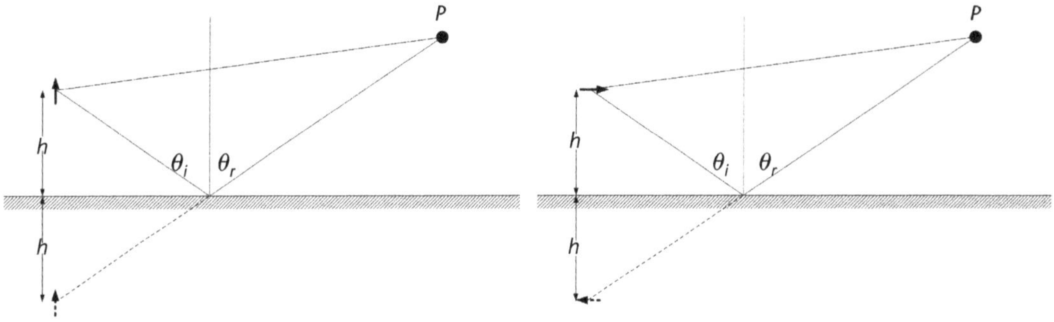

Figure 1.5 Image Theory for the two principal polarizations.

stress that the system has to be linear—the response of a system is unchanged when source and measurement are exchanged. That means that if I measure the fields at a location B of in the chamber for a given source A, I will get the same results if I measure the field at the location of the source A while placing the source at location B. In equation form the theorem takes the form $\iiint_v \vec{E}_B \cdot \vec{J}_A dv = \iiint_v \vec{E}_A \cdot \vec{J}_B dv$, where the v is the volume where the fields and sources exist, in this case the chamber or range.

This theorem implies that even if we analyze a range using the range antenna as the source, the result, or the performance, is the same as if the source was in the quiet zone and the range antenna was receiving.

1.4.3 Surface Equivalence Theorem

The surface equivalence theorem states that a radiator can be replace by a set of equivalent sources. This is an important theorem as it is the basis for a lot of numerical method formulations, and to a certain degree it is the basis of the near- to far-field antenna measurement methodologies.

The surface equivalence theorem implies that there is an imaginary surface can be defined around an object and that the fields outside that surface can be computed from a set of current densities that satisfy the boundary conditions.

1.5 Antenna Basics

It is a property of physics that accelerating charges radiate [4], and by acceleration we mean changes in the velocity vector of the charge. This means that any curve of an open transmission line will cause radiation even if the charge maintains a constant speed (i.e., constant velocity magnitude) because there is a change to the velocity vector by changing the direction. Time-varying currents, because of the change in velocity, also radiate as do time-harmonic currents. This property is the reason for electromagnetic compatibility (EMC) and making sure that the radiation from devices with electric circuits is minimized or canceled, or attenuated or shielded. Basically, any circuit trace with a curve on it will cause radiation as the charges accelerates.

Thus, everything can make an antenna and receive and transmit radio waves. The key is to make the structure efficient enough, so they efficiently radiate and receive (as is the case for antennas) or inefficient so that their radiation is minimized (as is the case in EMC).

Chapter 2 discusses the radiation pattern and other parameters of importance that are typically measured for antennas. Readers should refer to [2, 4] for further information on antennas.

1.5.1 Radiation Problems

Radiation problems in electromagnetics are typically solved using the magnetic vector potential $\vec{A}$. Using (1.2) we can represent $\vec{B}$ as the curl of another vector, so we define

$$\vec{H} = \frac{1}{\mu}\nabla \times \vec{A} \tag{1.40}$$

Balanis shows in [1] how to get field solutions from the magnetic vector potential as does Stutzman and Thiele [5]. The electric and magnetic fields can be computed from the magnetic vector potential. In general, for a volume source (such as an antenna) like the one depicted in Figure 1.6, the magnetic vector potential can be obtained by the following integral:

$$\vec{A} = \iiint_{v'} \mu\vec{J}\frac{e^{-j\beta R}}{4\pi R}dv' \tag{1.41}$$

where v' is the volume where the current density vector exists (i.e., the antenna).

It is clear that (1.41) is not a simple integral, and typically to come up with a solution in close form the point P is moved far away from the source volume such

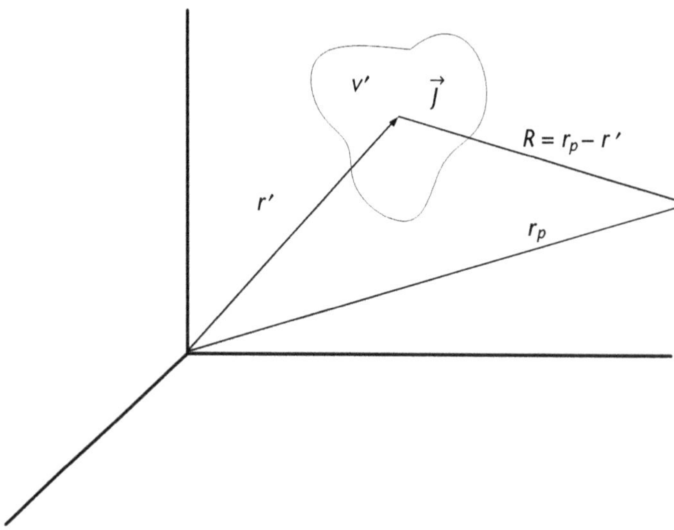

Figure 1.6 A general radiation problem.

that $R \approx r_p$. Moving the point P away gives us the far-field approximation, and the solution in the far field is independent of the distance and only dependent on the angular direction. This property of the radiated fields of antennas gives us the regions of the radiated field that will have an important effect on the design of ranges.

1.5.2 Radiation Regions

This book frequently mentions the far field and the near field. The radiation around the antenna can be divided into different regions. The most common approach [4] is to divide the radiated field regions of an antenna into three parts, or to paraphrase Julius Caesar regarding the Gallic Wars (*Gallia est omnis divisa in partes tres*). These parts are the reactive near field, the radiating near field, and the far field. The radiating near field is also known as the Fresnel region, while the far field is also referred to as the Fraunhofer region.

1.5.2.1 The Reactive Near-Field Region

In a reactive region, the energy around the antenna oscillates between the electric field and the magnetic field. That is, the electric fields are 90 degrees off-phase with respect to the magnetic fields. This region extends from the antenna to an outer boundary with radius R where the following inequality applies [4]:

$$R < 0.62\sqrt{\frac{D^3}{\lambda}} \tag{1.42}$$

Here, D is the largest dimension of the antenna, and λ is the wavelength as usual. When the dimensions of the antenna are dipole-like (i.e., $D \leq \lambda/2$), the outer boundary is taken as $R \leq \lambda/2\pi$.

1.5.2.2 The Radiating Near-Field Region

The radiating near-field is the region of interest for antenna measurements performed in the near field where the far-field behavior is obtained by means of a mathematical transform; also called the Fresnel region, it extends from the reactive near field to the inner boundary of the far field [4]. In electrically small antennas where $D \ll \lambda$, this region may not exist. In this region, the radiating fields predominate. However, the angular distribution of the fields varies with distance from the antenna, as opposed to the far field where the angular distribution of the fields is constant as the distance from the antenna changes. The outer boundary of the radiating near field is given by the inner boundary of the far-field region, as derived in Section 1.5.2.3.

1.5.2.3 The Far Field

Let us consider the sketch in Figure 1.7. Consider the largest dimension of the antenna to be D, and let us assume that we are a distance R from an observation point to the center of the antenna. If a point source is located at the point O, and it radiates a spherical wave at a distance $r = R$ from point O, there will be a spherical wavefront of equal phase, but the phase at point A will be lagging the phase at the center point of the antenna of size D. The phase delay associated with the distance δr is the phase variation across the antenna of size D. Using basic geometry it is clear that

$$\overline{OA} - \overline{OB} = \delta r = \sqrt{r^2 + \left(\frac{D}{2}\right)^2} - r \tag{1.43}$$

where (1.43) can be rewritten as

$$\delta r = \sqrt{r^2\left(1 + \left(\frac{D}{2r^2}\right)^2\right)} - r \tag{1.44}$$

and it can be further manipulated to give

$$\delta r = r\sqrt{\left(1 + \left(\frac{D}{2r^2}\right)^2\right)} - r = r\left(\sqrt{\left(1 + \left(\frac{dD}{2r^2}\right)^2\right)} - 1\right) \tag{1.45}$$

yielding

$$\delta r = r\left(\left(1 + \left(\frac{D}{2r^2}\right)^2\right)^{1/2} - 1\right) \tag{1.46}$$

Equation (1.46) can be approximated by Newton's binomial expansion

$$\delta r = r\left(\frac{1}{2}x - \frac{1}{8}x^2 + \frac{1}{16}x^3 - \frac{1}{128}x^4 + \ldots\right) \tag{1.47}$$

by making

$$x = \left(\frac{D}{2r^2}\right)^2 \tag{1.48}$$

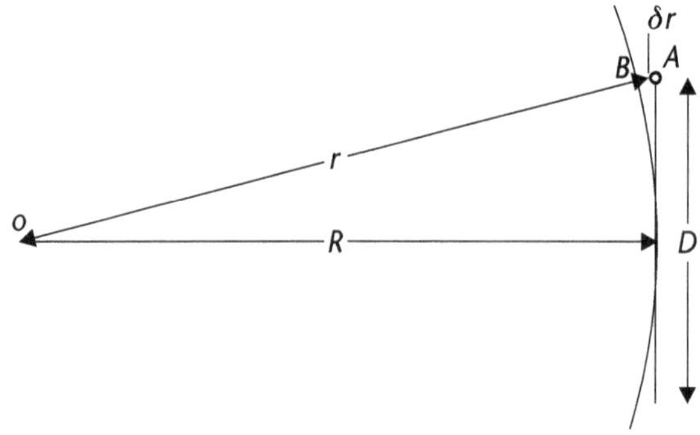

Figure 1.7 Derivation of the far-field inner boundary.

Using the assumption that $r \gg D$, we can drop the higher order terms and approximate (1.48) as

$$\delta r \approx \frac{D^2}{8R} \tag{1.49}$$

If a wavelength λ is equivalent to 2π, then $\lambda/16 = 22.5°$. If we choose $\delta r = \lambda/16$ then (1.49) becomes

$$R \approx \frac{2D^2}{\lambda} \tag{1.50}$$

which should be recognized by the reader as the traditional far-field distance. Notice that the assumption is that the distance is much larger than the antenna. Thus, that constraint must be met together with the distance being also much larger than a wavelength, so a more complete set of equations for the far field is given by

$$R \geq \frac{2D^2}{\lambda} \tag{1.50a}$$

$$R \gg D \tag{1.50b}$$

$$R \gg \lambda \tag{1.50c}$$

Thus, the Fresnel, or radiating near field region, is limited by

$$\frac{2D^2}{\lambda} > R > 0.62\sqrt{\frac{D^3}{\lambda}} \tag{1.51}$$

1.5.3 The Herzian Dipole Antenna

Let us assume that the current I exists on a line of length dl and that the line is coincident with the z-axis. Let us also assume that line is infinitesimal such that $dl \to 0$.

$$\vec{A} = \hat{a}_z \frac{\mu_o Idl}{4\pi} \frac{e^{-j\beta r}}{4\pi r} \tag{1.52}$$

If we express the vector potential in spherical coordinates, the following equations are obtained:

$$A_r = \frac{\mu_o Idl}{4\pi} \frac{e^{-j\beta r}}{4\pi r} \cos\theta \tag{1.53}$$

$$A_\theta = -\frac{\mu_o Idl}{4\pi} \frac{e^{-j\beta r}}{4\pi r} \sin\theta \tag{1.54}$$

From (1.40) the H field can be obtained as

$$\vec{H} = -\hat{a}_\phi \frac{Idl}{4\pi}\beta^2 \sin\theta\left[\frac{1}{j\beta r} + \frac{1}{(j\beta r)^2}\right]e^{-j\beta r} \quad (1.55)$$

and from the H field we can get the two components of the E field using (1.19):

$$E_r = -\frac{Idl}{4\pi}\eta_o\beta^2 2\cos\theta\left[\frac{1}{(j\beta r)^2} + \frac{1}{(j\beta r)^3}\right]e^{-j\beta r} \quad (1.56)$$

$$E_\theta = -\frac{Idl}{4\pi}\eta_o\beta^2 \sin\theta\left[\frac{1}{j\beta r} + \frac{1}{(j\beta r)^2} + \frac{1}{(j\beta r)^3}\right]e^{-j\beta r} \quad (1.57)$$

Notice that (1.55), (1.56), and (1.57) are independent of ϕ, so the fields are the same as we move around the dipole; this is what we call omnidirectional radiation, as discussed in Chapter 2. It should be mentioned that these equations are valid everywhere, even in the reactive near field. As we move in the far field where (1.50a–c) apply, then we can drop all the higher-order terms and we can approximate the far fields from the Herzian dipole as

$$H_\phi = j\frac{Idl}{4\pi}\beta\sin\theta\left[\frac{e^{-j\beta r}}{r}\right] \quad (1.58)$$

and

$$E_\theta = \eta_o H_\phi = j\frac{Idl}{4\pi}\eta_o\beta\sin\theta\left[\frac{e^{-j\beta r}}{r}\right]e^{-j\beta r} \quad (1.59)$$

Notice that for $\theta = 0$ and $\theta = 180$, the field goes to zero, so the Herzian dipole does not radiate the same for all values of θ. One of the reasons for covering the Herzian dipole is that it provides a close-form equation for the radiated fields everywhere, and from the general equations (1.55) to (1.57), the complexity of the field around the antenna can be seen.

Far more information on antennas can be found in [4, 5]. Chapter 2 revisits other antenna-related concepts.

1.6 About Numerical Methods

While it may appear incredibly simple that the entire EM phenomena can be described using four equations, with the exception of some canonical problems, there is not an easy way to solve Maxwell's equations. Solutions to problems can be obtained by using numerical techniques, but it is important to stress that any numerical technique is by definition an approximation whose accuracy depends on the assumptions made during the derivation. Unfortunately, the ease of use of commercially available computational electromagnetic software packages has caused people to use these tools without understanding some of their limitations. Thus, this section examines different types of numerical methods and highlights potential problems

There are a lot of numerical methods than can be used to solve numerical problems; some are just variations of existing methods. In general, we can separate numerical methodologies into full-wave analysis methods and asymptotic methods. The full-wave analysis methods include finite-element methods (FEMs) and finite-difference time domain (FDTD) or finite integral time domain (FITD) methods. In FITD, instead of starting from (1.1) to (1.4), we start the derivation from (1.5) to (1.8). Integral equation methods where the currents on the surfaces are solved using methods, such as the method of moments (MoM), do not constitute full-wave analysis methods since we are just solving for surface currents at interfaces.

All these methods require huge amounts of memory and time to perform the computations to solve large problems. Typically, we require 10 to 20 discrete points per wavelength where the field quantity or surface current is solved. A typical range can be easily 20 wavelengths long at a given frequency, so about 400 unknowns are required in that direction to get accurate solutions.

Full-wave analysis has been used to model chambers, but its application is limited in frequency range [6] or additional approximations are required to model the RF absorber in the range.

Overall, it is important to note that full-wave methods are not ideal for solving problems related to indoor ranges. They can be used, requiring large amounts of computer power and time, but the results may not be more accurate that a geometrical optics approximation or a simple ray-tracing exercise. As it will be shown, the absorber pieces used in a range are not identical, and the material properties vary slightly from batch to batch and from type to type, and the accuracy of the numerical model is not only related to the method, but to the input data used for the physical properties of the model. Asymptotic methods or approximations that treat wave propagation as we treat optical problems using ray tracing are better suited for electrically (meaning related to wavelength) large problems.

The full-wave analyses do have a place in range design and analysis. They can be used to model the absorber (as done in this text) to understand how a wall of absorber reflects energy. That data can be used with ray-tracing methods to study the potential reflected energy at a given location in the range.

Geometrical optics (GO), physical optics (PO), and other methods that include more accurate representation of the EM phenomena, such as the geometrical theory of diffraction (GTD) and the physical theory of diffraction (PTD), which extend the classical (GO) and (PO), are better suited to solving anechoic chamber design problems.

Work continues to develop methods that efficiently model anechoic chambers. For example, consider the simple approach I recently suggested, based on the absorber approximation discussed in Chapter 3 [7]. Other approaches have been introduced by Xiong et al. [8, 9].

References

[1] Balanis, C. A., *Advanced Engineering Electromagnetics*, New York: John Wiley & Sons, 1989.

[2] Chen, D. K., *Fundamentals of Engineering Electromagnetics*, Reading, MA: Addison-Wesley Publishing Company, 1993.

[3] Hunt, B. J., *The Maxwellians*, Ithaca, NY: Cornell University Press, 1991.

[4] Balanis, C. A., *Antenna Theory: Analysis and Design* (Second Ed.), New York: John Wiley & Sons, Inc., 1997.

[5] Stutzman, W. L., and G. A. Thiele, *Antenna Theory and Design* (Second Ed.), New York: John Wiley & Sons, Inc., 1998.

[6] Rodriguez, V., "Validation of the Polynomial for RF Absorber Reflectivity for the Prediction of Anechoic Chambers,"in *2017 International Symposium on Antennas and Propagations USNC/URSI National Radio Science Meeting*, San Diego, CA, July 9–14, 2017.

[7] Rodriguez, V., "Further Refining and Validation of RF Absorber Approximation Equations for Anechoic Chamber Predictions," in *12th European Conference on Antennas and Propagation (EuCAP 2018)*, London, United Kingdom, 2018.

[8] Xiong, Z., J. Chen, and Z. Chen, "Low Frequency Modeling for Electromagnetic Analysis of Arbitrary Anechoic Chambers," in *2016 IEEE International Symposium on Electromagnetic Compatibility (EMC)*, Ottawa, Canada, 2016.

[9] Xiong, Z., Z. Chen, and J. Chen, "Efficient Broadband Electromagnetic Modeling of Anechoic Chambers," in *2017 11th European Conference on Antennas and Propagation (EUCAP)*, Paris, France, 19–24 March 2017.

CHAPTER 2

Measurement Ranges

2.1 Indoor Ranges

Following the EM theory refresher in Chapter 1, this chapter introduces readers to the main subject of the book, providing a survey of the different antenna measurements ranges. In keeping with the idea of a basic survey, the introduction is very basic, leaving details in later chapters (Chapters 5 through 9) that discuss specific types of indoor ranges. However, the chapter continues with a review of antennas to help readers to determine which range approach is most suitable.

There are different types of indoor electromagnetic measurement ranges, commonly referred as anechoic chambers or anechoic ranges. Althouth this book mainly focuses on those ranges used for the measurement of antennas and antenna performance; Chapter 9 focuses on EMC measurement ranges. These can be anechoic, semi-anechoic, or partially anechoic ranges used to determine the levels of unintentional radiation radiated from electronic equipment, or the levels of EM radiation that electronic equipment can withstand. Tests for levels of RF radiation from a piece of equipment (such as a domestic appliance or an internal combustion vehicle) are called radiated emission tests. Tests in which a piece of equipment is operated in the presence of RF radiation levels are known as radiated immunity or susceptibility tests. EMC testing chambers have designs that are directed by a standard document.

Although there are other anechoic ranges used to test the functionality of equipment, and while one could argue that EMC chambers technically fall under this description, EMC chambers are concerned mainly with the equipment functioning within some legal limits or functioning in a given EM environment. The functionality tested in these other types of anechoic rooms is related to the performance of the device. Examples include guidance systems for missiles, transparency of radomes, and cell phone and wireless device operation.

In my opinion, CTIA, and other Wi-Fi, and multiple input and multiple output (MIMO) testing fall in this last category of anechoic chambers for testing equipment performance. These ranges, typically referred as hardware-in-loop (HWiL) testing ranges, are discussed in Chapter 10.

Another special type of range covered in this book, the RCS measurement range, is used to measure the radar signature of targets. Typically utilized for defense applications, RCS measurements may expand into the commercial world as more collision-avoidance systems are incorporated into vehicles. Chapter 8 discusses the design and performance of RCS ranges.

2.2 Antenna Measurements

An antenna is a device that receives and transmits radio waves. More specifically the IEEE 145-2018 standard defines an antenna as: "that part of a transmitting or receiving system that is designed to radiate or to receive electromagnetic waves" [1]. Antennas, therefore, are the part of a system—presumably electronic system—that transmits or receives a signal. An antenna may also receive *and* transmit for a given system. Antennas radiate waves, and we seek to determine how these waves propagate around the antenna. Chapter 1 shows that radiation changes as it leaves the antenna and that we can divide the radiation regions into a reactive near field, close to the antenna, a radiating near field and a far field. In the far field, the field distribution is a function of direction, not of distance, to the antenna. How is the power radiated as a function of the direction away from the antenna? The field can be computed or measured and then mapped. That map of the radiation as a function of direction is what we call the radiation pattern. This is what we are trying to measure in an antenna chamber. The radiation pattern is defined as "the spatial distribution of a quantity that characterizes the electromagnetic field generated by an antenna" [1].

The reason that we catalog the CTIA, Wi-Fi, and MIMO anechoic ranges as HWiL is that the antenna radiation is not what is being measured. While the pattern can be measured in those chambers, typically, sensitivity, total radiated power, data throughput, and error rates are the parameters being measured. It is true that the antenna is a key part of the system, but the overall system, or the whole hardware, is what is being tested.

The lines between antenna testing and HWiL are fading more and more. Active antennas are more common, and the line between where the actual antenna starts, and the electronics end, is disappearing. Still, it is my opinion that if the test involves finding the pattern as the beam is electronically steered or as multiple beams are switched on and off, the goal of the test is to find the pattern or parameters related to the pattern. Hence, these are antenna tests because the goal is to find the performance of the antenna as opposed to the entire system.

2.2.1 Pattern Parameters

There are a series of parameters related to the antenna pattern that may be of interest to the antenna designer or the overall electronic system designer. These parameters include directivity, gain, beamwidth, sidelobe levels, front-to-back ratio, and polarization (including cross-polarization). Sections 2.2.1.1–2.2.1.3 survey these parameters to help readers understand what are we trying to measure.

2.2.1.1 Gain

Gain relates how the antenna radiates to the input power at its port. In most areas of electrical engineering, gain implies amplification. Gain is typically defined as the power out of a system, divided by the power in.

Antennas are typically passive devices, so the output power should be less than the input power due to internal losses; hence all antennas should have negative gains using the definition for gain typically used in electrical engineering. However, in antenna engineering, gain (in a given direction, typically the direction of the maximum) is the ratio of the radiation intensity in a given direction to the radiation intensity if the power accepted by the antenna was radiated isotropically [1]. Thus, the gain, much like the radiated power, is a function of the direction. Typically, when talking about the gain of an antenna, we refer to the maximum gain or peak gain. Hence, if the direction is not mentioned, the gain refers to the direction of maximum radiation [1]. This gain is sometimes referred to as absolute gain. There is also a partial gain, which refers to the radiation intensity in a given polarization, still divided by the radiation intensity if the power accepted was radiated isotropically.

Another concept is the realized gain, the gain of an antenna reduced by the impedance mismatch. Thus, we are not looking at the power accepted but the incident power coming from the source. Figure 2.1 illustrates these concepts. The directivity is the radiation intensity in a given direction divided by the radiation intensity, assuming that the total radiated power is radiated isotropically. Notice the absolute gain is the same as the directivity if the efficiency is 1—that is, there are no losses. The realized gain takes the power available, not the power accepted by the antenna. The absolute gain and the realized gain are implicitly assumed to be in the direction of the maximum gain.

There are different approaches to measure gain, but these can be grouped into absolute gain measurements and gain-transfer measurements. Absolute gain measurements are based on the Friis transmission equation, whose simple form is given by

$$P_r = P_{\text{in}} G_{\text{A}} G_{\text{B}} \left(\frac{\lambda}{4\pi r} \right)^2 \tag{2.1}$$

In (2.1), P_r is the received power, and P_{in} is the power into the antenna that is transmitting during the test. If the two antennas are identical, meaning $G_A = G_B$, knowing the distance r and the wavelength λ, the gain of the antennas can be determined from the transmitted and received powers.

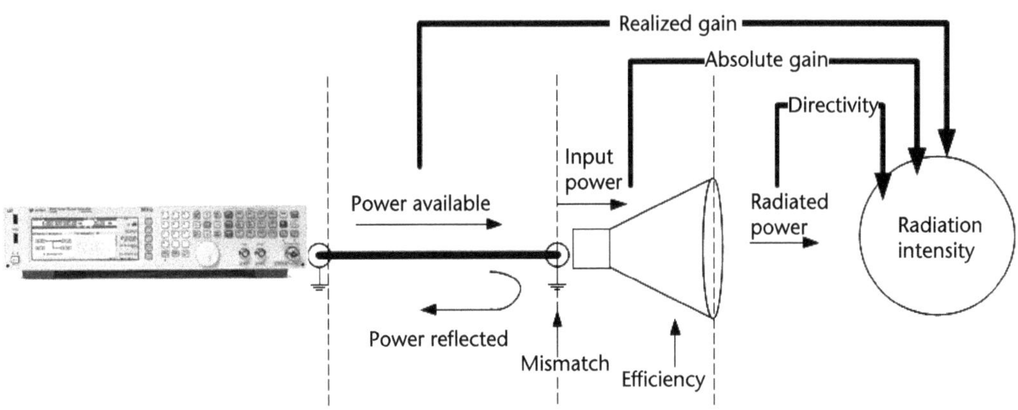

Figure 2.1 Illustration of the concepts of directivity, realized gain, and absolute gain.

It is more common for the two antennas not to be identical. Even if the transmit and receive antenna are the same models, variations in manufacturing will add to the uncertainty. To avoid this issue of having two identical antennas, a variation to the two-antenna absolute gain measurement is the three-antenna method, where the gain for the three antennas can be obtained by solving a system of equations once the three measurements are obtained. For three antennas A, B, and C, the equations are

$$\begin{aligned}(G_{\mathrm{A}})_{\mathrm{dB}} + (G_{\mathrm{B}})_{\mathrm{dB}} &= 20\log_{10}\left(\frac{4\pi R}{\lambda}\right) - 10\log_{10}\left(\frac{P_{\mathrm{in}}}{P_r}\right)_{\mathrm{AB}} \\ (G_{\mathrm{A}})_{\mathrm{dB}} + (G_{\mathrm{C}})_{\mathrm{dB}} &= 20\log_{10}\left(\frac{4\pi R}{\lambda}\right) - 10\log_{10}\left(\frac{P_{\mathrm{in}}}{P_r}\right)_{\mathrm{AC}} \\ (G_{\mathrm{B}})_{\mathrm{dB}} + (G_{\mathrm{C}})_{\mathrm{dB}} &= 20\log_{10}\left(\frac{4\pi R}{\lambda}\right) - 10\log_{10}\left(\frac{P_{\mathrm{in}}}{P_r}\right)_{\mathrm{BC}}\end{aligned} \tag{2.2}$$

Because there are three separate measurements, there are several sources of uncertainty including misalignment, connection repeatability issues, and variations in temperature.

One important fact is that these methods are based on the Friis equation. The Friis equation describes the transmission between two antennas in free space. Thus, modifications to the equation are required when using the methodology in cases where there is a ground plane present. This methodology cannot be used in ranges where there is no true free-space condition, such as tapered ranges, and in ranges where the range antenna cannot be switched because of the measurement approach. For example, near-field methods and compact ranges are not suitable for absolute gain measurements since there are specific constraints on the type of range antenna that is used.

Most antenna measurement ranges rely on reference gain methods, where the gain of the antenna under test G_T is obtained by comparing its received power to the power received by a known antenna with known gain G_S. That antenna is what is called a "gain standard." (These are typically calculable antennas like standard gain pyramidal horns or dipoles.) In those cases, the equation to find the gain is given by

$$G_T(\mathrm{dB}) = G_S(\mathrm{dB}) + 10\log_{10}\left(\frac{P_T}{P_S}\right) \tag{2.3}$$

If the gain of the gain standard is not calculable, then it is calibrated using an absolute calibration in a range where the equations in (2.2) apply. For most applications, calculable antennas such as pyramidal horns are used [2]. Extrapolation ranges are another method for calibrating gain standards [3, 4]. In these extrapolation ranges, the antennas are moved away from each other, and the data collected versus distance is then approximated with a polynomial curve. The limit of the polynomial as the distance goes to infinity is computed, and that provides the far-field gain for the antenna under tests.

As discussed, the gain is usually the peak gain, or the gain in the direction of highest radiation. Accordingly, it is important to determine that direction, which requires mapping the radiation pattern, as discussed in Section 2.2.1.2.

2.2.1.2 Radiation Pattern

Like many other areas of human research, we like to classify things in order to study them. Radiation patterns do not escape this human need. The first division in this classification is how the energy is radiated. We divide antennas into omnidirectional and directional categories. Before we dig further into those two types of patterns, let us look at little bit more into the pattern definition.

2.2.1.2.1 Graphical Representation of the Pattern

A pattern is a map of the radiation intensity as a function of direction. Hence, the pattern is a three-dimensional map, where we can map the direction as spherical coordinates θ and ϕ and use the r direction to map the intensity of the radiation. Figure 2.2 illustrates a 3D pattern, intensity versus θ and ϕ, on to a Cartesian set of axes, further reduced to a 2D representation on the pages of this book. Today, it

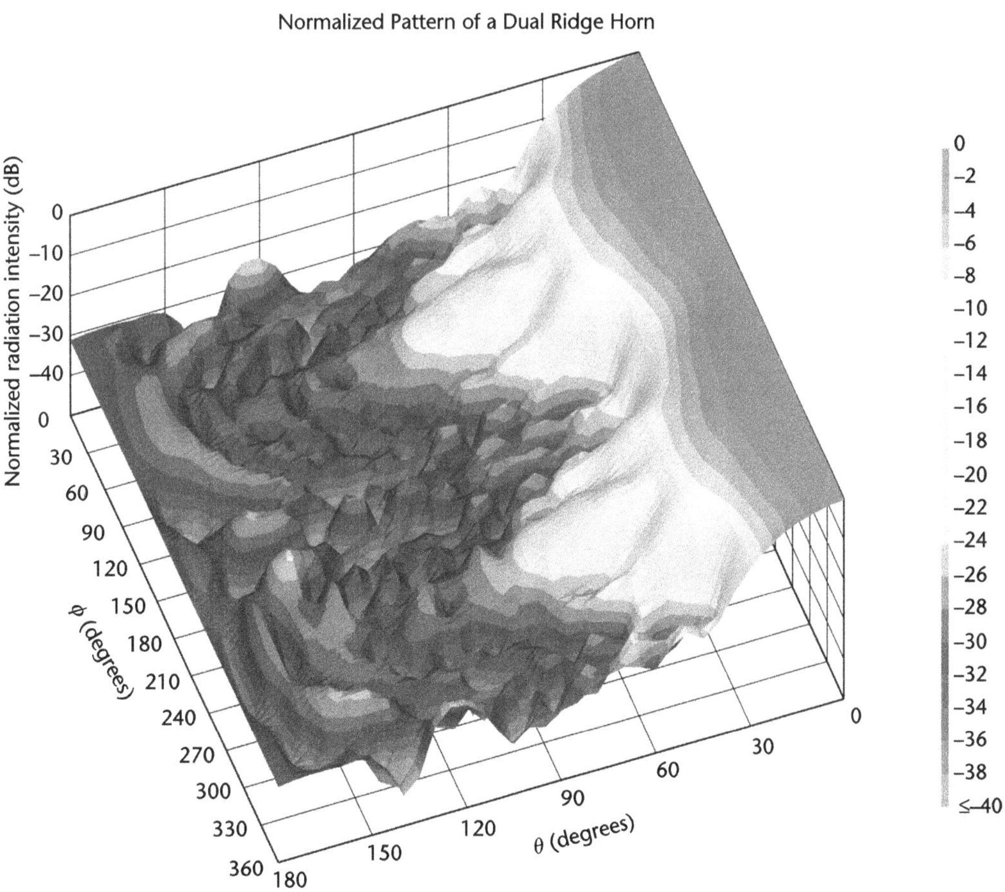

Figure 2.2 A representation of a 3D pattern onto a 2D page.

is easy to create these 3D plots, to rotate them, and visualize them on a computer screen. However, 30 years ago this required significant computer power, and 60 years ago, creating these plots was a very involved process. Engineers solved this problem by looking at plane cuts for a constant ϕ or θ to visualize the antenna in a simpler way.

Therefore, antenna engineers defined the principal E-plane as the plane containing the electric field vector and the direction of maximum radiation [1]. It should be noted that this definition is, of course, for linearly polarized antennas. Similarly, antenna engineers defined the principal H-plane as the plane containing the magnetic-field vector and the direction of maximum radiation. The principal plane that we call the H-plane is orthogonal to the principal plane referred as the E-plane. This is the case because in the far field, where plane waves propagate, the E-field vector is perpendicular to the H-field vector. Hence, for the 3D pattern plotted in Figure 2.2, we can obtain a set of two plots that show the radiation pattern on the principal plane. Figure 2.3 illustrates the representation of a radiation pattern as two principal plane cuts. Typically, these plots are shown in Cartesian or polar coordinates.

While it has become the norm to compute and measure 3D patterns and visualize them on a computer screen, antenna engineers still use the E- and H-plane representation for the radiation patterns. Now that we have introduced the concept of the principal planes, we continue with the classification of directional and omnidirectional radiation patterns.

2.2.1.2.2 Omnidirectional

Omnidirectional, whose Latin prefix "omni" means all or every, describes radiation patterns that radiate in "all" directions or "every" direction. This is confusing,

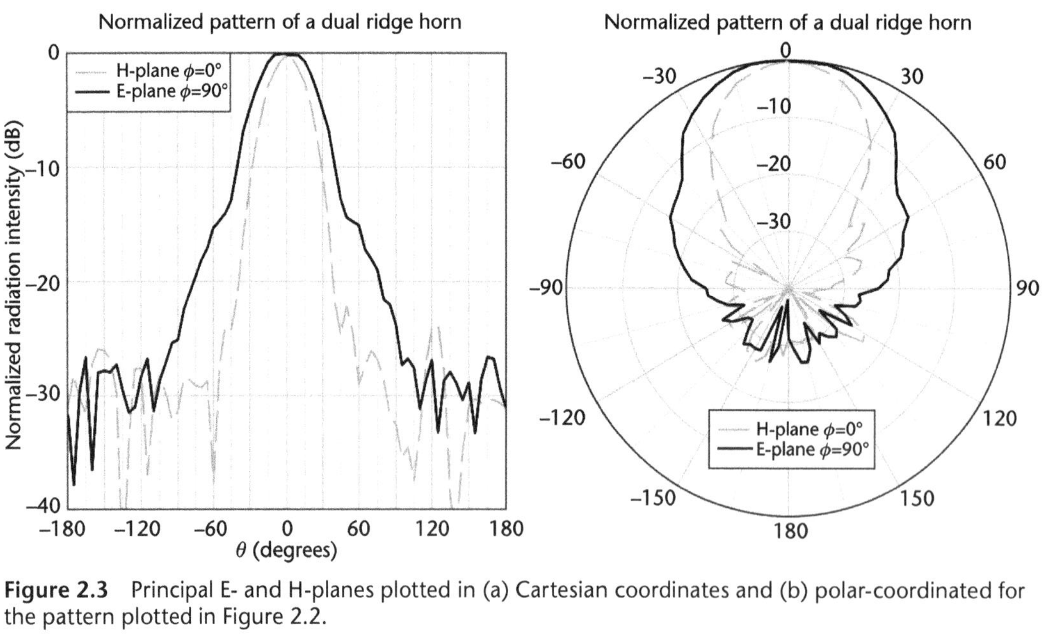

Figure 2.3 Principal E- and H-planes plotted in (a) Cartesian coordinates and (b) polar-coordinated for the pattern plotted in Figure 2.2.

because it seems to imply that the antenna radiates toward every point around it, but the most classic of the omnidirectional antennas, the humble dipole, has two nulls where no energy radiates. Omnidirectional antennas radiate in all directions on one of the two principal planes; on the orthogonal plane, the radiation has certain directivity since there are nulls or areas where there is little to no radiation.

While it may come as a surprise to the antenna engineer reading this book, there is confusion between isotropic and omnidirectional antennas. However, antenna textbooks are very clear on the differences. In fact, there are no isotropic antennas. The isotropic radiator is a mere mathematical concept and not a real implementable antenna. An isotropic antenna is lossless. That by itself relegates isotropic radiators to Plato's realm of ideas. Isotropic radiators do radiate equally in all directions. Unlike the word "omnidirectional," which comes from Latin, "isotropic" is constructed from the Greek origin prefix "isos" (ἴσος), meaning equal, and the suffix "tropos" (τρόπος), meaning direction or way. Hence, isotropic antennas radiate equally in all directions. We can find the power density radiated from an isotropic antenna by simply dividing the input power by the surface of a sphere of a given radius r. The gain of any other antenna can be compared to this ideal isotropic radiator. Thus, we compare the power density (at the peak) to the power density from an isotropic antenna to find the gain, which in decibels is given in decibels-isotropic (dBi), where the "i" stands for isotropic or gain related to an isotropic antenna.

Interestingly, antennas are still sold as isotropic antennas. These are typically arrangements of three dipoles orthogonal to each other. They may have a degree of "isotropicity," but they are not isotropic and are usually intended to measure field levels in a given location and are a collection of linear sensors or antennas placed orthogonally, and in close proximity, to each other. These are typically calibrated to equate the responses of the different sensors. However, such antennas should not be confused with the concept of isotropic radiators.

2.2.1.2.3 Directional Patterns and Features of the Pattern

Directional antennas are the other main family in which we can divide antennas by their pattern. These antennas have their radiation propagating mainly in one direction. The patterns shown in Figure 2.3 show a directional antenna. Oriented in the $\theta = 0$ direction is the main lobe. The main lobe is where the main radiation from the antenna is directed. The gain versus isotropic reported for any given antenna is in the main lobe direction. Moving away from the peak of the beam are the –3-dB points. At these points the power density radiated by the antenna is half compared to the peak value. This half-power beamwidth (HPBW) provides some information about the shape of the beam, and there is a HPBW reported for each of the principal planes. Hence the E-plane will have one value of HPBW, and the H-plane could have, and typically does, a different value of HPBW.

The HPBW may determine the accuracy of the positioning equipment used in an antenna measurement. Very large apertures and arrays may have an HPBW of a few degrees. Figure 2.4 illustrates such an antenna. This antenna is an 18-inch (45.7-cm) diameter flat array, where the array elements are slots cut out on a waveguide. This type of antenna is commonly used in weather radar applications for commercial aircraft. The radar operates at 9.375 GHz; hence it is 14.3λ in size ($\lambda = 3.198$ cm at 9.375 GHz). This is not an extremely large antenna, yet it has a 3-dB beamwidth on its E-plane that it is only 5.6°, and 5.1° on its H-plane. The pattern in Figure 2.4 was measured on a compact range as reported by Fordham and D'Agostino in [5]. Antennas with an HPBW of less than a degree are possible, and the accuracy of the positioning system is important in measuring the HPBW.

To the sides of the main beam are the sidelobes and nulls. These are areas of lower radiation, so the levels being measured are much lower than the main beam levels. For the antenna illustrated in Figure 2.4, the highest sidelobe level (SLL) is at −25.93 dB at $\theta = -11.8°$. That is the first sidelobe on the E-plane. The secondary sidelobes are much lower, with some of them as low as −39 dB, and as discussed in Chapter 5, the reflections from the range walls cannot be too high, or the error on the measured SLL quantity can be significant to the point of making it impossible to measure these features of the pattern.

When measuring an antenna, the question that the designer must ask is, "What is the acceptable error for each of the parameters being measured?" This error budget or uncertainty estimation should be done to select the proper approach to measure the antenna and the parameters of interest. This book provides rules of design to minimize the effects of the range on the measurement. Similarly, these rules can be used to estimate the effects of the range on measurements being done on existing ranges or to choose from potentially available facilities.

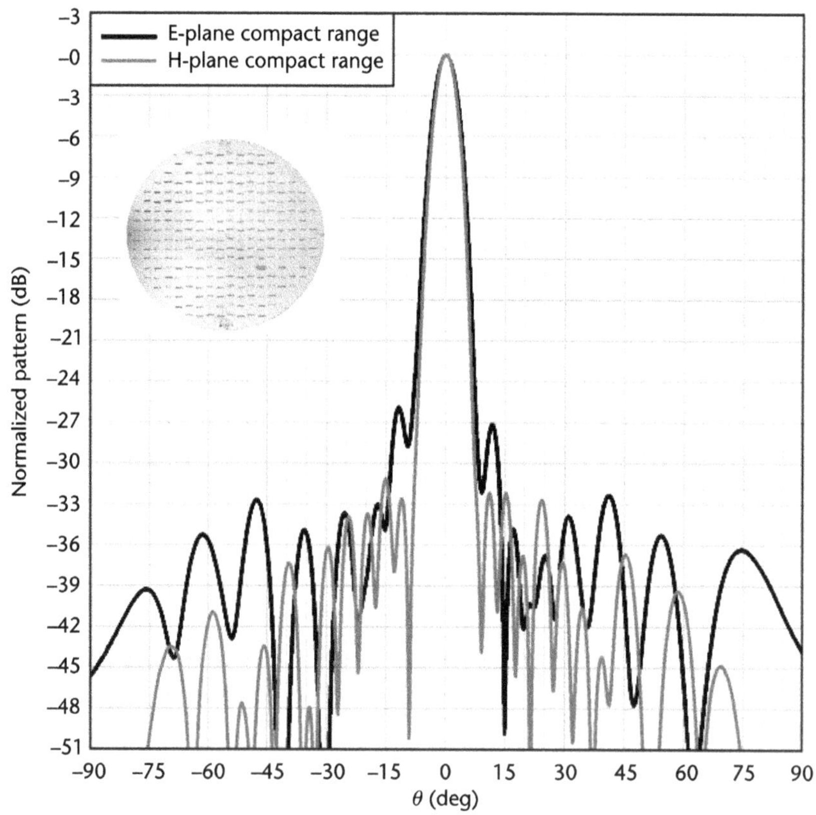

Figure 2.4 A highly directive flat-slotted array antenna pattern in the two principal cuts.

2.2.1.3 Polarization

Microwave Antenna Measurement [6] dedicates a complete chapter to polarization. Polarization is an important aspect of a radiating antenna that must be understood. Not all ranges are equal when it comes to measuring polarization. On some, like in compact ranges, the polarization measurements are limited by the cross-polarization of the range itself. The limit of what levels of polarization can be measured must be understood. In most ranges the range antenna plays a part in the level of cross-polarization that can be measured; however, it is not the only source of error or limiting factor. The range itself has an effect on the cross-polarization, as in tapered ranges where cross-polarization measurements beyond a certain level cannot be conducted.

2.3 Antenna Ranges

While this book focuses on indoor antenna measurement ranges, it also covers other ranges for other applications. Figure 2.5 shows a decision tree that can help decide which range type to use.

Using the decision tree in Figure 2.5, assume that we have an antenna to be measured. The first question is whether we need a shielded facility (a topic detailed

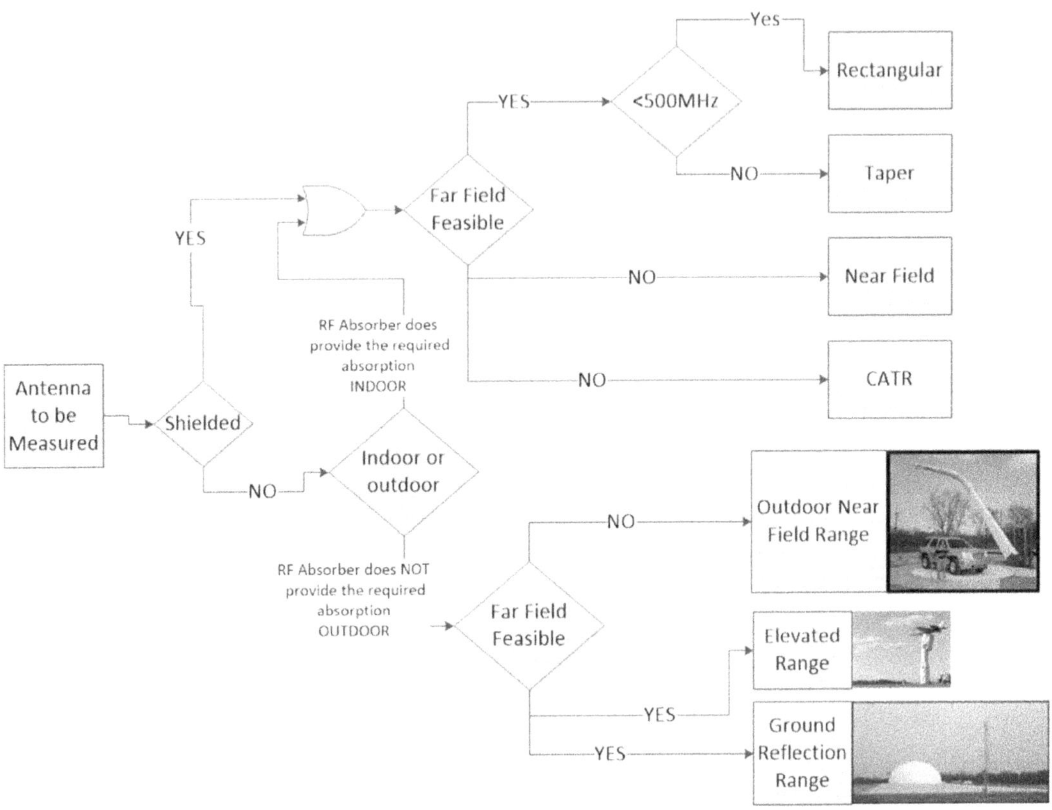

Figure 2.5 The antenna range decision tree.

in Chapter 4). If a shielded facility is necessary, then an indoor range is a must; however, notice that an indoor facility may be chosen for reasons other than the need for shielding, hence the "OR gate" symbol used in Figure 2.5. Indoor ranges may be chosen for environmental-control reasons (concerning temperature, for example), meaning that an indoor range does not have to be shielded.

The next decision on Figure 2.5's tree is to decide between indoor and outdoor facilities. The factor that weighs heavily on that decision is the available RF absorber technology (covered in Chapter 3). It should be mentioned that, currently, accurate antenna measurements in indoor spaces are not easily achievable below 80 MHz. The antenna test engineer and the antenna engineer must decide if the potential uncertainties of performing indoor measurements of antennas at really low frequencies (i.e., under 100 MHz) are acceptable for the given antenna, and that will depend on the parameters of interest.

Regardless of which of the two branches (indoor or outdoor) of Figure 2.5's decision tree we are following, the next decision is related to the far field. Is the far field feasible? Let us look at the far-field requirements to attempt to answer this question.

2.3.1 Far-Field Ranges

Far-field ranges are the simplest types of ranges. The range antenna is placed at the far field of the antenna under test (AUT), and then the AUT is rotated to measure the field intensity as a function of direction, which is the definition of the radiation pattern. While simple in theory, there are some limitations to far-field ranges, mainly related to the size and the losses. Chapter 1 defines the lower limit of the far field region as [1]

$$R = \frac{2D^2}{\lambda} \tag{2.4}$$

where D is the physical size of the antenna, and λ is the wavelength at a given frequency. Let us now express the size of the antenna D as a function of the wavelength. It should be noted that (2.4) is one of the constraints of the far field, an approximation that is be used for antennas that where the dimension D is larger than λ [1]. In addition, it is shown in [7, 8] that when testing at the distance given by (2.4), the first sidelobe may have a significant error, and the first null may be "filled. "In general, it is better to test at a longer distance if the sidelobe level and the first null are important quantities. This book defines the far field as starting at a distance given by (2.4), but the reader should bear in mind that for some antennas it will be desirable to test at distances such that $R = 4D^2/\lambda$.

Continuing with the lower limit distance given by (2.4), for an antenna of size $D = n\lambda$, we can see that the lower limit of the far field becomes

$$R = 2n^2\lambda \tag{2.5}$$

So as the antenna increases in size in terms of wavelengths (the electrical size), the far field increases as twice the square of the electrical size. For example, a 5-λ antenna has the lower limit of the far field at 50λ, while a 6-λ antenna will have the far field move to 72λ. The antenna illustrated in Figure 2.4 has the far field at 409λ, which is 13.08m (42.91 ft.).

As the test distance increases, so does the path loss. We define the propagation loss as

$$L(\text{dB}) = 20\log_{10}\left(\frac{4\pi R}{\lambda}\right) \tag{2.6}$$

Substituting (2.5) into (2.6) we get

$$L(\text{dB}) = 20\log_{10}\left(8\pi n^2\right) \tag{2.7}$$

Thus, the antenna in Figure 2.4, if measured at the far field, has losses of 74.22 dB. That level of over 70 dB is a significant amount of loss, and the high losses that occur in far-field ranges are one of the reasons for going to alternative methods to measure antennas. The losses on a far-field range as given by (2.7) are plotted in Figure 2.6.

The far-field distance is not necessarily a critical issue. While it is true that a large indoor far-field facility may be economically unfeasible, when money is not an issue such large facilities can be built. Open areas and elevated ranges can have the range antenna far away from the antenna under test. The ground reflections can be handled through the use of diffraction fences or by sufficiently elevating the range. The ground reflection can be used to our advantage, and outdoor ground reflection ranges do use ground reflection, as the name suggests. The rules for designing outdoor ranges are detailed in [7, 8]. Chapter 5 describes the theory behind ground reflection ranges in the discussion of tapered ranges. Readers should also refer to [9] where the famous *Microwave Antenna Measurements* book [6] is available. Rules

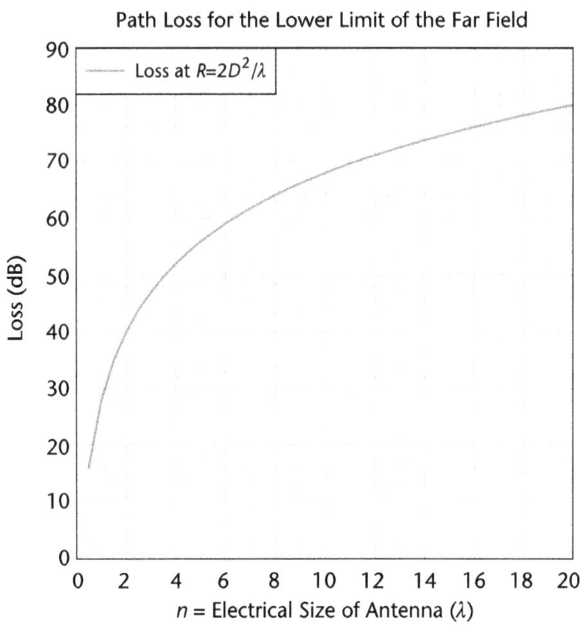

Figure 2.6 Far-field range losses for an antenna of a given electrical size.

Figure 2.7 An outdoor antenna far-field range. The picture of the range antenna tower is taken from the AUT room, and the picture of the building housing the AUT room is taken from the base of the AUT tower. The inset shows the AUT positioner in the AUT room. The door on the side of the building is open during measurements. (*Source:* Author's private collection.)

for the design of outdoor antenna ranges are discussed in [6]. These outdoor test ranges were once extremely popular, and several of them are still in use around the world. Figure 2.7 illustrates one of these ranges, a hybrid range, named as such since the AUT positioner is in an indoor room that opens to the outside.

While there are cases where an outdoor range is the proper solution, there are drawbacks to outdoor ranges. One of the drawbacks of outdoor ranges can be seen in Figure 2.7. Notice the beautiful sky in the picture of the range antenna tower; in the meantime, the sky is cloudy in the picture of the AUT room. Outdoor ranges are susceptible to weather. That susceptibility as well as the legal limits to the radiated power over a given band, is a limitation that pushed the development of indoor ranges. Indeed, the range in Figure 2.7 uses an indoor room to keep the AUT out of the weather. Figure 2.8 illustrates another approach to weatherizing outdoor ranges. In Figure 2.8, a large radome is used to cover the AUT. This is a ground reflection range. The area between the range antenna tower location and the AUT has a metallic ground plane so that the reflection is known and used in the measurement.

Let us go back to the discussion of far-field ranges and their losses. Let us assume that we have an antenna with a circular aperture that is 5λ in diameter. The gain of that antenna, assuming 100% efficiency, can be approximated by $g(\text{dB}) = 10\log_{10}[(\pi D/\lambda)^2]$ as shown in [10]. For the assumed 5-λ antenna, the gain is 24 dB, and the path loss to the far-field distance is about 56 dB, as shown in Figure 2.6. The losses are, in part, counteracted by the gain of the antenna. Thus, the range antenna will receive a signal that is not 56 dB down but only 22 dB down; hence, if the range antenna has significant gain, then the path loss may not be an issue.

Consider now the antenna shown in Figure 2.4 as our AUT, which has far-field propagation losses of 74.2 dB. The antenna has, in the best case, a gain that is roughly 30 dBi. Now let us use the 5-λ aperture antenna of the previous example to measure the flat-plate array. The flat-plate array with its 30-dBi gain will put

Figure 2.8 A ground reflection range using an RF-transparent radome to cover the AUT. (Photo courtesy of NSI-MI Technologies.)

at the far field in front of the 5-λ range antenna a level that is −44.2 dB, but with the 24-dB gain of the 5-λ range antenna the power level at the port will be −20 dB compared to the input at the flat-plate antenna. However, as we rotate the flat array, and we try to measure the sidelobes, the levels will be as low as −60 dB below the input signal at the AUT. Depending on the dynamic range of the receiver and the other losses on the system, it may not be possible to measure those levels, as we may run into system noise.

Figure 2.9 shows the path loss with the gain of the antenna subtracted. The plot shows that path loss still increases with frequency. The plot, however, does not show that, as the gain increases, the sidelobes will be much lower, and the losses may be too high to accurately measure them. In Figure 2.9, a 20-λ antenna will put power at the far field that is 44.1 dB lower than its input power. (We are still assuming no losses on the antenna.) However, a −40-dB SLL will be at −84.1 dB versus the input power to the antenna.

This process of adding gains and losses on a system is called the link budget, and it is a process that should be followed in any range geometry to ensure that the different parameters can be measured. Section 2.3.3 discusses the link budget further. First, Section 2.3.2 examines the typical antenna measurement system configurations.

2.3.2 The Antenna Measurement System

Let us consider a typical antenna measurement system geometry. This is a general geometry with remote mixing and a separate RF source and receiver. We will use this system to illustrate the link budget methodology and determine the losses in a

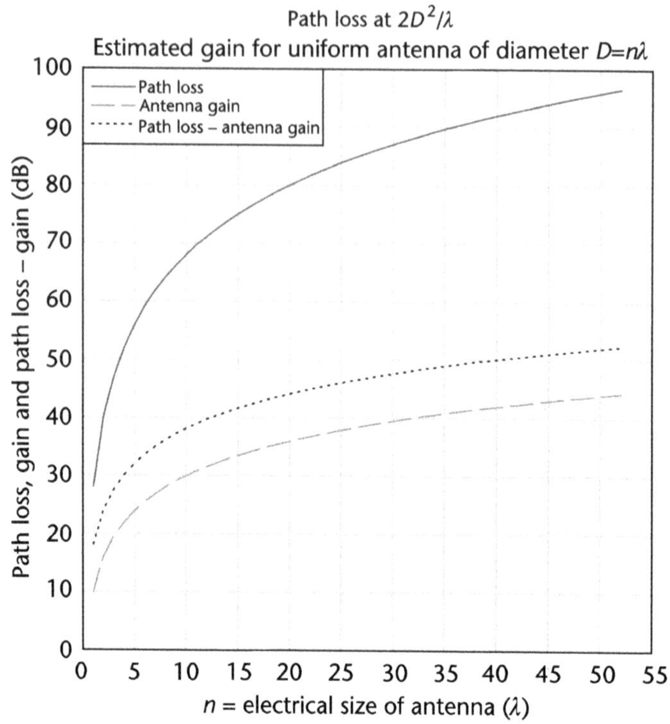

Figure 2.9 Path loss at the far field minus the gain of a circular aperture of that size.

system. The system illustrated in Figure 2.10 has a receiver with an internal local oscillator (LO) source, but an additional external source can also be used as the LO. At the mixers the RF signal is downconverted to an intermediate frequency (IF) so that there is lower loss on the cables. The mixers can be internal to the receiver, as can the RF source. A typical example of this is a vector network analyzer, but this configuration, although popular, is not recommended, as the cable losses reduce the system dynamic range significantly and increase the uncertainty. Vector network analyzers (VNAs) can be purchased with the option of using external mixing and LO, but this approach is typically more costly than dedicated receivers like the system shown in Figure 2.10.

Walking through the system, we start with the RF source. This device will provide the excitation to one of the antennas in the range. This could be the range antenna or the AUT. The signal out of the source is sampled via a directional coupler. That sampled signal will serve as the reference and be fed back to the reference channel on the receiver. The reference is required to measure phase. Phase is a must when measuring polarization as well as in near-field measurements where the magnitude and phase are required to mathematically obtain the far field. Having the reference phase also helps in rejecting some of the ambient noncoherent noise, reducing the need for shielding.

In the system shown in Figure 2.10, the sampled signal is fed to a mixer to downconvert to IF. The mixers have internal diplexers to separate the LO traveling from the receiver internal LO source to the mixer from the IF traveling to the receiver reference channel.

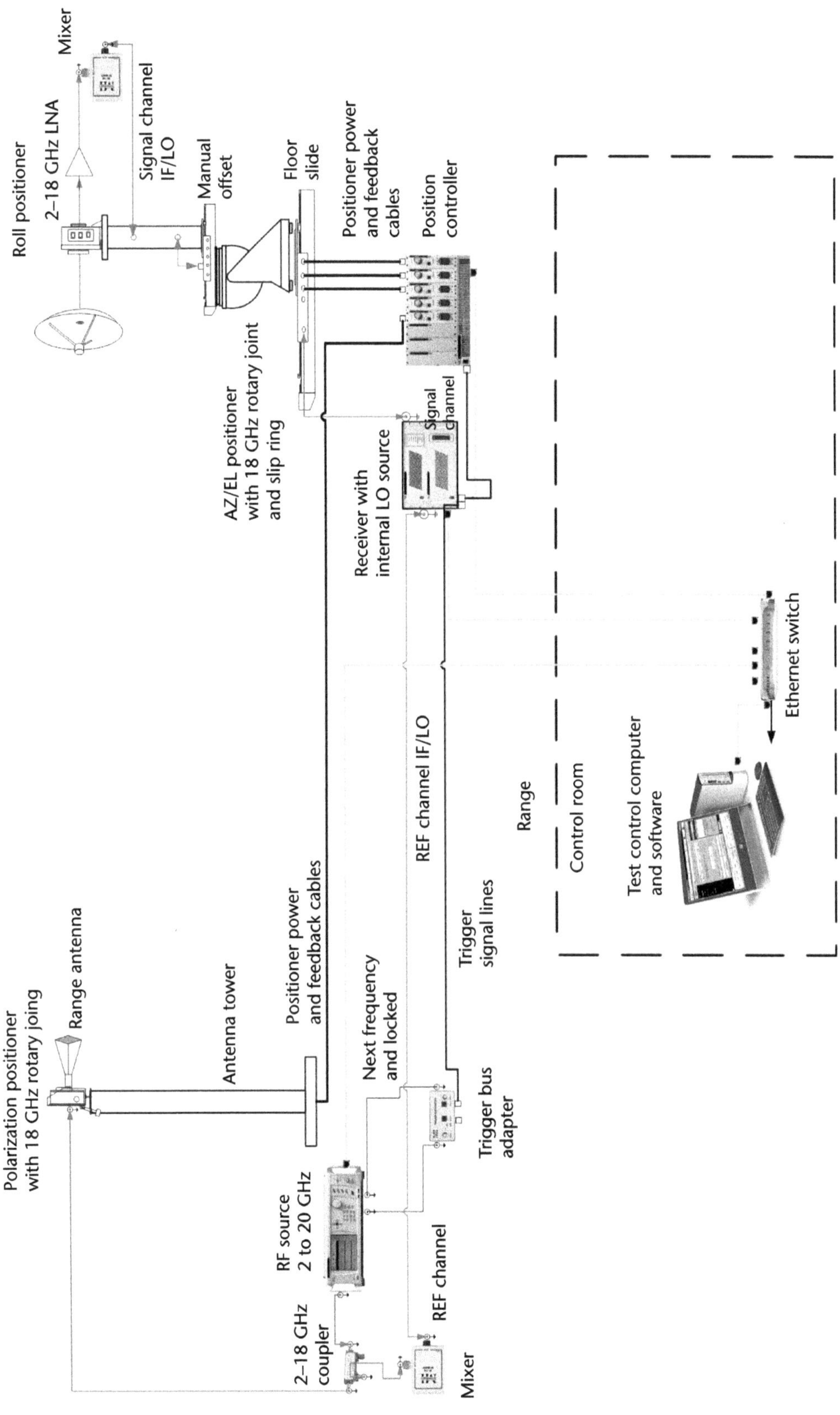

Figure 2.10 A common antenna measurement system.

In Figure 2.10, the RF is fed via cables to the range antenna, and it radiates toward the AUT. The antenna is mounted on a polarization positioner that rotates the range antenna about its axis. The positioner is controlled by position-controller equipment.

On the other side of the range is the AUT, in this case, in receive mode. The AUT is positioned by a mechanical system controlled by the position controller, so that all aspects or angles of arrival of the AUT may be tested. The positioning equipment may allow for rotation of the AUT on two orthogonal axes or about a single axis. In some cases, the range antenna is rotated about the AUT, as is the case in some spherical near-field systems. The power received by the antenna may be amplified using a low-noise amplifier (LNA). Care should be taken that the signal from the range antenna does not saturate the LNA. This is one of the drawbacks of outdoor ranges as LNAs may be easily saturated by ambient noise. To avoid this, instead of amplifying the received power, the transmit power is amplified. However, here there may be legal limitations as to how much power can be transmitted without a license and over which bands, and that may limit use of far-field outdoor ranges.

If the far-field outdoor range is not feasible, near-field outdoor ranges can be constructed, as well as outdoor compact ranges like the one at Ft. Huachuca in Arizona [11].

Whether the signal is amplified or not, it has to be taken via cables to the receiver. In the sample system shown in Figure 2.10, the signal is downconverted at a mixer close to the AUT to minimize the cable losses.

The entire measurement system is controlled by a computer that communicates with the different devices in the system. Communication used to be done via IEEE-488, but this approach had limited range. Extenders can be used for large ranges, but nowadays, communication via Ethernet is more common.

The instruments have to work together as well, so there is a series of trigger signals that command the source to switch frequencies. The source also sends a trigger to let the receiver know when it has switched to the next frequency to allow the receiver to start measuring. The position controller is also commanded by triggers to go on to the next position when all the data is acquired at a given position, but it may be set to continuously rotate, with the measurements performed on-the-fly.

2.3.3 Link Budgets

Now that we have taken an overall look at the system, let us create a link budget. For efficiency, our link budget will be a very simplistic one, making assumptions about the losses of components. A more thorough analysis can be done using actual insertion loss measurements for the different components or using manufacturers' specifications. In this simplistic approach, we will typically assume a 1-dB loss for most of the inserted components on the RF path.

We also assume that there is no LNA at the output of the AUT. In this example we can see the importance of remote mixing. Remote mixing does require additional analysis to make sure that the LO signal is high enough to drive the mixer properly. If it is not, amplification along the LO line may be required to properly drive the mixers.

Table 2.1 shows the link budget for a range operating from 2 to 18 GHz. The cable losses can be obtained from a manufacturer as either decibels per meter or decibels per foot. The free-space path loss is obtained from (2.6). The noise floor of the receiver for a given IF bandwidth (IFBW) is also obtained from the manufacturer's specifications.

Notice that there is a 3-dB attenuator located at the input of the TX antenna and at the output of the receive antenna. It may seem like a waste to have these in place and lose 6 dB of power; however, these pads reduce the mismatch losses in the system and reduce the uncertainty as different antennas are used, as is the case in three-antenna methods or when using reference gain standards.

Regarding the range space loss, Chapter 7 shows that since the wave from the reflector (or the plane-wave generator) on a compact range to the AUT is a plane wave, it does not attenuate; hence, there is no space loss on those antenna ranges, except for the loss between the feed antenna and the reflector. For near-field systems where the separation between the transmit and receive antennas is between 3λ and 10λ, the loss between the AUT and the sensing probe is approximated as a difference of gains $G_{AUT} - G_{Probe}$.Once the link budget is laid out we can start considering

Table 2.1 A Power Link Budget for an Antenna Range

Frequency (GHz)	*Cable Type or Model*	*Length (ft.)*	*2*	*4*	*8*	*12*	*16*	*18*
Range Source Output (dBm)			10.0	10.0	10.0	10.0	10.0	10.0
Cable Loss to Ref Coupler	UFB142A (F)	2	−0.24	−0.32	−0.54	−0.72	−0.82	−0.90
Ref Coupler Insertion Loss			−1.0	−1.0	−1.0	−1.0	−1.0	−1.0
Cable Loss to Rotary Joint	UFB142A (F)	20	−2.40	−3.20	−5.40	−7.20	−8.20	−9.00
Rotary Joint Loss			−1.0	−1.0	−1.0	−1.0	−1.0	−1.0
Cable Loss to TX Antenna	UFB142A (F)	1	−0.12	−0.16	−0.27	−0.36	−0.41	−0.45
Matching Pad/Isolator Loss			−3.0	−3.0	−3.0	−3.0	−3.0	−3.0
TX Antenna Gain (dBi)			24.0	24.0	24.0	24.0	24.0	24.0
Power Transmitted			26.2	25.3	22.8	20.7	19.6	18.7
Range Loss (dB)	Space	42.91	−60.9	−66.9	−72.9	−76.4	−78.9	−79.9
Power @ RX Antenna (dB)			−34.6	−41.6	−50.1	−55.7	−59.3	−61.3
RX Antenna Gain (dBi)			20.0	22.0	22.0	23.0	23.0	24.0
Matching Pad/Isolator Loss			−3.0	−3.0	−3.0	−3.0	−3.0	−3.0
Cable Loss to Rotary Joint	UFB142A (F)	1	−0.12	−0.16	−0.27	−0.36	−0.41	−0.45
Rotary Joint Loss			−1.0	−1.0	−1.0	−1.0	−1.0	−1.0
Cable Loss to Signal Mixer	UFB142A (F)	4	−0.48	−0.64	−1.08	−1.44	−1.64	−1.80
Signal at Mixer Input			−19.2	−24.2	−33.5	−38.5	−42.4	−43.5
Sample Rate for Each Frequency (IFBW)			10 kHz	10 kHz	10 kHz	10 kHz	10 kHz	10 kHz
RF Noise Floor for that Sample Rate			−106	−106	−106	−106	−106	−106
Dynamic Range			**86.8**	**81.6**	**72.5**	**67.5**	**63.6**	**62.5**

some potential cases. It is a good idea to use a spreadsheet to set the link budget and then change specific cells to see the effects on the receive power and dynamic range. Table 2.1 assumes typical standard gain horns for the AUT gain and 24 dBi of gain for the range antenna. The link budget in Table 2.1 places the range antenna and the AUT boresighted to each other (i.e., we are considering the peak gain). At these peak gains the dynamic range of the system is very good, exceeding 62.5 dB above the noise floor at the highest frequency. As the AUT is rotated towards null or sidelobes, is a good SNR still preserved? That question can be answered by reducing the gain of the AUT by 20 or 30 dB to simulate the performance when measuring SLL and to determine if the dynamic range of the system is sufficient.

2.4 Selecting the Range

When selecting a range approach to testing a specific antenna we have look at the available approaches and determine which one is preferred. The decision tree illustrated in Figure 2.5 leads us to ask a series of questions. The first question regards the need for shielding as well as the level of shielding, and that question is answered in Chapter 4. The next question is whether RF absorber technology can meet the reflectivity requirements, which is addressed in Chapter 3.

Subsequently, the decision tree focuses on the feasibility of a far-field range. The previous sections have shown that as the antenna becomes electrically larger, the far-field condition is further away, and the range losses are more significant. A link budget can let us know if measurements with a given dynamic range can be achieved. However, there are practical constraints as well; it is still not the same 50-λ far-field distance for an antenna operating at 2 GHz (7.5-m far-field distance) than at 500 MHz (30-m far-field distance). Table 2.2 illustrates the limits of the physical size of the range for indoor ranges.

Table 2.2 Far-Field Distance for Different Frequencies and Antenna Sizes

Frequency (MHz)	*Wavelength (m)*	*Antenna Size (λ)*	*Far Field (m)*
100	3	2	24
100	3	3	54
200	1.5	2	12
200	1.5	3	27
300	1	2	8
300	1	3	18
500	0.6	3	10.8
500	0.6	4	19.2
750	.4	4	12.8
750	.4	5	20
1000	.3	5	15
3000	.1	6	7.2

At 100 MHz, the path length for even electrically small antennas is extremely large. Indoor ranges can be much more expensive than outdoor ranges. For indoor ranges I do not recommend ranges longer that 25 m unless absolutely necessary because of security or shielding requirements.

2.4.1 Indoor Far-Field Ranges

If an indoor far-field range is the chosen approach, there are two options available: the rectangular range and the tapered range. Some authors group compact ranges as far-field ranges, and to a certain degree they are, since it can be argued that the compact range will create a far-field condition at the AUT. Although the compact range is a far-field range is a true statement, I prefer to define far-field ranges as those where the two antennas are placed physically in the far field of each other.

2.4.1.1 Rectangular Anechoic Chambers

These ranges are easy to implement and allow the user to do real-time measurements of radiation patterns. They are the most popular approach to setting up an antenna range, but they are limited by the reflectivity of the lateral walls. They potentially need to be very tall and wide to achieve measurements. As explained in Chapter 5, they are limited in the lower frequency, and limited on the physical size of the AUT as the frequency increases. These two limitations are the reasons behind the development of the other ranges and measurement approaches.

2.4.1.2 Tapered Anechoic Chambers

Tapered ranges try to solve the low-frequency issues of the rectangular far-field range. Their main limitation is that they are not a free-space condition, as required for the measurement of absolute gain. Thus, they require the use of gain standards. Finding accurate gain standards at the typical frequencies of use for tapered chambers is not a trivial task. Standard gain horns, if available, are physically very large; thus dipoles are potentially the best gain standards. As detailed in Chapter 5, tapered chambers are extremely useful, particularly in the 100–700-MHz range where standard rectangular ranges become extremely physically large as covered in [12, 13].

2.4.2 Near-Field Ranges

If the far field is physically far away, the near-field measurement is potentially the most effective. In addition, because the far field is extracted mathematically, the pattern is obtained at the true far field, that is, at infinity. The issues noted in the Standard 149 [7] when doing measurements at the lower limit of the far field are not seen, and better definition is obtained for nulls and sidelobes.

Near-field systems come in different types. The simpler is the planar near field, and the most complex is the spherical near field. Recently, work has been conducted on moving away from these canonical solution approaches (i.e., planar, spherical, and cylindrical surfaces) to arbitrary surfaces that conform better to the AUT, and in this case, instead of the traditional scanner, robotic arms are being used to position the measurement probe around the AUT. (See Figure 2.11.) Near-field methods are where the newest advances are taking place in antenna measurements.

2.4.2.1 Planar Near Field

Planar near-field systems are limited to directive antennas, usually with directivities higher than 15 dBi. They do not provide a full 3-D pattern, the PNF measured the pattern around the high-gain beam and the adjacent sidelobes. If the scan plane is sufficiently large, they can be used to measure out to 70°.

Planar near-field systems are very powerful for those specific highly directive antennas, and they can be installed in relatively small areas. However, they cannot be used to measure the back lobe or front-to-back ratio or sidelobes pointing in the $\theta = 90°$ direction.

2.4.2.2 Spherical Near Field

A more complex solution, the spherical near-field approach, can be used to measure all types of antennas. These systems have been implemented both indoors and outdoors. The outdoor SNF systems are popular in measuring antennas on vehicles.

They are not trivial systems to integrate. Alignment and correction for the range antennas or probes used are critical, and like all the near-field approaches, they can

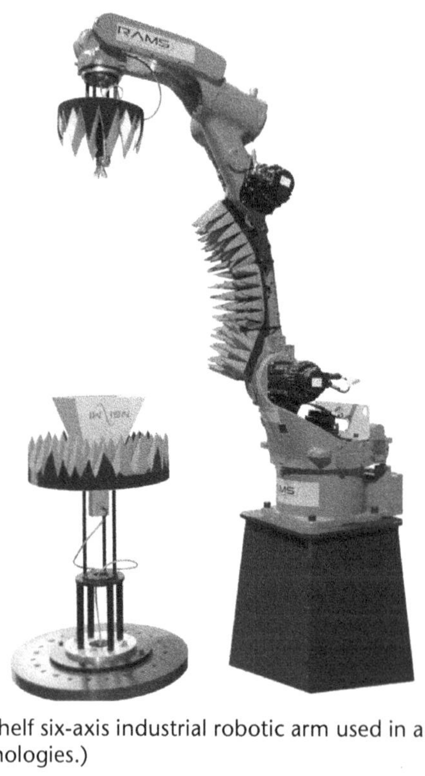

Figure 2.11 An off-the shelf six-axis industrial robotic arm used in antenna measurements. (Courtesy of NSI-MI Technologies.)

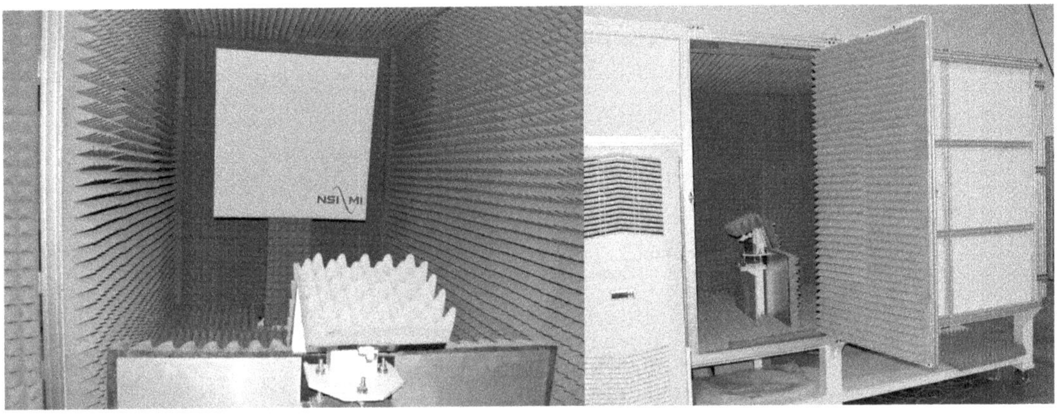

Figure 2.12 A compact range for operation in the 21- to 42-GHz frequency range for 5G testing. (Source: NSI-MI Technologies.)

be time consuming as the field must be measured at a large number of locations on a given surface prior to doing the mathematical transformation.

2.4.2.3 Cylindrical Near Field

The cylindrical near-field approach system is the least common of the near-field canonical approaches. Like the planar near field, it limits the beamwidth of the antenna on one of the principal planes. It assumes the there is little radiation in two opposite directions of the pattern. These constraints limit this approach to high-gain omnidirectional antennas and to sectoral horns.

2.4.3 Compact Ranges

Compact ranges have grown to be extremely popular since their introduction in the late 1960s. They do not require the mathematical post-processing step that the near-field systems require, and because of that, they can be faster in measuring antennas. In addition, because they reduce the overall size of a far-field range as well as the losses, these ranges have caught the attention of system designers for 5G devices. This upcoming technology utilizes large arrays that can steer high-gain beams to the required directions to establish high-gain links with the base stations. These arrays operate at frequencies in the 24–42-GHz range and need to be tested in the far field. The compact range (Figure 2.12) allows for real-time measurements on a reduced footprint with lower losses.

Compact ranges can get extremely large as the samples in Chapter 7 show, and they do not necessarily need to be indoors as shown by the Ft. Huachuca outdoor range [11]. However, they do provide a reduced footprint for testing antennas indoors, and an indoor facility provides the environmental constraints that the reflector requires to avoid deformation due to thermal effects. There are other methods to reduce the test distance by using plane waves, but the compact range is, by far, the most common.

References

[1] IEEE, IEEE STD 145-2013 IEEE Standard for Definitions of Terms for Antennas, New York: The Institute of Electrical and Electronic Engineers, 2013.

[2] Slayton, W. T., "Design and Calibration of Microwave Antenna Gain Standards," Naval Research Office Report 4433, Washington, D.C., November 9, 1954.

[3] Newell, A. C., R. C. Baird, and P. Wacker, "Accurate Measurement of Antenna Gain and Polarization at Reduced Distances by an Extrapolation Technique," *IEEE Transactions of Antennas and Propagations*, Vol. 21, No. 4, 1973, pp. 418–431.

[4] Repjar, A. G., A. C. Newell, and D. T. Tamura, "Extrapolation Range Measurements for Determining Antenna Gain and Polarization," National Bureau of Standards NBS, Technical Note 1311, United States Department of Commerce, U.S. Government Printing Office, Washington, D.C., August 1987.

[5] Fordham, J., and F. D'Agostini, "Spherical Spiral Spcanning for Automotive Antenna Measurements," in *37th Annual Antenna Measurement Techniques Association Symposium AMTA 2015*, Long Beach, CA, 2015.

[6] Hollis, J. S., T. J. Lyon, and L. Clayton, Jr., *Microwave Antenna Measurements* (Third Ed.), Atlanta, GA: Scientific Atlanta, 1985.

[7] IEEE, IEEE Std 149-1979(R2008) IEEE Standard Test Procedures for Antennas, the Institute of Electrical and Electronic Engineers, 2008.

[8] Lyon, T. J., J. S. Hollis, and T. G. Hickman, Chapter 14, "Antenna Range Desgn and Evaluation," in *Microwave Antenna Measurements*, Atlanta, GA: Scientific Atlanta, Inc., 1970, pp. 14-7–14-10.

[9] Hollis, J. S., T. J. Lyon, and L. Clayton, Jr., "Microwave Antennas Measurements," NSI-MI Technologies, [Online]. Available: https://www.nsi-mi.com/images/PDFs/Microwave_Antenna_Measurements.pdf.

[10] Stutzman, W. L., and G. A. Thiele, Equation (7-77) of Chapter 7, "Aperture Antennas," in *Antenna Theory and Design* (Second Ed.), New York: John Wiley Sons, Inc., 1998, p. 294.

[11] Francis, M. H., and R. C. Wittman, "Evaluating the Performance of the Ft. Huachuca Compact Range," in *2014 IEEE International Symposium on Antennas and Propagation and USNC-URSI National Radio Science Meeting*, Memphis, TN, July 6, 2014.

[12] Rodriguez, V., and J. Hansen, "Evaluate Antenna Measurement Methods," *Microwave and RF*, Vol. 49, No. 10, October 2010, pp. 62–67.

[13] Rodriguez, V., "On Selecting the Most Suitable Range for Antenna Measurements in the VHF-UHF Range," in *2018 IEEE Conference on Antenna Measurements and Applications*, Vasteras, Sweden, September 2018.

CHAPTER 3

Radio-Frequency Absorber

3.1 Absorber Family

The 1950s and 1960s saw the beginnings of RF absorber development, which is described in books such as *Radar Man: A Personal History of Stealth* by Edward Lovick [1]. In [1], Lovick describes the earlier materials under the brand name Echosorb by Emerson Cumming and the Hairflex brand manufactured by B. F. Goodrich. I consider myself lucky to have seen samples of these earlier absorbers. While I was not aware of their names at the time, the descriptions by Lovick clearly picture the materials I have seen in my career. The early Echosorb was an expanded polystyrene foam block, with a series of conical shaped holes on one side. The side with holes was the one placed against the wall of the range. The interior of those holes were coated with a carbon-loaded paint. I saw samples of this material at the University of Mississippi. While I do not recall the exact reflectivity level, I do remember my doctoral coadvisor, Dr. Charles E. Smith, measuring the reflectivity of a sample during an experiment as part of the advanced microwave measurements class. The other material described by Lovick [1], B. F. Goodrich's Hairflex, is also described by Gillespie [2]. Gillespie reports the reflectivity for Hairflex fabricated as 3.75 inches (9.5 cm) total height pyramidal absorbers with a pyramidal section being 2.5 inches (6.4 cm) in height and a square base measuring 3 inches by 3 inches (7.6 cm). Gillespie reports the reflectivity improving from –20 dB at 2 GHz to –30 dB at 25 GHz. Lovick describes "Hairflex" as a "tangle of graphite-coated animal hair resembling a black pot scrubber" and goes on to mention that the absorber was better performing than the Echosorb that was previously used in his range. He correctly attributes the reason for the improved reflectivity performance to its pyramidal shape. I was able to see these early animal fiber absorbers at the ETS-Lindgren factory in Durant, Oklahoma. The fibers were horsehair, and the graphite coating had rendered them stiff. Clearly the mass of hair had been placed in a pyramid-shaped mold to give it a pyramidal shape.

These two materials are the ancestors of our current RF absorbers. The substrates may have changed, but graphite or carbon black powder remains the key ingredient to provide absorption. It is a common approach in science to classify the items being studied in certain families or subclasses. The subclasses are usually defined by some common key characteristics shared by the items placed in a given class or family. Following that approach, we can classify absorber materials as members of three main families. The key characteristics for their classification are the types of losses that are being used to accomplish the absorption of electromagnetic waves. Following that, we divide the absorber technologies into three families:

- Electric losses;
- Magnetic losses;
- Hybrids.

Electric losses, or electrically lossy absorbers, are the largest family, including the bulk of the materials used in antenna measurements. In general, there is a substrate onto which the electrically lossy material is impregnated or coated. The early materials follow either of these approaches. Hairflex, with its tangle of coated fibers, is a volumetrically loaded absorber. Most of the typical absorbers used in ranges follow this approach where the substrate is loaded with the lossy substance. The early Echosorb described in [1] is a coated material. In these coated materials, a substrate supports the lossy substance on its surface.

Magnetically lossy materials comprise a smaller family. Here we have ferrite materials shaped as a thin layer, or tile or plastics, that have been loaded with ferrite powders. The high relative permeability of the material makes these absorbers physically small given the frequencies at which they operate. However, the frequency dependency of the materials is such that the ferrite materials are limited in frequency band. This limitation reduces their usability at microwave and millimeter-wave frequencies. These materials find an excellent niche in EMC measurements as they can provide adequate absorption from 20 MHz to about 2 GHz. When injected into plastics and molded as pyramidal shapes, their absorption is about –20 dB to frequencies beyond 6 GHz. These levels, however, are not enough for general antenna pattern measurements.

Hybrid absorbers, as the name indicates, describes a family that shares the characteristics of the previous two families. They provide the low-frequency absorption and reduced physical size of the magnetically lossy materials and the high-frequency performance of the electrically lossy materials. However, the resultant absorbers never achieve the absorption levels that the traditional electrically lossy materials exhibit for reasons that are explained later in this chapter.

The following sections of this chapter describe the different families in more detail and present some theory behind the operation of the different absorbers.

3.2 Electrically Lossy Absorbers

Members of this family of absorbers serve as the workhorses of antenna and RCS measurement ranges in the world. As described in [2], the 1960s and the 1970s saw growth in the availability of RF absorbers with improved electrical characteristics. It is in this period that the archetypical blue pyramidal absorber entered the market. Most of the absorbers manufactured today are not different from the ones available almost 50 years ago in the late 1960s. Because of their wide use, we devote this section of the chapter to describing the operation, manufacture, and characteristics of this family of absorbers.

3.2.1 Absorber Theory

3.2.1.1 The Reason for Pyramids

The key to RF absorber design lies in the penetration of the energy. The RF energy must enter the absorber where it will be transformed into a different kind of energy. The absorber takes the radiated energy and turns it into thermal energy, which will be dissipated into the surrounding air as heat. This concept was known to the pioneers of absorber technology for broadband operation. Lovick mentions that the original Echosorb was not ideal because the high difference in material properties between the absorber and air caused too much reflection at the surface of the material; hence, not much energy entered the material where it was absorbed. Emerson [3] describes the use of a broadband material for reducing the radar signature of periscopes developed by the Germans during WWII. This absorber comprised layers of plastic and resistive sheets where the resistance decreased at an exponential rate toward the back surface of the material. Clearly the goal of the resistance taper was to allow for penetration of the wave into the material where it was attenuated as it propagated through it.

Cutting the material into pyramidal shapes creates a similar effect as the resistive layers without having to manufacture a lot of different materials with different losses. Consider a propagating electromagnetic wave approaching a lossy material. Let us assume that that material is a loaded foam similar to the ones used in absorber manufacturing and that the complex relative permittivity is $\varepsilon_r = 1.8 + j1.7$. The conductivity of the material is given by $\sigma = \omega\varepsilon''$. Assume that the complex permittivity was measured at 5 GHz. Given the frequency and the complex permittivity, the conductivity is 0.4729 S/m. (Recall that $\varepsilon = \varepsilon_0\varepsilon_r$, where ε_0 is the permittivity of free space.) As it can be seen, this makes the radio-frequency absorber neither a good dielectric nor a good conductor since the good dielectric condition $(\sigma/\omega\varepsilon') \ll 1$ is not true; nor is the good conductor condition, $(\sigma/\omega\varepsilon') \gg 1$. Let us continue with the propagating wave. As the wave hits the boundary between the free space and the material, some of it will be transmitted and some will be reflected. The reflection coefficient is given by (3.1), whose derivation can be found in [4]:

$$\Gamma = \left(\frac{\eta_1 - \eta_o}{\eta_1 + \eta_o}\right) \tag{3.1}$$

where η_0 is the impedance of free space ($\eta_0 = 120\pi\,\Omega$), and η_1 is the wave impedance inside the absorber foam media. The impedance in the media is given by (3.2) [4] as

$$\eta_1 = \sqrt{\frac{j\omega\mu}{\sigma + j\omega\varepsilon'}} \tag{3.2}$$

Notice that we are using the exact equation and not the good dielectric or good conductor approximation. Using the values given above, the reflection coefficient can be computed. The reflection coefficient is $\Gamma = -0.2309 + j0.181$. This is a reflectivity of -10.64 dB. Notice that we are looking at propagating electric fields, so the factor 20 is used in obtaining the decibels. This reflectivity level is very poor. The

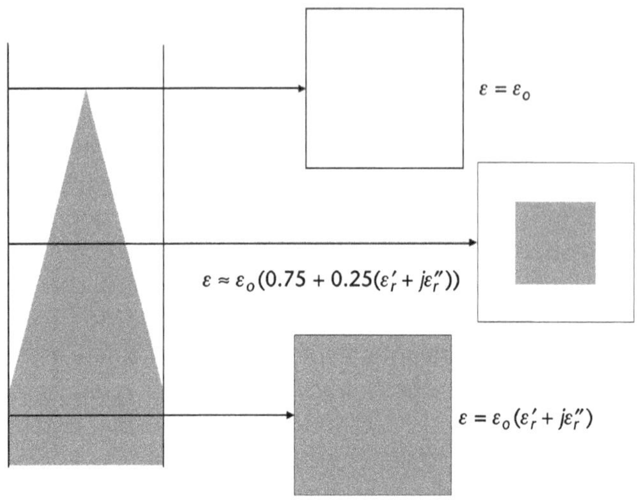

Figure 3.1 Reasoning behind the pyramidal shape as a method for obtaining an impedance taper.

pyramid shape reduces this first reflectivity and allows for more power to enter the material. As Figure 3.1 shows, as the wave enters the material, the wave impedance changes from air to an effective wave impedance that is a weighted average of the permittivity of the material and the permittivity of the surrounding free space.

Therefore, we can approximate the pyramid of the absorber as a set of multiple layers where every layer is made up of a permittivity that is between the absorber material and the surrounding air. Looking at Figure 3.1, we can do a very coarse approximation to the pyramid where we have three media layers. For simplicity we will assume that the base extends to infinity, which is a mathematical way of saying that the base is electrically very thick. Then our absorber model for this pyramid at 5 GHz becomes something similar to what we see in Figure 3.2.

This is a very rough approximation, but the purpose is not to be exact, but to help illustrate the concept of the pyramidal shape. The reflection coefficient at the input can be computed from (3.3) presented in [4]:

$$\Gamma_{\text{in}} = \frac{(1 - Z_{12})(1 + Z_{23}) + (1 + Z_{12})(1 - Z_{23})e^{-2\gamma_2 d}}{(1 + Z_{12})(1 + Z_{23}) + (1 - Z_{12})(1 - Z_{23})e^{-2\gamma_2 d}} \tag{3.3}$$

where the different media are free-space, the average of air and lossy material and the third layer is lossy material. The wave impedances of the material are given by

$$Z_{ij} = \frac{\mu_i \gamma_j}{\mu_j \gamma_i} \tag{3.4}$$

and the propagation constants are given by

$$\gamma_k = \pm\sqrt{j\omega\mu_k(\sigma_k + j\omega\varepsilon_k)} \tag{3.5}$$

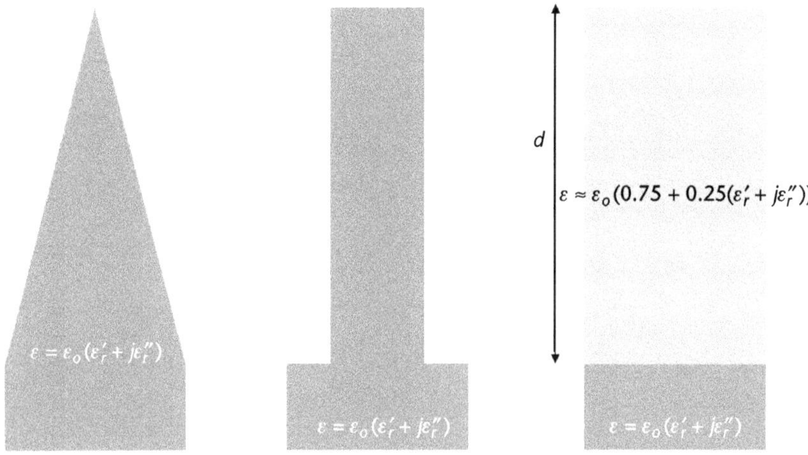

Figure 3.2 Approximation of a pyramid as a layer of average permittivity.

Let us assume that the pyramid height d is 12.7 cm (5 inches) and that the base extends for several wavelengths. The relative permittivity of the layer that approximates the pyramid is $\varepsilon_r = 1.2 + j0.425$. The compued reflectivity is now −19.7 dB—a significant improvement from the −10.64 dB that was computed for a solid layer of material. If we continue adding layers to model the pyramidal shape, the advantage of the pyramidal shape becomes more apparent.

The input reflection coefficient for multiple layers of equal thicknesses, sandwiched between two semi-infinite spaces can be approximated, provided that the reflection between layers is small, by (3.6) [4]:

$$\Gamma_{in} \cong \Gamma_0 + \Gamma_1 e^{-2\gamma_1 d} + \Gamma_2 e^{-2d(\gamma_1+\gamma_2)} + \ldots + \Gamma_N e^{-2d(\gamma_1+\gamma_2+\ldots+\gamma_N)} \tag{3.6}$$

where each reflection coefficient for each layer is given by

$$\Gamma_n = \left(\frac{\eta_{n+1} - \eta_n}{\eta_{n+1} + \eta_n}\right) \tag{3.7}$$

For $n = 0$, $\eta_0 = 120\pi\,\Omega$, and the reflection coefficient of the last layer, Γ_N is given by

$$\Gamma_N = \left(\frac{\eta_L - \eta_N}{\eta_L + \eta_N}\right) \tag{3.8}$$

where η_L is the wave impedance of the foam material, in our case, $\varepsilon_r = 1.8 + j1.7$. Let us split the 12.7-cm pyramid into five layers. Let us also assume that the base of the pyramid is 5.08 cm (2 inches). Let us now approximate the materials for each of the layers of this approximated pyramid. Figure 3.3 shows how the pyramidal shape is approximated. Table 3.1 provides the values of epsilon for the different layers.

We can apply these values to get the reflection coefficient at each layer, based on the wave impedances that can be computed from these material properties. Then,

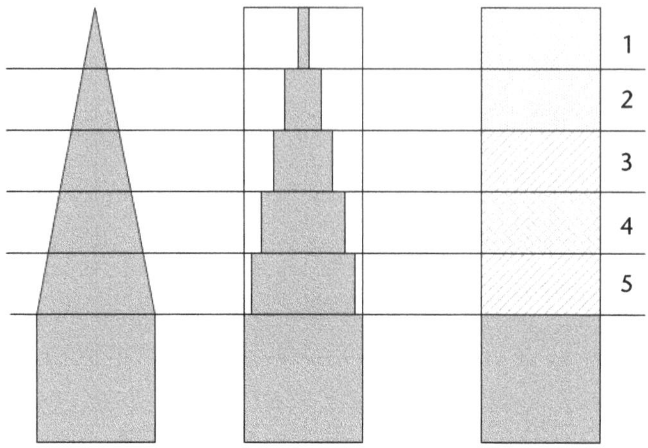

Figure 3.3 Approximation of the pyramidal shape as a set of effective layer permittivities.

we perform the summation using the complex propagation constant γ for each layer and arrive at a reflectivity of –22 dB. Clearly the approximation in Figure 3.3 is closer to a pyramid than the approximation in Figure 3.2. Appendix 3A provides a MATLAB® script for the reader to try different numbers of layers. It should be noted that as the number of layers increases, the reflectivity approaches a limit, and increasing the layers does not change the reflectivity. Figure 3.4 shows the reflectivity plotted against the inverse of the number of layers chosen in the calculation. If we compare the –58-dB reflectivity for the pyramid-shaped piece of material with $\varepsilon_r = 1.8 + j1.7$ to the –10.64-dB reflectivity of a flat slab of the same material, it is clear that the pyramidal shape provides a much reduced reflectivity compared to the flat slab.

It should be noted from the results of Figure 3.4 that to achieve a reflectivity better than –50 dB, a sandwich of at least 10 layers is required, with each of those layers having different material properties. The manufacturing cost of such an absorber is very high. As discussed in Section 3.2.3, there is a type of flat absorber made up of layers, usually three, that can be used in certain applications. As expected, its reflectivity does not match the levels of pyramid-shaped pieces.

Table 3.1 Material Properties for the Different Layers

Layer	*Real Relative Permittivity*	*Imaginary Relative Permittivity*
1	1.008	0.0170
2	1.072	0.153
3	1.2	0.425
4	1.392	0.833
5	1.648	1.377
base	1.8	1.7

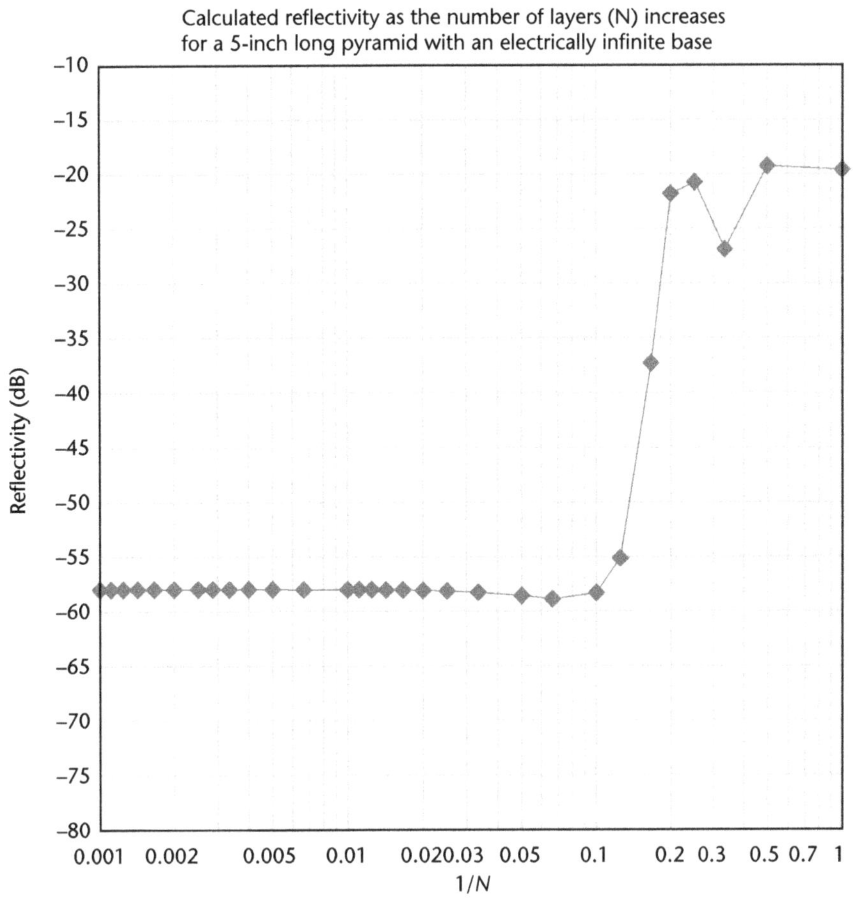

Figure 3.4 Reflectivity of *N* layers used to model a pyramidal shape.

Imagine now the permittivity of these levels being optimized to achieve a reduced overall reflectivity and to minimize the reflectivity between the layers. This would yield a different set of effective permittivity in each layer. In order to get the different effective permittivity at each layer, the average between air and loaded foam has to be changed leading to something that is no longer pyramidal in shape but has a more complex curve. These curvilinear absorbers are available on the market. At the lower end of their operational frequency range, these absorbers exhibit improved performance compared to standard pyramidal absorbers of similar overall dimensions.

3.2.1.2 Different Sizes, Different Loadings

If we look at the any manufacturer of pyramidal absorbers, we notice that there is a close to constant ratio for the base of the pyramid and the height. This height-to-base ratio is about 2.5. This ratio is mainly for mechanical reasons. With the exception of some more rigid substrates like polyethylene and expanded polystyrene, most absorbers use a flexible polyurethane foam for substrate. This flexible foam requires a minimum thickness for it to hold its shape when mounted horizontally as is the case with the walls of an anechoic range. Some manufacturers build longer pyramids using what is called a twisted pyramid shape. These shapes usually have a slightly reduced reflectivity when compared to standard pyramids [5]. In general, the longer the pyramid, the wider the base. These thicker absorber pyramids will have an effect on the carbon-loading and permittivity of the foam.

Again, the key to absorber design is penetration. A highly loaded foam will have a higher conductivity, and a higher conductivity implies a smaller penetration. That is, the skin depth is smaller. We define the skin depth as

$$\delta = \frac{1}{\alpha} = \frac{1}{\omega\sqrt{\mu\varepsilon}\left(\frac{1}{2}\left(\sqrt{1+\left(\frac{\sigma}{\omega\varepsilon}\right)^2}\right)-1\right)^{\frac{1}{2}}} \tag{3.9}$$

Figure 3.5 shows the skin depth for media made up of the typical complex permittivity used in manufacturing polyurethane absorbers that are 12 inches (30.5 cm) in height. Figure 3.5 also shows the skin depth for the material used for manufacturing pyramids that are 72 inches (1.83m) in height. The base of a pyramid in a piece of 12-inch absorber is usually 4 inches by 4 inches (10.16 cm), while for a 72-inch piece of absorber, the base is 24 inches by 24 inches (61 cm). A piece of 72-inch absorber manufactured out of the same foam used in making

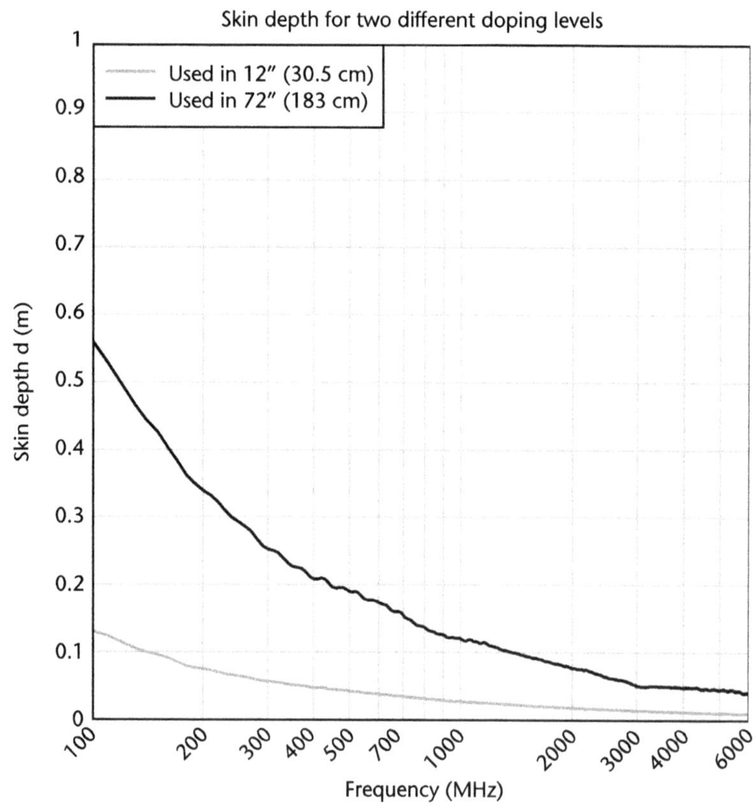

Figure 3.5 Skin depth versus frequency for two different loaded polyurethne foam types.

a 12-inch absorber will not have a better absorption even if we have more losses. This is related to having lower levels of energy penetrating the lossy medium. This is the mechanism behind the optimal loading described by Hemming [6]. In addition, in examining the real part of the permittivity, one observes that the higher the loading of carbon back, the higher the permittivity at a given frequency, as shown in Figure 3.6 and previously reported in the literature [7]. Based on the skin depth shown in Figure 3.5, we see that at 100 MHz for a 72-inch pyramid, little energy will enter more than 10 cm into the foam. For such a physically large pyramid, this is a lot of unused lossy material.

3.2.1.3 Approximations for Pyramidal Electrically Lossy Absorbers

We have so far considered the reasons for shaping the electrically lossy RF absorber into pyramids and the loading variations for different sizes. It should be clear that the pyramid shape optimizes the performance for a given angle of incidence. To be more precise, the angle of incidence should be such that, the wave propagation is normal to the absorber wall. As it will be shown in subsequent chapters, for range design, we are interested in incidence at oblique angles as shown in Figure 3.7. For these angles, where $\theta > 0°$, the reflectivity is higher, given that the component of the wave traveling parallel to the absorber wall (the positive x direction in Figure 3.7)

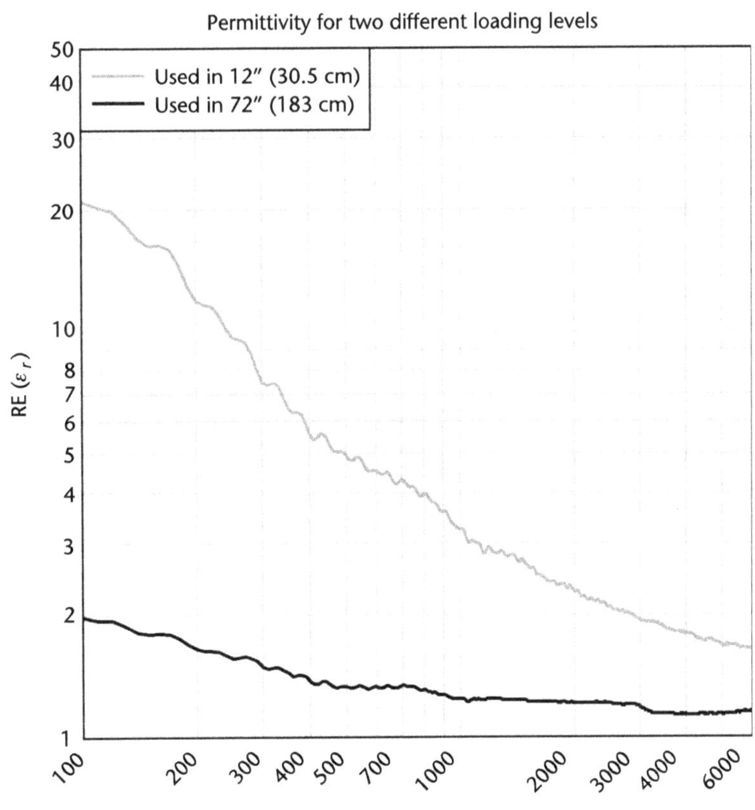

Figure 3.6 Real part of the complex permittivity of two different polyurethane loaded foams.

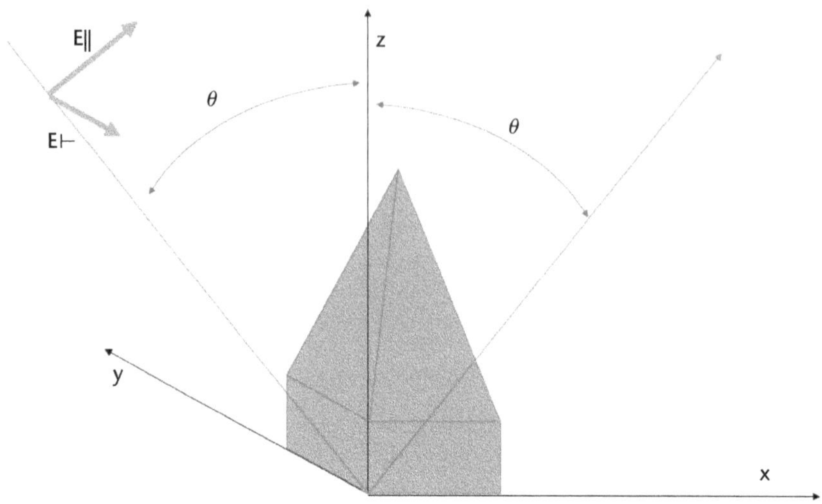

Figure 3.7 Off-normal incidence and the two polarizations of the incident field.

is not being attenuated since it does not travel into the absorber. Additionally, there will be a difference between the two possible polarizations of the electric field as shown in Chapter 1.

Despite its importance, it is rare to find oblique incidence performance of pyramidal absorbers in manufacturer's specifications, the reason being that it is very costly to perform these measurements. Furthermore, the frequency band at which the measurement is performed is lower as the area required for the absorber sample becomes larger and larger. The antennas used to make the measurement in the VHF band will have very broad beams, or otherwise be physically extremely large. At UHF, the situation does not improve much. Another issue that appears when performing oblique incidence measurements is that it becomes difficult to minimize the coupling between the antennas used to measure the reflectivity of the sample. Figure 3.8 shows the direct path between the measurement antennas in this oblique incidence measurement. The direct coupling path and the reflected path are very close in length, making it nearly impossible to use time-gating to cut out the direct path.

The best approach is to use a numerical method to simulate the pyramidal absorber. A series of simulations can be performed using commercially available packages, and periodic boundaries or the approach described by Holloway et al. [5] can be relatively easily programmed. Simulating a wide variety of pyramidal sizes and foam loadings produces a lot of data. This data can then be translated to reflectivity at a given angle for a given electrical size of absorber, that is, the size of the absorber in wavelengths. For the different electrical sizes, we can choose the worst-case reflectivity. The result of these manipulations is a set of curves that shows the reflectivity of the pyramidal absorber versus the angle of incidence for different electrical sizes of pyramids (Figure 3.9). This approach, presented in [7], is extremely useful since it provides an estimated worst-case reflectivity for pyramidal

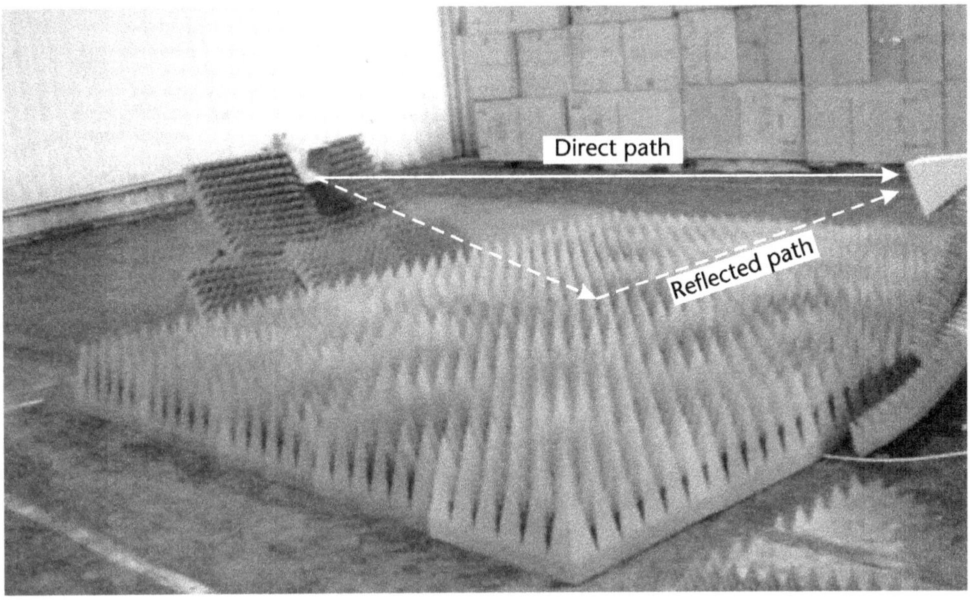

Figure 3.8 Oblique incidence measurement setup. Notice that the direct and reflected paths are very close in length. (*Source:* Author's private collection.)

absorbers of different electrical sizes at different angles of incidence. From this data, a series of coefficients and polynomials to estimate the reflectivity of the absorber at an oblique incidence were developed for typical polyurethane-based foams. The so-called Rodriguez's polynomials are given by the following series of equations.

The general equation for reflectivity is as follows:

$$R(\text{dB}) = C_0(t) + C_1(t)\theta + C_2(t)\theta^2 + C_3(t)\theta^3 + C_4(t)\theta^4 + C_5(t)\theta^5 \tag{3.10}$$

where t is the thickness of the material in wavelengths. For normal incidence, when $\theta = 0°$, the independent coefficient is given by (3.11), which represents an approximation to normal incidence reflectivity:

$$C_0(t) = -54.98 + 55.21e^{-0.7744t} \tag{3.11}$$

The other coefficients of (3.10) are given by polynomials of the seventh order. In general, the coefficients have the following form:

$$C_n = A_0 + A_1 t + A_2 t^2 + A_3 t^3 + A_4 t^4 + A_5 t^5 + A_6 t^6 + A_7 t^7 \tag{3.12}$$

There are two sets of coefficients A_n for (3.12). One set is used for cases where the thickness of the absorber is smaller than 1.8λ but larger than 0.25λ. The other set is for cases where the absorber is larger than 1.8λ. Once the absorber is longer than 18λ, the performance is calculated as if the size seventh-order polynomial in (3.12) is that it can describe all the coefficients for (3.10). There is, however, one

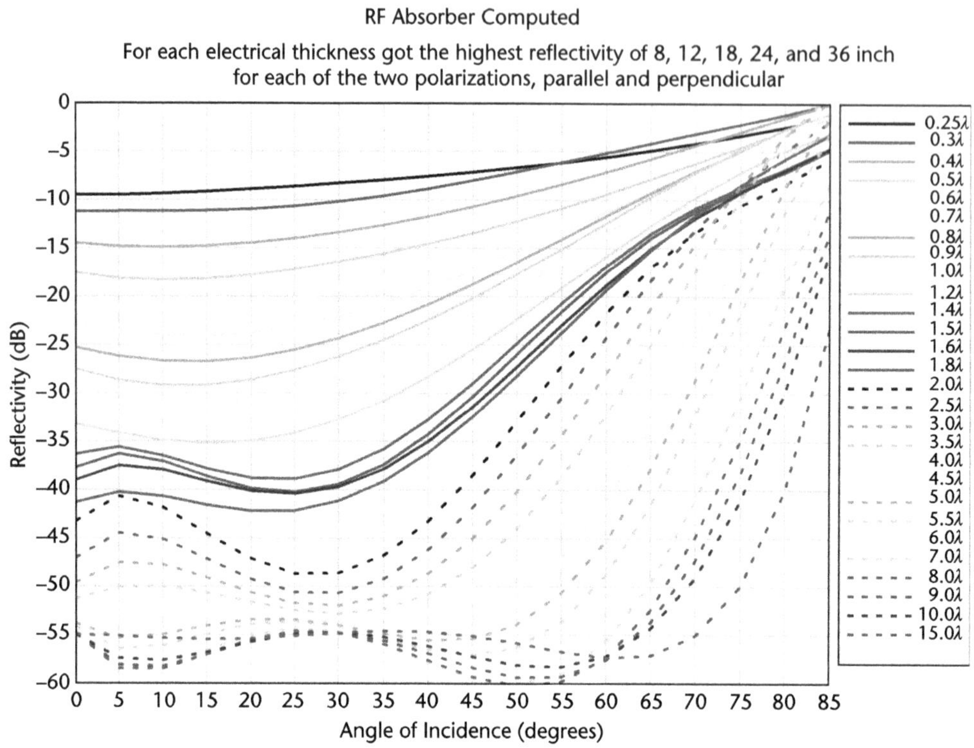

Figure 3.9 Worst-case reflectivity of pyramidal material versus angle for different pyramid heights.

disadvantage: that a lot of significant figures are required to get the correct coefficient and solution. Tables 3.2 and 3.3 show the coefficients A_n for (3.12).

Other approaches for estimating the oblique incidence reflectivity are given in [6], and some manufacturers provide some information based on those approaches. The approach based on the polynomials was shown to provide a good estimate compared to computed results and to manufacturer estimates [7]. Additionally, this approach has been shown to be able to predict anechoic range performance [8].

3.2.2 Absorber Manufacturing

The main substrate has moved from the animal fibers of the Hairflex to polyurethane foam, which is used by most the manufacturers worldwide. The majority of these absorbers are volumetrically loaded materials. The typical process involves the dipping of raw polyurethane blocks into vats of aqueous solutions of latex and carbon black. Carbon black is any finely divided form of carbon that it is not crystalline-like graphite. The blocks of polyurethane are dried and weighed. The difference in weight is an indication of the amount of carbon and latex solids that have been added to the foam. A second dipping into a salt solution for limiting the flammability of the absorber usually follows. After drying, the large foam block is cut into the desired pyramidal shape. Other manufacturers cut the raw polyurethane into the pyramidal shape and then proceed to the impregnation.

Table 3.2 Coefficients for (3.12) for $0.5\lambda \leq t \leq 1.8\lambda$

	C_1	C_2	C_3	C_4	C_5
A_0	-6.3210690199E+00	7.6880225853E-01	−2.6912146493E-02	3.5113868215E-04	−1.5529824891E-06
A_1	7.2267938615E+01	−8.6597762712E+00	3.0168480772E-01	−3.9373084837E-03	1.7444914212E-05
A_2	−3.2278068371E+02	3.8364405955E+01	−1.3300774967E+00	1.7369236066E-02	−7.7111307085E-05
A_3	7.3099338658E+02	−8.6725914370E+01	2.9993264140E+00	−3.9206538788E-02	1.7441418132E-04
A_4	−9.1764369957E+02	1.0884507395E+02	−3.7572593998E+00	4.9162133561E-02	−2.1912441417E-04
A_5	6.4244485883E+02	−7.6286796335E+01	2.6301952632E+00	−3.4448414528E-02	1.5380783527E-04
A_6	−2.3361509236E+02	2.7811126567E+01	−9.5847779882E-01	1.2566670885E-02	−5.6195960853E-05
A_7	3.4280598868E+01	−4.0964212067E+00	1.4122528124E-01	−1.8537524098E-03	8.3014688640E-06

Table 3.3 Coefficients for (3.12) for $1.8\lambda < t \leq 1.8\lambda$

	C_1	C_2	C_3	C_4	C_5
A_0	−1.391039779E+00	2.033775097E-02	7.631876394E-04	−2.420064589E-05	1.626120944E-07
A_1	3.070364757E+00	−2.454851890E-01	7.308066068E-03	−8.179930282E-05	3.076624295E-07
A_2	−1.233311882E+00	1.170747811E-01	−3.857386866E-03	4.793188957E-05	−2.013685407E-07
A_3	1.773626849E-01	−1.862915801E-02	6.463419255E-04	−8.335949232E-06	3.612733560E-08
A_4	−1.001908712E-02	1.227464256E-03	−4.527085011E-05	5.985008184E-07	−2.623393236E-09
A_5	4.917586556E-05	−2.285563716E-05	1.051384196E-06	−1.455306267E-08	6.344426436E-11
A_6	1.507549767E-05	−8.659380032E-07	2.005956450E-08	−2.356371032E-10	1.097898307E-12
A_7	−3.956411780E-07	3.079188422E-08	−9.254596248E-10	1.181444440E-11	−5.354257560E-14

Painting usually follows the cutting of the material. Traditionally the absorber was painted a shade of blue. The reason for the blue color was a purely aesthetic one. The polyurethane open-cell foam was difficult to paint. Once painted, the black open cells make the color a shade darker. Whites look like grays, red like burgundy, and yellows like a light brown. Blue used to be the most eye-pleasing color. Nowadays, more flashy colors are available in rubberized coatings that make the typical polyurethane material less susceptible to damage, as shown in Figure 3.10. These thicker paints do not appear to have a significant effect on the absorption of the pyramidal absorber. The National Institute of Standards and Technology (NIST) conducted a series of experiments on the reflectivity of these heavily latex-coated pyramidal absorbers and found no degradation in performance for an 8-inch pyramidal absorber in the 5–15-GHz range [9]. A more recent study by Zhong [10] showed the effects of paint on the reflectivity at the W-band (75–110 GHz). It showed that the effects of the paint may depend on the loading of the absorber; it also showed little difference between black tips and painted absorber, but the results showed that at those bands, it is better to have a fully unpainted absorber.

Other manufacturers use different substrates, including polyethylene. More rigid than polyurethane foams, polyethylene is a hard foam that does not break apart with repeated friction from handling or rubbing against it. This rigid foam seems to have some limitations regarding power-handling, as discussed in Section 3.5. It should be noted that while rigidity is seen as a positive characteristic for absorbers versus flexible substrates, there are situations where rigidity is not a positive attribute. A rigid absorber is more difficult to conform to curved surfaces like those of positioners. Also, in some ranges, like tapered ranges, the flexible materials are easier to work with when lining critical areas that have curved surfaces.

Another popular material is expanded polystyrene. Cutting pieces of both polyurethane foam and polyethylene foam absorbers, and performing a visual inspection, shows a uniform loading of the carbon black, the lossy material added to the foam. Expanded polystyrene materials show the white interior of the substrate

Figure 3.10 Rubber-like painted absorber showing its flexibility.

beads surrounded by a coating of the lossy material. This may reduce the losses and absorption at high frequencies (millimeter-wave range) as the density of losses per wavelength is much lower than in other substrates. Recent measured data for polystyrene absorbers have shown that the performance up to 18 GHz is comparable to that of the traditional polyurethane absorbers, at least for 18-inch-tall pyramidal material. Figure 3.11 shows examples of cuts of polystyrene- and polyurethane-based absorbers.

Expanded polystyrene foam absorbers are rigid and impervious to humidity. Therefore, polystyrene is a popular material in hybrid absorbers, as discussed in Section 3.4.

There are other types of substrates used because of their mechanical properties and ability to have air flowing through the absorber. These are discussed in Section 3.5.

3.2.3 Types of Electrically Lossy Absorbers

Section 3.2.2 describes different types of electrically lossy absorbers. The classification is based on the substrate material that is used for the manufacturing of the absorber. This section examines different types of electrically lossy absorbers but uses the shape as the classifying parameter, expanding our scope beyond the traditional pyramid-shaped absorbers.

3.2.3.1 Flat-Laminate Absorber

Another type of electrically lossy absorber is the flat-laminate type, a sandwich of, usually, three layers of foam. These layers, which go from a lightly loaded one to a heavily loaded one, effectively create a stepped pyramid like the ones shown in Figure 3.3. Their absorption is limited to about 20 dB. At some frequencies phase cancelation provides a higher level of absorption. These absorbers are a good choice for smaller boxes, masking positioners, and other structures present in the chamber. One manufacturing technique is to make the bases of pyramidal absorbers out of layers to try to improve the lower frequency performance for a given overall size.

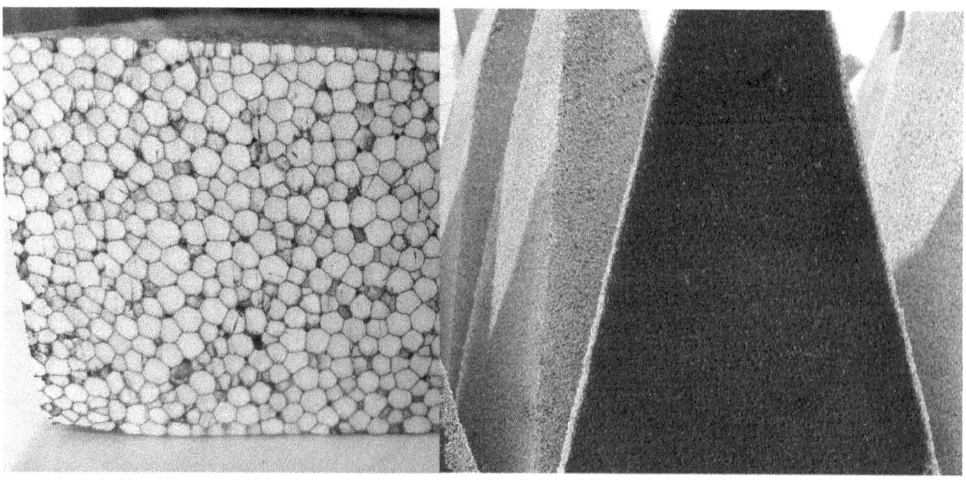

Figure 3.11 Cuts of polystyrene-based absorbers and polyurethane-based absorbers.

3.2.3.2 Lossy Foam Block

Flat laminate should not be confused with lossy foam block (LFB). As the name suggests, LFB is a block of foam that has been loaded. Commonly, it is used for corner treatments in chambers or to treat the access doors at the location of the door handle. Users should be wary of LFB, as manufacturers usually do not track the loading of the foam from which it was cut. Additionally, there are no performance requirements, so the foam is not tested. If the need arises for range technicians to use LFB, I suggest that they request from the manufacturer a lightly loaded foam (with lower permittivity where there is a lower reflection from the surface of the material). For areas of the range that may be critical, flat laminate is a better choice.

3.2.3.3 Convoluted Absorber

Convoluted absorber can be classified in the same group as LFB and Flat-Laminate absorber. (See Figure 3.12.) Convoluted foam is well known in the furniture manufacturing world. Also known as egg-crate foam, it has multiple uses, including as a mattress topper to improve comfort. At some point, somebody decided that this polyurethane foam should be loaded and used as absorber. It is difficult to model numerically because of its complex shape. Because of the physically smaller sizes in which the raw foam material is available [typically between 1.5 inches (3.8 cm) and 4 inches (10.2cm)], it is used in the upper microwave and the millimeter-wave range. Convoluted foam is a good option for covering positioners due to its flexibility, and it is slightly better in performance than flat-laminate types.

3.2.3.4 Walk-on Absorber

At some point in the operation of a range, it is necessary for the range operator to enter the range and approach the range antenna, the AUT positioner, or some other equipment that resides inside the range. Clearly, it is not possible to walk on top of

Figure 3.12 Convoluted absorber.

pyramidal foam pieces, and even more rigid substrates are liable to break under the weight of an adult human. For that reason, a walk-on absorber was created. For many years, walk-on absorber was made by taking a regular pyramid and nesting it into a piece of expanded polystyrene foam. This approach creates something similar to some of the absorber described by Lovick [1]. In addition to the polystyrene, a higher density polystyrene layer was added to increase the rigidity of the block. Finally, it was capped by a thick layer of rubber, as shown in Figure 3.13.

All these additional layers of materials added to the reflectivity of the absorber. The relative dielectric constant of rubber is listed as high as 3.0 [4]. Adding that layer of rubber on top of the pyramids can increase the reflectivity by up to 20 or 25 dB compared to the original pyramidal absorber. The MATLAB script used to generate Figure 3.4 can be used, forcing the first layer out of a total of 20 layers (hence, the top layer is 0.3175 cm) to be made out of rubber (assuming no losses on the rubber). The reflectivity changes from −58 dB to −10.6 dB. That extreme example serves to illustrate the problem with a walk-on absorber.

That said, walk-on absorber is a necessary evil inside chambers. Fortunately, it has improved recently. Instead of the heavy rubber layer, new materials with honeycomb structures with lower effective permittivity are being used to provide the rigidity required for an adult human to stand on top of the material. (See Figure 3.14.) Nevertheless, care should be taken to minimize the use of walk-on absorber and to restrict it to noncritical areas of the chamber.

3.2.3.5 Wedge Absorber

Wedge absorbers, which represent a special type of pyramidal absorber, are used to treat specific areas of anechoic ranges where backscattering may be a problem.

If we think of typical pyramidal absorber, it has a given height and a square base. If we take the base and start to make it rectangular in shape and continue

Figure 3.13 Top of a classical walk-on absorber.

Figure 3.14 A modern walk-on RF absorber.

to reduce the length of the smaller side of the rectangular base, we end up with a wedge when the smaller side of the base becomes zero. Figure 3.15 illustrates this transformation from pyramid to wedge.

The reasons for turning pyramids into wedges can be found in the literature. In [11], there is an analysis of periodic structures. In general, as long as the spatial period (D) for the pyramids—that is, the distance from tip to tip—is such that $D < \lambda$, the reflected wave will travel in the bistatic direction. That is the vector pointing

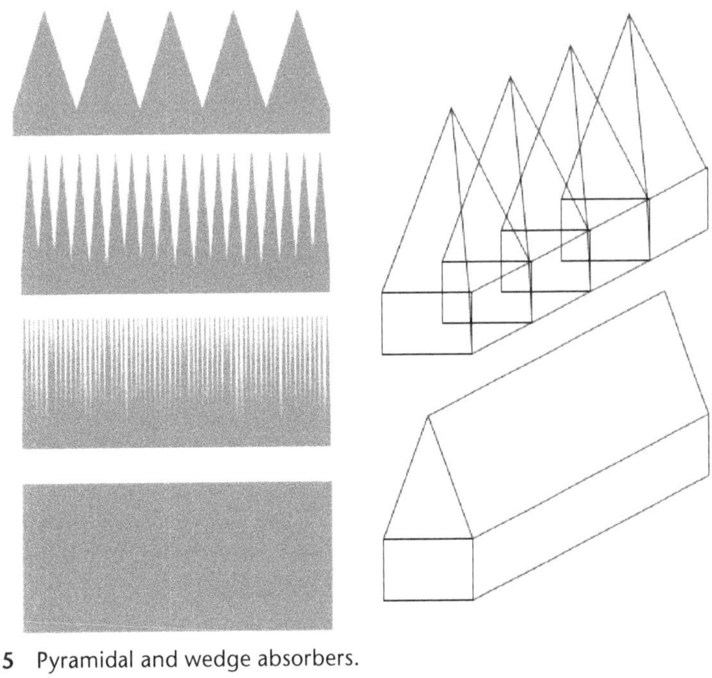

Figure 3.15 Pyramidal and wedge absorbers.

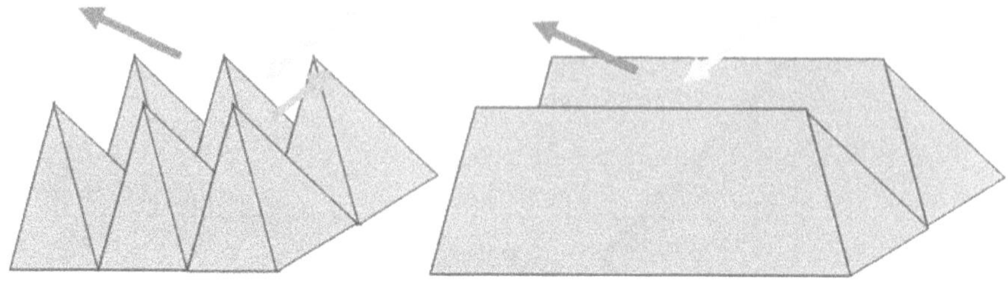

Figure 3.16 Backscattering in pyramidal absorbers and wedges.

away from the foam and opposite the direction of the incoming vector in Figure 3.16. This assumes a very large area of absorbers, ideally infinite. Once $D > \lambda$, then other possible modes will propagate and some of the reflected waves will travel back toward the origin in the incident field shown as the smaller vector in Figure 3.16. By shaping the absorber as a wedge, we have reduced D to zero in one of the dimensions. This means that for $D > \lambda$ to be true, the wavelength must be smaller than zero. Since this cannot happen, then there is no backscattering from the wedge absorber as long as we are incident onto the direction of the wedge.

Wedge absorbers are used to treat specific areas of anechoic ranges where the backscattering may be a problem.

3.3 Magnetically Lossy Absorbers

Magnetically lossy absorbers are based on the magnetic losses of the materials. These materials have high relative permeability; however, their permeability is not on the order of several thousand as in ferromagnetic materials. Hence these substances fall in the class of ferrimagnetic materials [4]. Ferrites, a type of ferrimagnetic material, have low electric conductivities and a complex permeability. Ferrites, commonly used in the manufacturing of microwave components such as isolators and circulators, are complex materials that are very frequency-dependent. Ferrites are ceramic materials that are manufactured by heating powders under pressure at 1,000 to 1,500°C. Because they are ceramics, ferrites are brittle; they are also heavy since they are made of iron, manganese, and other metal oxides. Their applications are known in many areas. For the purpose of RF absorbers, ferrites are usually manufactured in thin tiles, between 4 mm and 7 mm. Ferrite powders have also been used suspended into other materials such as silicon rubbers and plastics.

3.3.1 Ferrite Tiles

Ferrite tiles are predominantly a 10-MHz to 1.5-GHz absorber. This is in part related to the frequency dependency of ferrite materials [4]. Ferrite tiles are to be used with a metal backing, where the reflection from the bottom of the ferrite tile combines with the reflection from the air to ferrite boundary to help in the reflectivity of the tile. As discussed in [5], analysis of ferrite tile absorber is straightforward using the equations for the reflectivity of a slab. In this case, the slabs are free-space, the

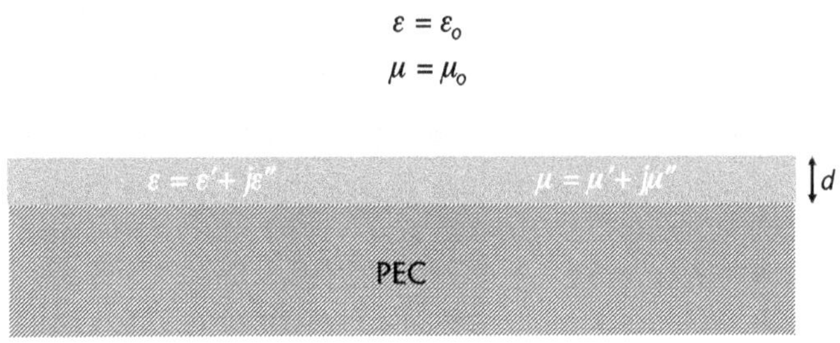

Figure 3.17 Ferrite tile model.

ferrite tile, and the metal backing. Figure 3.17 shows the geometry of a ferrite tile of thickness d over a metal backing simplified as PEC.

The reflection coefficient at the air-to-ferrite boundary can be obtained from

$$\Gamma_{in} = \frac{\Gamma_{12} + \Gamma_{23}e^{j2\gamma_2 d}}{1 + \Gamma_{12}\Gamma_{23}e^{j2\gamma_2 d}} \tag{3.13}$$

where Γ_{12} is the intrinsic reflection coefficient between the air (or vacuum to be more precise) and the ferrite material, and Γ_{23} is the reflection coefficient between the tile and the PEC backing, which happens to be −1. The propagation constant γ_2 is the propagation in the ferrite, which can be computed from

$$\gamma_2 = \omega\sqrt{\varepsilon\mu} \tag{3.14}$$

where ε and μ are the permittivity and permeability of the ferrite. If we take the values for the ferrite listed in [5] for 30 MHz we have that $\varepsilon = 10.88\varepsilon_0 + j0.16\varepsilon_0$ and $\mu = 52.31\mu_0 + j236.17\mu_0$. The reflectivity at the top tile boundary is $\Gamma_{in} = -0.0138 - j0.1024$, which, expressed in decibels, is a level of −19.717 dB for a tile thickness of 6.4 mm. Appendix 3A provides a Matlab script that performs this calculation. The script provides the reflectivity at different frequencies and plots them versus frequency. Figure 3.18 shows the results for the reflectivity of a tile at normal incidence. It should be noted that this thin tile provides reflectivity levels better than 20 dB from 30 MHz to 400 MHz. However, the material properties change with frequency, and the reflectivity at 1 GHz is reduced to −13 dB. At 2 GHz most ferrite tile is as reflective as the metal backing behind the tile. It should also be noted that ferrite tile must have a metallic backing for it to operate correctly.

If we analyze the tile replacing the PEC layer shown in Figure 3.17 with vacuum, then the reflectivity changes. Γ_{23} is not −1 as in the PEC-backed case, but its value is given by

$$\Gamma_{23} = \left(\frac{\eta_0 - \eta_{\text{ferrite}}}{\eta_0 + \eta_{\text{ferrite}}}\right) = -\Gamma_{12} \tag{3.15}$$

The reflectivity of this tile of thickness d placed between two infinite half-spaces of air (or vacuums to be precise) is −9.76 dB at 30 MHz and only −7.03 dB at 1 GHz. Figure 3.18 shows how much poorer the reflectivity of the ferrite tile is without the metal backing. Sometimes the tiles are fixed to a piece of particle board and then fixed to the wall of the chamber. The thickness of the wood backing is chosen to provide a broader band performance of the tile compared to when it is fixed directly to the metal. In addition, gluing the tiles to the backing board (which is usually about 60 cm by 60 cm in area) reduces the time of installation on site. Typical tiles are manufactured in sizes of 10 cm by 10 cm so their installation can be very time-consuming.

Another type of tile is the so-called waffle or grid ferrite. The presence of square holes on the tile create an effective permeability and permittivity. Ferrite grid tends to have less of a null compared to the ferrite tile and better performance at 1 GHz.

3.3.2 Ferrite "Cones"

As mentioned at the beginning of Section 3.3, there are also plastic conical and pyramidal shapes that have been loaded with ferrite powders. Mostly manufactured in Japan, these materials, which appeared at the turn of the century, have to be used on top of ferrite tiles. They do not provide any advantage at lower frequencies without the ferrite tile underneath. Additionally, the ferrite tile must be placed on top of a metal backing. These requirements make these ferrite cones, as they are frequently called, a high-cost solution. They are mainly used in EMC chambers. Measuring their reflectivity shows that in the microwave range they are limited to between −30 and −25 dB of normal incidence reflectivity, as shown in Figure 3.19.

3.4 Hybrid Absorbers

Hybrid absorbers hit the market in force in the last decade of the twentieth century. However, their idea is much older; in fact, a U.S. patent was awarded to Kunihiro Suetake in 1971 [12]. Hybrid absorbers are basically an attempt to merge the

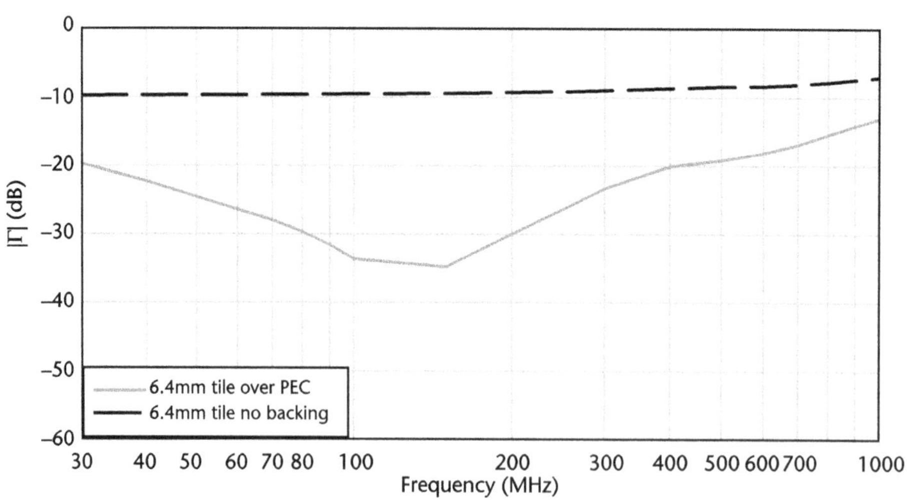

Figure 3.18 Computed results for ferrite tile with and without metal backing.

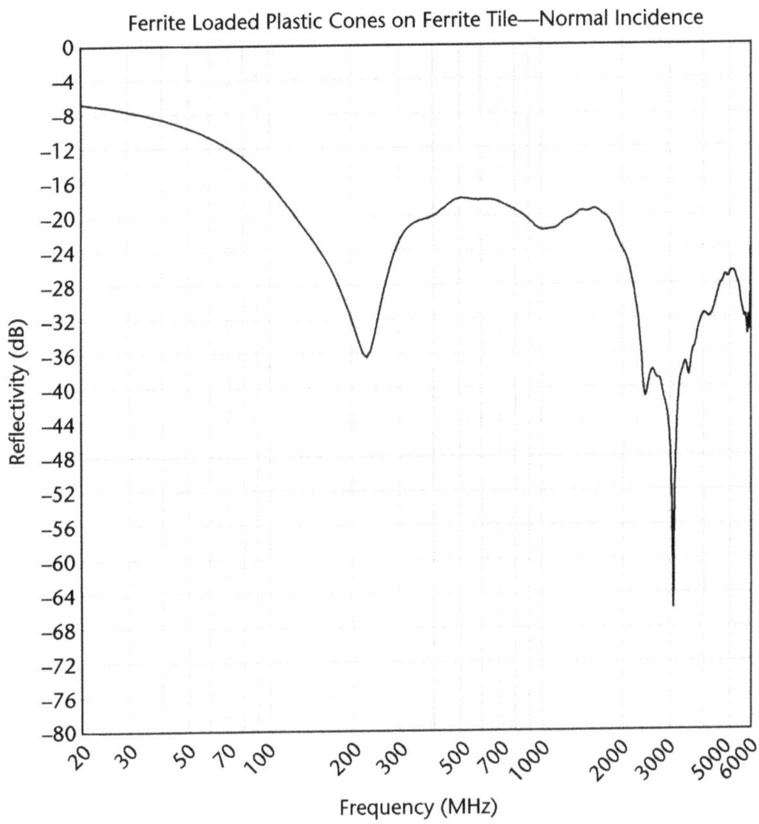

Figure 3.19 Measured data for a type of ferrite powder-injected plastic pyramid.

high-frequency performance of electrically lossy absorbers with the low-frequency performance of magnetically lossy absorbers.

3.4.1 The Mismatch Issue

The idea behind hybrid absorbers appears to be very simple: take a piece of pyramidal electrically lossy absorber and place it on top of ferrite. Unfortunately, that will not work. As the wave penetrates the lossy material of the pyramidal absorber, it travels in a medium with a wave impedance that is different from the wave impedance of air. The wave impedance is also different from the wave impedance inside the ferrite tile. This causes the wave to reflect at the interface of the electrical absorber and the ferrite tile, and effectively it hides the tile from the wave. Hence, no benefit is achieved from the ferrite tile at the base of the electrically lossy absorber.

Figure 3.20 shows computed data presented in [13] that illustrates the issue in matching the parts that make up hybrid absorbers. The computed data shows the reflectivity of ferrite tile over metal backing and the reflectivity of an 18-inch pyramidal absorber over metal backing. When the 18-inch pyramidal absorber is placed over the ferrite tile over metal, we notice that reflectivity of this new "hybrid" absorber leaves much to be desired. Clearly the ferrite has not added anything to the absorption, except for changing the phase of the reflected wave from the surface of the ferrite and causing the null at 200 MHz to shift to 290 MHz.

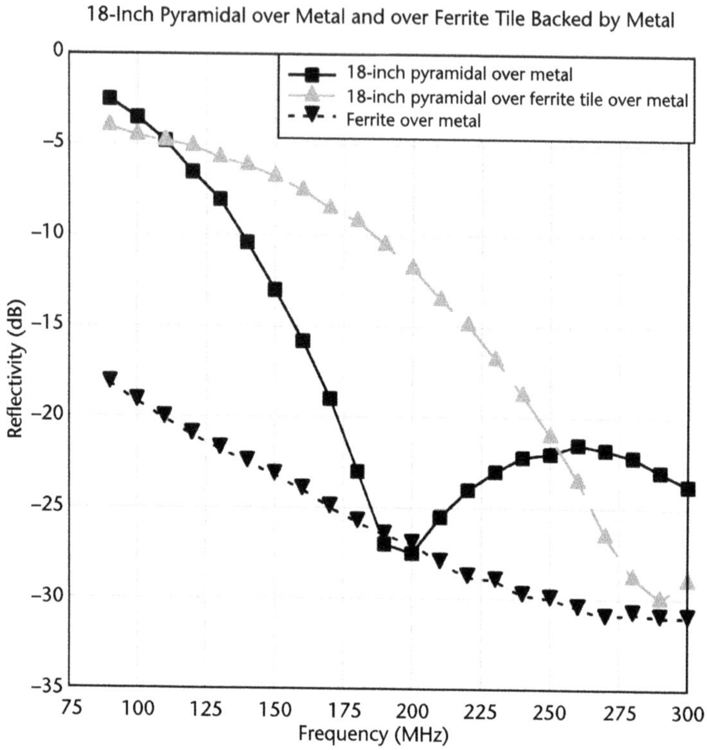

Figure 3.20 Computed effects of placing highly loaded polyurethane foam on ferrite.

Measurements of absorbers on metal and on ferrite tile confirm that no advantage is apparent when placing traditional electrically lossy absorber onto ferrite tile. In Figure 3.21, the reflectivity at 200 MHz for the absorber on ferrite tile is higher than –20 dB; the same absorber on metal, measured better than –20 dB at 200 MHz. Only above 500 MHz is the response of the absorber the same, but the

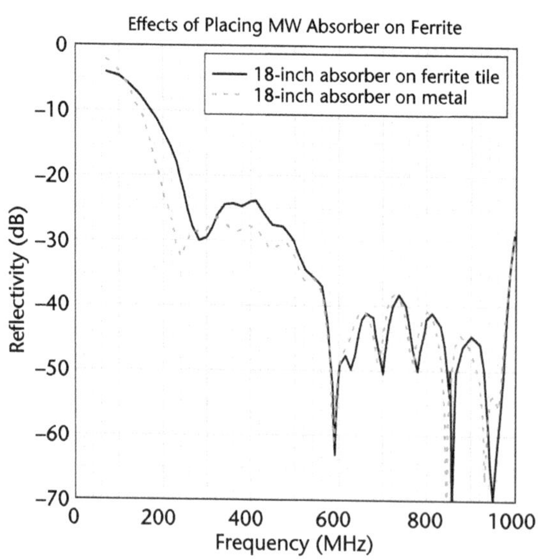

Figure 3.21 Measured results for an 18-inch pyramidal absorber on metal and on ferrite.

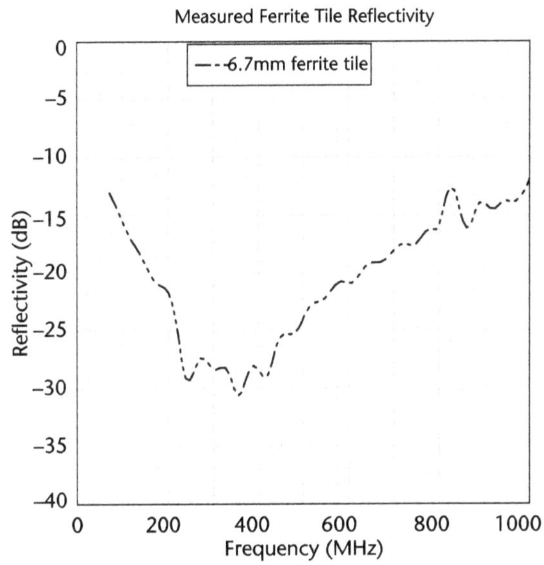

Figure 3.22 Measured results for ferrite tile.

ferrite, whose reflectivity is shown in Figure 3.22, does not add any advantage and is basically hidden by the foam absorber.

Hence, matching the electrically lossy material to the ferrite tile is the critical part of hybrid absorber design. One approach is to lower the carbon loading of the foam that is placed on top of the ferrite. In some cases the carbon black levels are about one-tenth or even one-hundredth the levels in standard polyurethane absorbers.

Other approaches consist of making hollow pyramids to be placed on top of the ferrite, while maintaining higher levels of carbon on the material. By hollowing the pyramidal shape, the overall permittivity at lower frequencies is an average of the air and the material, making it appear as if there are lower permittivity and lower losses at the lower frequencies where the ferrite operates.

One of the drawbacks of these special shaping of pyramids or special lower loading of materials is that at the microwave range, hybrid absorbers usually never exceed –40 dB of normal incidence reflectivity.

3.5 Power Handling

It is a known physical law that energy cannot be created nor destroyed. However, the energy in the electromagnetic wave incident onto the absorber is "absorbed." Where does it go? The answer is heat. The electromagnetic energy is transformed to thermal energy and dissipated to the surrounding air. The more power on the incoming wave, the more energy, since power is energy per time. The more energy onto the absorber, the more heat. Here is where the problem resides. Most absorber substrates are poor heat conductors, and the heat accumulates inside the absorber. At some point the temperature increases to a point where the substrate will ignite.

For most antenna measurement applications, the power handling of the absorber was not a problem. Most measurements were conducted at very low power levels.

There were exceptions where the antenna under test had a very high gain, and the power was concentrated on a small area, causing a high power density on the absorber. EMC measurements were the other application where power handling was a concern. This was mainly on military and automotive EMC testing where immunity to very high field levels was required. Over the past 10 years, the power handling of the RF absorber has become a hot topic, as newer antennas are very sophisticated integrated systems. Phase shifters, amplifiers, switches, multipliers, and other nonlinear components are integrated into the elements of large arrays and due to the nonlinearity of these components, it is not possible to conduct antenna testing at low powers and then scale up the measured results to the operational power level. The antennas have to operate during testing at high power levels radiating high amounts of power.

In addition, these active arrays can steer their beams so it is no longer a single area of absorber that will be illuminated with high power levels. Large areas of the absorber treatment are susceptible to be illuminated with a high-power density beam.

Traditional polyurethane-based absorber is rated to 0.5 W/in^2, which is equivalent to 775 W/m^2. Where these levels came from is lost in the archives of absorber manufacturer facilities. Recent testing suggests that 1,000 W/m^2 is a safe level for polyurethane-based absorbers. Different companies conducted tests or had tests conducted for them by third parties. There is, however, not a clearly established method for testing the power handling of absorbers. Some approaches like the ones used in [14, 15], are loosely based on EMC susceptibility testing as described by MIL-STD-461F [16]. This seems to be a good approach. After all, the susceptibility test is intended to irradiate some equipment under test (EUT) with different electric field levels to see if there is a failure. Finding the power at which a failure occurs on the absorber is the goal of the power-handling test.

The absorber test chamber shown in [17] is a modified system where the test area is more enclosed so that there is no need for a larger chamber. The enclosed test box concentrates the power into the sample; however, because of the reflections from the walls, the illumination is not very uniform over the sample. In another approach used in the past, the absorber was placed in a reverberation chamber, an enlarged microwave oven, to see how the absorber responded to being illuminated by high fields. In some of these tests, a volume of water was placed in the location of the absorber being tested to get a measurement of the power levels by comparing the temperate increase in the water before and after the test.

Polyethylene absorbers appear to have lower power handling than polyurethane, but by a very small amount. Manufacturers claim levels of 500 W/m^2. That level is close to 0.65 times the levels that polyurethane can handle. Polystyrene absorber does not seem to have been thoroughly tested for power requirements. Some manufacturers just specify that it can handle 200 V/m, a common level used in automotive EMC testing. Other polystyrene absorber manufacturers mention power handling levels of 1,500 W/m^2.

Figure 3.23, which appeared in [14, 15], shows how the RF absorber gets hot as it is illuminated by RF energy. In this particular case, the frequency of the test was at 12.4 GHz, and the aim was to generate 700 V/m, which is equivalent to 1,299.7 W/m^2. It can be seen in the infrared camera picture how the surface of the absorber has heated up to up to 62.7°C. It is a very interesting experience to touch the absorber with your fingers after one of these tests and feel how hot it is. Notice that the tips are cooler. Being smaller, the heat buildup in the tips is quickly dissipated into the surrounding air. It is interesting to think that the RF absorber is acting as some sort of frequency multiplier since it is absorbing electromagnetic waves at microwave frequencies and then radiating some of the energy as electromagnetic waves at infrared frequencies.

As power is increased, the temperature on the absorber continues to rise until the temperature reaches a point at which the materials that make up the absorber ignite. Most absorbers have combustion-limiting compounds added to their materials, so that in the event of reaching a temperature at which the material will ignite the material will smolder, rather than burn with an open flame. The added materials are typically salts that will hydrate. As the temperature increases, the salts will dehydrate and the water vapor will displace the oxygen, limiting the combustion. There are different tests that are conducted on the materials to ensure they meet certain flammability standards. Some tests commonly quoted in manufacturers' marketing materials are flammability tests that are more general, that is, not specific to anechoic ranges or to RF absorbers. The UL-94 is a test that applies to polymers used in the manufacturing of appliances and other devices [18], while the often quoted DIN 4102 is a test developed in Europe by the German Deutsches Institut für Bautechnik for testing construction materials, including fabrics [19]. The Naval Research Laboratories (NRL) test for flammability addresses specifically the testing of RF absorber materials. The testing tries to simulate the type of ignition that will be experienced in a piece of RF absorber illuminated by an electromagnetic wave [20]. It looks at increasing the temperature inside a sample of the foam to check different parameters including time for self-extinguishing after the heat source is removed and oxygen levels required for open flame.

Temperature increase is the problem, so part of the key is to increase the heat exchange from the absorber to the surrounding air. This approach to cooling is one of the most common to obtaining higher power handling from the RF absorber.

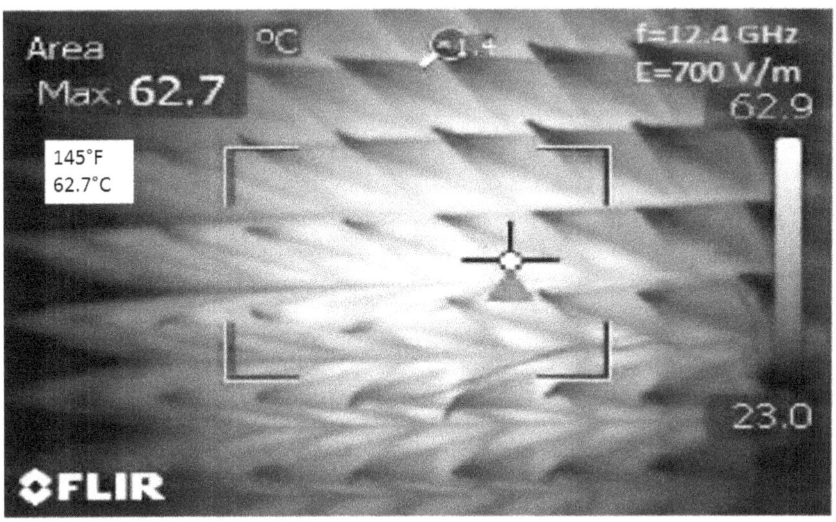

Figure 3.23 Infrared picture of absorber illuminated by high power.

Increasing the surface area of the material to increase the heat exchange is a common approach and is the basis for the medium-power absorber introduced in [16].

The other approach to increase the power handling of absorber is to change the substrate; however, sometimes, the change in substrate relates to the increased area. Polyurethane foam comes in different forms. Some foams have larger cells. These foams are usually used to cover ventilation holes or apertures in anechoic ranges. However, because the larger cells create a larger surface, these foams are known to handle a higher power density than common polyurethane foam. Figure 3.24 shows a close-up of this type of foam, which in general, can handle power densities that are a factor of 4 to 6 times the power handling of regular polyurethane absorbers. The engineer reading these lines will ask, "What is the trade-off?" One is cost. These materials are higher-cost than standard foam, in part because of the smaller amount of this absorber that is manufactured. Lower amounts of the raw material translate in a higher cost compared to the standard foam. The other trade-off associated with this absorber is that because of the larger cells, there is a larger volume of air per volume of material, which translates into lower overall losses. Hence these foams have higher reflectivity than similar absorbers manufactured with the regular foam.

Because air can flow through these foams (one of the reasons they are used to cover vents), air flow can be used for cooling the foam (thus exchanging the heat

Figure 3.24 So-called filter foam polyurethane absorber.

with the surrounding air and taking the hotter air out of the range). This forced-air cooling increases the power handling of the absorber.

Honeycomb absorber is the better known of the high-power materials (see Figure 3.25). The honeycomb structure allows for better flow of air through the material and better exchange of heat with the surrounding air. It also allows for better flow under forced air conditions. In addition to the structure, the substrate is a phenolic paper that can handle much higher temperatures. In general these honeycomb absorbers have a power handling capability that is nine to 20 times the level of standard polyurethane materials, depending on the manufacturer. These levels are without forced air; increasing the air flow will increase the power handling of these absorbers.

There are other substrates like ceramics or some of the ferrite-injected plastics that can handle higher powers; however, the reflectivity is much lower. Some of the new materials marketed by some manufacturers can handle levels over 50 times higher than standard polyurethane foams and about 2.5 times the levels of honeycomb absorbers. As the need for higher power absorbers increases as more antennas need to be tested at high power, it is expected that we will see newer materials developed in the coming years.

It is important to ventilate the range. Unless forced air is used through the absorber, where the heated air is exhausted out of the range, the entire ambient temperature of the range will increase. This increase in the room temperature may not be good for the equipment inside the room, such as receivers, and sources; also it may cause distortion of some mechanical components such as compact range reflectors. Furthermore, the absorber temperature reaches a balance between the heat being generated by the incoming wave and the heat lost to the surrounding air. Heating of the surrounding air decreases the amount of heat exchanged from the absorber, which can alter the balance and increase the interior temperature of

Figure 3.25 Honeycomb high-power absorber.

the absorber. An air conditioning system that continuously pumps cooler air into the chamber is a way of ensuring that the absorber will handle the power according to its specifications.

There are testing approaches that can be used in some cases to reduce the power requirements of the absorber. One of them is to pulse the radiated signal from the antenna at some given duty cycle. In Figure 3.26, the temperature inside a 24-inch piece of pyramidal absorber is plotted versus time. A thermocouple probe used to measure the temperature was inserted inside the pyramid at about 4 inches (10.2 cm) from the tip of the pyramid. As it can be seen in the plot, the rise in temperature is not instantaneous; it took almost six minutes for the temperature to rise to 48°C from the starting temperature of 24°C. The plot also shows that immediately after shutting down the incident field, the absorber piece started to cool. In about three minutes the absorber internal temperature had dropped by about 10°C. Pulsing the source while testing is a way of bringing the average power density to a level that will not make the absorber exceed the temperature to levels at which it will ignite and smolder.

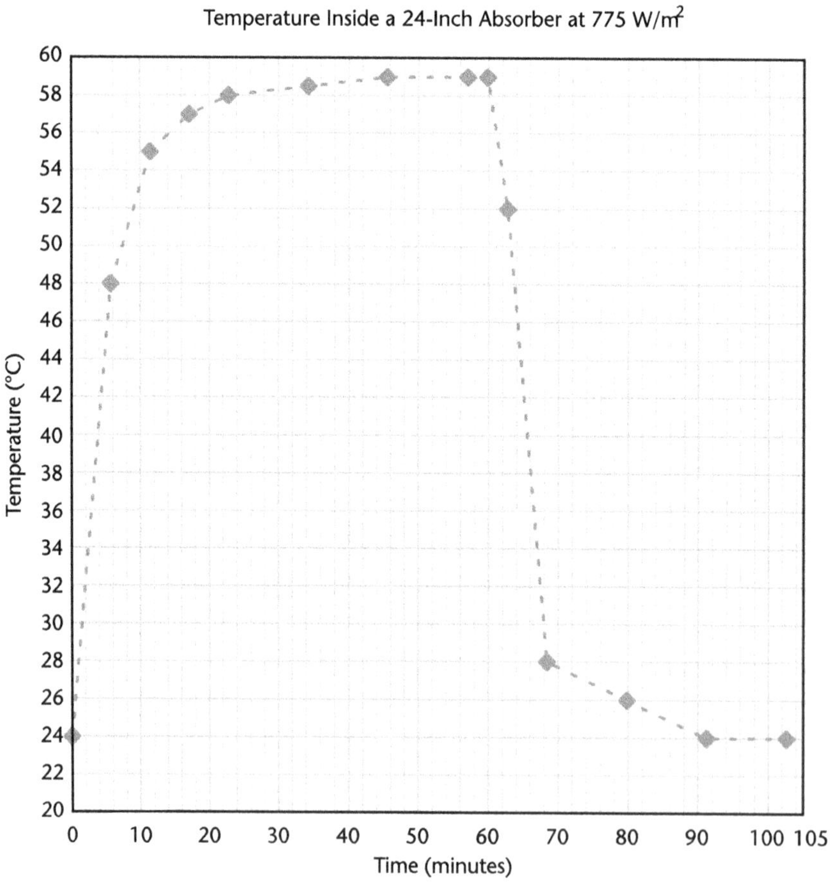

Figure 3.26 Internal temperature versus time with the source on for one hour.

References

[1] Lovick, E., *Radar Man: A Personal History of Stealth*, New York: iUniverse, Inc., 2010.

[2] Kummer, W. H., and Gillespie, E. S., "Antenna Measurements—1978," *Proceedings of the IEEE*, Vol. 66, No 4. April 1978.

[3] Emerson, W. H., "Electromagnetic Wave Absorbers and Anechoic Chambers Through the Years," *IEEE Trans. on Antennas and Propagation*, Vol. 21, No. 4, July 1973.

[4] Balanis, C. A., *Advanced Engineering Electromagnetics*, New York, NY: John Wiley & Sons, 1989.

[5] Holloway, C. L., et al., "Comparison of Electromagnetic Absorber Used in Anechoic and Semi-Anechoic Chambers for emissions and Immunity Testing of digital Devices," *IEEE trans. on Electromagnetic Compatibility*, Vol. 39, No. 1. Feb 1997. pp. 33–47.

[6] Hemming, L., *Electromagnetic Anechoic Chambers: A Fundamental Design and Specification Guide*, Piscataway, NJ: IEEE Press Wyley-Interscience, 2002.

[7] Rodriguez, V., and E. Barry "A Polynomial Approximation for the Prediction of Reflected Energy from Pyramidal RF Absorbers," *Proceedings of 38th Annual Antenna Measurement techniques Association Symposium AMTA 2016*, Austin, TX, Oct. 31, 2016, pp. 155–160.

[8] Rodriguez, V., "Validation of the Polynomial for RF Absorber Reflectivity for the Prediction of Anechoic Chambers," *IEEE AP-S Symposium on Antennas and Propagation and USNC-URSI Radio Science Meeting*, San Diego, CA, July 9–14, 2017.

[9] Guerrieri, J., et al., "RF Characterization of Latex-Coated Pyramidal Absorber," *Proceedings of 34th Annual Antenna Measurement techniques Association Symposium AMTA*, Seattle, WA, 2012.

[10] Chen, Z., "Common Microwave Absorber Evaluations in the W-Band (75–110 GHz)," *Proceedings of the 39th Annual Antenna Measurement techniques Association Symposium AMTA 2017*, Atlanta, GA, Oct. 16, 2017, pp. 20–24.

[11] Sun, W., and C. Balanis, "Analysis and Design of Periodic Absorbers by Finite-Difference Frequency-Domain Method," Report No. TRC-EM-WS-9301, Telecommunications Research Center, Arizona State University, Tempe, AZ, 1993.

[12] Suetake, K., "Super Broadband Absorber," U.S. Patent 3 623 099, Nov. 23, 1971.

[13] Rodriguez-Pereyra, V., "A Study of the Effect of Placing Microwave Pyramidal Absorber on Top of Ferrite Tile Absorber on the Performance of the Ferrite Absorber," *19th Annual Review of Progress in Computational Electromagnetics (ACES 2003), Symposium*, Monterey, CA, March 2003.

[14] Chen, Z., and V. Rodriguez, "Surface and Internal Temperature Versus Incident Field Measurements of Polyurethane-Based Absorbers in the Ku Band," *35th Annual Antenna Measurement Techniques Association Symposium*, AMTA 2013, Columbus, OH, Oct. 6–11, 2013.

[15] Chen, Z., and V. Rodriguez "Surface and Internal Temperature versus Incident-Field Measurements of Polyurethane-based Absorbers in the Ku-Band," *[AMTA Corner] IEEE Antennas and Propagation Magazine*, Vol. 56, No. 1, 2014, pp. 248–254.

[16] Mil Std. 461 F, Department of Defense Interface Standard: Requirements for the Control of Electromagnetic Interference Characteristics of Subsystems and Equipment, Dec. 10, 2007.

[17] Rodriguez, V., G. d'Abreu, and K. Liu, "Measurements of the Power Handling of RF Absorber Materials: Creation of a Medium Power Absorber by Mechanical Means," *31st Annual Antenna Measurement Techniques Association Symposium AMTA 2009*, Salt Lake City, UT, Nov. 1–,6 2009.

[18] Underwriters Laboratories, "UL Standard for Safety Tests for Flammability of Plastic Materials for Parts in Devices and Appliances," Oct. 29, 1996.

[19] DIN 4102—Part 1, B2, Reaction to Fire Tests—Ignitability of Building Products Subjected to Direct Impingement Of Flame, 1998.

[20] Tatem, P. A., P. D. Marshall, and F. W. Williams, "Modified Smoldering Test of Urethane Foams Used in Anechoic Chambers," NRL Report 8093, Naval Research Laboratory, Washington D.C., March 9, 1977.

Appendix 3A: MATLAB® Scripts

3A.1 Simulation of a Pyramidal Shape

Reflection from multiple layers. This script provides the reflectivity of a pyramid. The size of the pyramid can be specified, as well as the material properties. The script estimates the average permittivity for each layer and provides the reflectivity at the top layer.

```
% set up the variables and constants
j=1i;
mu0=4*pi*1E-07;
e0=8.85418782E-12;
f=5.E9;
lambda=3E8/f;
thickness=5*0.0254;
base=2*0.0254;
N=1000;%number of layers
d=thickness/N;
omega=2*pi*f;
eprimeL=1.8;
edprimeL=1.7;
anglepyr=atan((base/2)/thickness);
%set up the layers
%middle of layer
md=d/2;
sigmaL=omega*e0*edprimeL;
etaL=sqrt((1i*omega*mu0)/(sigmaL+(1i*omega*e0*eprimeL)));

for ii=1:1:N;
 location=md+((ii-1)*d);
 layerbase=location*tan(anglepyr);
 factor(ii)=((2*layerbase)^2)/base^2;
ecomplex(ii)=(1-factor(ii))+factor(ii)*(eprimeL+(j*(edprimeL)));
eprime(ii)=real(ecomplex(ii));
edprime(ii)=imag(ecomplex(ii));
sigma(ii)=omega*e0*edprime(ii);
eta(ii)=sqrt((j*omega*mu0)/(sigma(ii)+(j*omega*e0*eprime(ii))));
end
%get the reflection coefficient at each layer
```

```
biggamma0=(eta(1)-(120*pi))/(eta(1)+(120*pi));
for jj=1:1:N-1;
biggamma(jj)=(eta(jj+1)-eta(jj))/(eta(jj+1)+eta(jj));
end
%set the propagation constants
for kk=1:1:N;
biggamma(N)=(etaL-eta(N))/(etaL+eta(N));
k1=(omega*sqrt(mu0*e0*eprime(kk)));
kk1=sigma(kk)/(omega*e0*eprime(kk));
gamma(kk)=k1*sqrt(0.5*(sqrt((1+(kk1)^2))-1))+j*(k1*sqrt(0.5*(sqrt((1+
  (kk1)^2))+1)));
end
%being calculating the reflectivity
biggammain=biggamma0;
for nn=1:1:N;
 gammatot=0;
 for m=1:1:nn;
 gammatot=gammatot+gamma(m);
 end
 biggammain=biggammain+biggamma(nn)*exp(-2*d*gammatot);
end

resultsin=20*log10(abs(biggammain))
```

3A.2 Normal Incidence Reflectivity of Ferrite Tile

Normal incidence reflectivity of a ferrite tile. This script provides the reflectivity of a tile of ferrite. The parameters loaded are those shown in [5].

```
% set up the variables and constants
%for ferrite layer
j=1i;
mu0=4*pi*1E-07;
e0=8.85418782E-12;
c=299792458.0;
thickf=0.00638; %6.mm thick ferrite
tile=[
 30 10.88 0.16 52.31 236.17;
 40 10.93 0.37 31.26 181.36;
 50 11.04 0.50 21.05 147.02
 60 11.24 0.51 15.59 123.51;
 70 11.39 0.27 12.32 106.32;
 80 11.41 0.09 9.96 93.56;
 90 11.37 0.07 8.16 83.57;
 100 11.19 0.04 6.74 75.52;
```

```
 150 11.29 0.53 3.39 51.64;
 200 10.97 0.07 1.85 38.77;
 250 11.37 0.26 1.40 31.38;
 300 10.93 0.18 0.52 26.23;
 400 11.00 0.17 0.12 19.82;
 500 11.03 0.02 0.47 15.66;
 600 10.95 0.19 0.59 12.93;
 700 10.99 0.28 0.59 11.07;
 800 11.18 0.23 0.48 9.69;
 900 11.31 0.14 0.38 8.55;
 1000 11.36 0.04 0.29 7.69
 ];

freqn=length(tile);

f=tile(:,1)*1e6;
eprime=tile(:,2)*e0;
edprime=tile(:,3)*e0;
muprime=tile(:,4)*mu0;
mudprime=tile(:,5)*mu0*1;

lambda=c/f;
% we have three layers, ferrite, spacer, and metal backing
% we will assume the metal backing as PEC
% need to create the eta's for the air, ferrite
eta0=sqrt(mu0/e0);
for nn=1:freqn;
 eta1(nn)=sqrt((muprime(nn)+j*mudprime(nn))/
   (eprime(nn)+j*edprime(nn)));
end
% now we need the capital gamma for the air to ferrite, the
   ferrite to metal and the

for nnn=1:freqn;
 Bgamma12(nnn)=(eta1(nnn)-eta0)/(eta1(nnn)+eta0);
 Bgamma23(nnn)=-1.; % reflection off the PEC backing make = to %
   %Bgammma12(nnn)* -1 for air backed tile.
end
%set up omega
omega=f*2*pi;
% set Propagation constant for each layer lower case gamma
for k=1:freqn;
 muf=muprime(k)+j*mudprime(k);
 ef=eprime(k)+j*edprime(k);
 gamma1(k)=omega(k)*sqrt(muf*ef);
 end

%now let us compute the reflectivity at each frequency and plot
   the results
for kk=1:freqn;
```

```
    BigGamma(kk)=(Bgamma12(kk)+Bgamma23(kk)*exp(j*2*gamma1(kk)*thi
      ckf))/(1+Bgamma12(kk)*Bgamma23(kk)*exp(j*2*gamma1(kk)*thickf));
    end
    BigGammadB=20*log10(abs(BigGamma));
    plot(f,BigGammadB)
```

3A.3 Computation of Reflectivity Using Rodriguez's Equations

This function will return the reflectivity of a given thickness of absorber in inches and a given frequency in gigahertz. This approach is described in [7].

```
function Reflect = RodriguezReflec(Freq, Thick, Inc)
% this function provides the bistatic reflectivity of an
  electrically large
% treatment of absorber the are a series of inputs
% Freq is the frequency of interest in GHz
% Thick is the thickness of the absorber in inches
% Inc is the angle of incidence on to the absorber
% Reflect is the reflection in the bistatic direction in dB
% the domain of the function is 0.25<Thick<18 and 0=<Inc=<85
%
% NSI-MI Technologies February 2017
%
% the function was written by Vince Rodriguez and it is based on the
% Rodriguez Equations reported in
% Rodriguez and Barry "A Polynomial Approximation for the
  Prediction of
% Reflected Energy from Pyramidal RF Absorbers" AMTA 2016, Austin TX.
%%%%%%%%%%%%%%%%%%%%%%%%%%%%%%%%%%%%%%%%%%%%%%%%%%%%%%%%%%%%%%%%%%%%%%
  %%
Reflect = 0;
R = 0;
c = 299792458.; %speed of light in m/s
lambda = c /(Freq*1.E9) ; % wavelength in meters
Lambin = lambda * 39.36 ; % wavelength in inches
T= Thick / Lambin; % electrical height of absorber
if T>= 18 ;
 %for very large absorber set it to 18 lambda
 T = 18;
 fprintf('electrically very large absorber, setting its thickness to
  18 lambda')
end
if T < 0.25 ;
 fprintf('the electrical thickness of the absorber')
 fprintf(' must be larger than 0.25 lambda')
end

if T >= 0.25 && T < 2. ;
 C(1)=-54.98 + 55.21 * exp (-0.7744 *T);
```

```
% rodriguez coefficients for electrically smaller absorber
A = [-6.32106901985E+00 7.68802258525E-01 -2.69121464934E-02
  3.51138682149E-04 -1.55298248909E-06;
7.22679386149E+01 -8.65977627121E+00 3.01684807722E-01
  -3.93730848369E-03 1.74449142124E-05;
-3.22780683714E+02 3.83644059547E+01 -1.33007749667E+00
  1.73692360659E-02 -7.71113070847E-05;
7.30993386584E+02 -8.67259143697E+01 2.99932641399E+00
  -3.92065387879E-02 1.74414181320E-04;
-9.17643699566E+02 1.08845073951E+02 -3.75725939979E+00
  4.91621335614E-02 -2.19124414169E-04;
6.42444858829E+02 -7.62867963346E+01 2.63019526316E+00
  -3.44484145284E-02 1.53807835270E-04;
-2.33615092361E+02 2.78111265669E+01 -9.58477798823E-01
  1.25666708854E-02 -5.61959608529E-05;
3.42805988676E+01 -4.09642120674E+00 1.41225281240E-01
  -1.85375240980E-03 8.30146886404E-06];
for ii = 1:1:5;
C(ii+1) = A(1,ii)+A(2,ii)*T+A(3,ii)*T^2+A(4,ii)*T^3+A(5,ii)*T^4+A(6,ii
  )*T^5+A(7,ii)*T^6+A(8,ii)*T^7;
end
R = C(1)+C(2)*Inc+C(3)*Inc^2+C(4)*Inc^3+C(5)*Inc^4+C(6)*Inc^5;
if R < -55;
 R=-55;
end
if R > 0;
 R=0;
end
Reflect = R ;
end

if T >= 2 && T < 20. ;
C(1)=-54.98 + 55.21 * exp (-0.7744 *T);
% Rodriguez coefficients for larger absorber
A = [-1.3910397785E+00 2.0337750966E-02 7.6318763944E-04
  -2.4200645893E-05 1.6261209444E-07;
3.0703647571E+00 -2.4548518904E-01 7.3080660682E-03 -8.1799302820E-05
  3.0766242948E-07;
-1.2333118819E+00 1.1707478107E-01 -3.8573868662E-03 4.7931889567E-05
  -2.0136854065E-07;
1.7736268491E-01 -1.8629158011E-02 6.4634192554E-04 -8.3359492323E-06
  3.6127335603E-08;
-1.0019087120E-02 1.2274642557E-03 -4.5270850115E-05 5.9850081842E-07
  -2.6233932363E-09;
4.9175865563E-05 -2.2855637162E-05 1.0513841962E-06 -1.4553062674E-08
  6.3444264364E-11;
1.5075497675E-05 -8.6593800325E-07 2.0059564496E-08 -2.3563710316E-10
  1.0978983071E-12;
-3.9564117797E-07 3.0791884224E-08 -9.2545962479E-10 1.1814444399E-11
  -5.3542575601E-14];
```

```
for ii = 1:1:5;
C(ii+1) = A(1,ii)+A(2,ii)*T+A(3,ii)*T^2+A(4,ii)*T^3+A(5,ii)*T^4+A(6,ii
  )*T^5+A(7,ii)*T^6+A(8,ii)*T^7;
end
R = C(1)+C(2)*Inc+C(3)*Inc^2+C(4)*Inc^3+C(5)*Inc^4+C(6)*Inc^5;
if R < -55;
 R=-55;
end
if R > 0;
 R=0;
end
Reflect = R ;
end
if Inc > 85. ;
Reflect = 0
fprintf ('angle of incidence is larger than 85 Rodriguez Eq. do
  not apply')
end
end
```

CHAPTER 4

RF Shielding

4.1 To Shield or Not To Shield?

The necessity of RF shielding is one of the key questions that a user of an indoor range needs to ask and answer. As discussed in Chapter 2, if shielding is required, an indoor range is necessary, but not all indoor ranges need to be shielded. The main reference for shielding is without any doubt the book by Hemming [1]. Hemming himself agrees with the statement given above regarding the necessity of shielding is required, and in Chapter 2 of his book, he states that, "once the shielding effectiveness requirements have been defined ..." then "a cost-effective shielding system" can be chosen or designed.

There lies the key: The shielding effectiveness needs must be decided. It is common to immediately specify a high level of shielding when, as Hemming points out, the shielding is intended primarily to provide a constant reflective backing to the RF absorber, in which case a simple metal foil glued to the walls of the range will suffice [1]. It should be noted that a shielded anechoic range will always be better, as the ambient noise will be extremely low and the IFBW may be wider, making for faster measurements. However, given the additional cost, is it necessary to have a high-level shielded enclosure for the indoor range?

Shielding performs three functions: It blocks external sources from affecting the measurement; it blocks fields generated internally from radiating outside the range; and, it provides a consistent backing to the RF absorber. This is important, as shown in Chapter 3, since at certain frequencies, the absorption is in part improved by the phase cancelation between the reflection from the tip of the material and the reflection from the base of the material. Having a wood backing or any other material with a permittivity that changes with humidity can cause differences in measurements. This is the case when the absorber is less than one wavelength in thickness. There are some absorbing materials, such as ferrite tiles, that do need a metal backing (Chapter 3). This metal backing, however, does not need to have very high shielding performance.

4.1.1 Phase-Locked Measurements

In general, in antenna measurements the frequency being transmitted is known, and typically continuous wave (CW) measurements are performed. In addition, the measurement is typically a phase-locked loop measurement. In a typical antenna measurement system, an RF source generates a signal at a given frequency. That signal is fed to the range antenna (or the antenna under test, but one of the two antennas in the antenna range). Prior to the signal going to the antenna port, we can sample the signal with a directional coupler (as shown in Figure 2.10). This signal becomes our reference signal, which we need to measure phase. The antenna being measured (the AUT) will receive the same signal (attenuated, due to losses, and with a different phase due to the different paths) from the range antenna.

The phase between the reference and the received signal at the AUT will be coherent. Any sources of noise will be noncoherent. With the proper averaging and IFBW, the effects of the noise will be minimized. Thus, it is possible to do antenna measurements in outdoor ranges or in ranges with minimal shielding (i.e., less than 60 dB of shielding effectiveness at the frequencies of interest).

4.1.2 Reasons for Shielding

4.1.2.1 Measuring Very Small Signals

In EMC measurements, the frequencies at which noise may appear are unknown. Typically, a device that is a nonintentional radiator is being measured. The levels being measured are typically very small [2]. Take the Federal Communication Commission (FCC) field strength limits shown in Table 4.1.

If we take the values in Table 4.1 and express them in terms of power density, we get that above 960 MHz we are trying to measure signals that are lower than $2.38 \cdot 10^{-10}$ w/m^2. The levels assume that the measurement is performed at a 10-m distance from the equipment being tested. The effective isotropic radiated power (EIRP) is defined as the input power in watts for a lossless isotropic radiator to give the same maximum power density radiated by the actual radiating antenna at a distance from the radiator, hence $EIRP$ (dBw) = G (dBi) + P_{in} (dBw). Using the EIRP definition, the isotropic radiator input power will be 0.3 μW or −65.23 dBw for a 10-m radius sphere (with a surface of 1256.64 m^2). Consider that a typical antenna measurement is done using RF sources that put a maximum power between 10 and 15 dBm, or −20 to −15 dBw. That difference between the power levels for antenna measurements and for EMC emissions puts in perspective the typical levels measured in an EMC emission measurement. It should be obvious that in EMC, emission applications shielding is necessary to provide a quiet environment where such small signals can be measured.

4.1.2.2 Shielding High Field Levels

The other side of EMC is immunity. In these cases, an electromagnetic field is being generated to check if the EUT will continue to operate while illuminated by such

Table 4.1 FCC Limits When Measured at 10m from the EUT

Frequency of Emission (MHz)	*Field Strength (μV/m)*
30–88	90
88–216	150
216–960	210
Above 960	300

a high field. Shielding the exterior of a range from high fields radiated inside the range is a must since most countries have laws that prohibit radiating across broadbands without a license. More important is the safety of the personnel managing the range or test facility.

The IEEE C.95 standard [3] provides levels for the exposure of humans. There are different levels for general public or unrestricted areas and for technicians and range operators or restricted areas. Figure 4.1 shows the limits as provided in the IEEE C 95.1 standard [3].

EMC measurements were known for generating high fields, but in recent years the power levels have increased for antenna measurements with the proliferation of active electronically steered array (AESA) antennas. These antennas are nonlinear given that active circuit components are included as part of the antenna. The presence of these nonlinear elements makes it impossible to test these antennas at low power and scale the results to obtain the performance parameters at high power. It is not rare for some of these antennas to have very high EIRPs; in my experience, some customers have approached me with EIRP levels as high as 65 dBw. If the highest field level or power density that will be generated inside the range is known, it is possible to estimate the level of shielding required to ensure that the legal and safety limits of radiation are met.

4.1.2.3 Measuring Modulated Signals

In HWiL tests, where a device is being tested for performance, shielding may be necessary. For example, cellular phone devices or Wi-Fi device tests for throughput may need to be tested in a quiet environment where there is no interference from surrounding cellular telephony systems and surrounding Wi-Fi networks. Additionally, it is not desirable for these tests to affect the cellular communications and

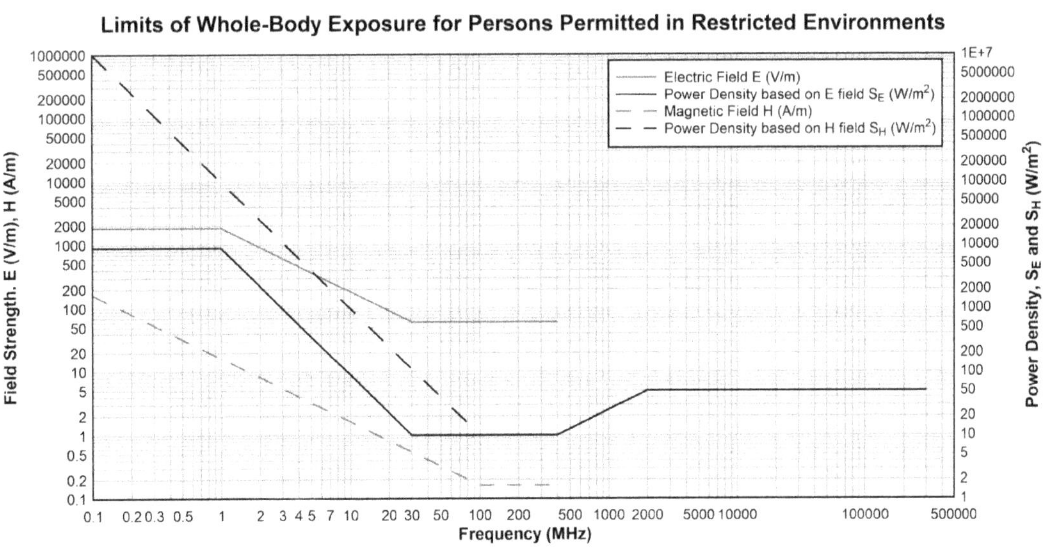

Figure 4.1 Limits for human whole body average field exposure to RF in restricted areas up to 30 minutes [3].

Wi-Fi services in the adjacent areas to the test range. It should be noted that for these cases, there is no need to shield down to a few kilohertz; the main band for the test being performed in the range may be from 450 MHz to 6 GHz. Shielding in that range is easily accomplished by aluminum foil 10 mils (0.254 mm) in thickness.

4.1.2.4 Secrecy

Another reason for shielding is to avoid eavesdropping. This is typically the case for defense-related ranges owned both by government or private contractors. Some government agencies have their own internal standards and specifications for shielded enclosures [4].

4.1.3 What Level of Shielding Is Required?

Typically, manufacturers offer shielding levels or shielding effectiveness in the 100-dB level from 10 kHz to 40 GHz for modular-type shielding and 100 dB from 10 kHz to 94 GHz for welded enclosures. These levels of shielding may only be required under some conditions. The questions to be asked, based on the reasons for shielding already discussed, are listed as follows:

- Is the enclosure to perform EMC testing, both emissions and immunity?
- Do power densities and field levels generated in the range exceed legal and or safety limits?
- Are the frequencies and modulations used in the test classified?
- Is there an HWiL test being conducted that may interfere with outside communication systems and vice versa?

If the answer is affirmative for any of these questions, then shielding of some type is required. The user should be careful in specifying the level of shielding. Shielding is usually tested prior to the RF absorber installation, so the attenuation through the absorber is not accounted for. This means that enclosures with 60 dB of shielding attenuation performance may be sufficient to meet most requirements.

The highest levels of shielding and the broadest frequencies of operation for the shielding should be left to EMC and secured applications. The other applications, while still requiring shielding, may require lower levels.

Let us go over an example. Let us assume that we want to measure an antenna at 10 GHz with an expected gain of 40 dBi and an expected EIRP of 100 dBm. Let us also assume that we have no losses between the RF source and the antenna input; hence the EIRP equation can be used to calculate the input power to the antenna:

$$EIRP\left(\mathrm{dB_W}\right) = P_t\left(\mathrm{dB_W}\right) - L_c(dB) + G_a(\mathrm{dBi}) \tag{4.1}$$

where P_t is the power into the antenna, L_c represents the cable losses to the antenna, and G_a is the gain of the antenna. Since the losses are assumed to be zero, then input power on the antenna is about 911.8W. At this point, we need another piece of information: the aperture size of the antenna, which we can estimate from the gain and the frequency of operation. Let us assume that we know that the antenna is 0.7853m^2 (assuming a 1-m diameter antenna). We can estimate that if this is a uniform amplitude aperture, such as an array, that the power density at the aperture is 1161.05 W/m^2. This level is much higher than the exposure levels of 50 W/m^2 for 10 GHz shown in Figure 4.1. Let us further assume that we are testing this antenna in a chamber that is approximately 8m wide by 8m tall by 16m long with the end wall closest to the antenna being 4.5m from the antenna. The elevation geometry of this hypothetical range is shown in Figure 4.2. What level of field do we have in the proximity of the antenna? We can approximate using the equation used by Baggett and Hess in their AMTA paper [5], which approximates the power density for a circular aperture. Chapter 7 provides a more in-depth discussion of the near-field estimation for different apertures. Another approach to estimate the near-field power densities is to use commercial antenna design numerical electromagnetic packages that can also solve for the field at a given location around the antenna. The equation provided by Baggett and Hess [5] is

$$P_d\left(\frac{w}{m^2}\right) = P_{da}\left(\frac{w}{m^2}\right)\cdot\left[1-\left(16\cdot\frac{r}{\frac{2D^2}{\lambda}}\cdot\frac{\sin\left(\frac{\pi}{8\left(\frac{r}{\frac{2D^2}{\lambda}}\right)}\right)}{\pi}\right)+\left(\frac{128\left(\frac{r}{\frac{2D^2}{\lambda}}\right)^2}{\pi^2}\left(1-\cos\left(\frac{\pi}{8\left(\frac{r}{\frac{2D^2}{\lambda}}\right)}\right)\right)\right)\right] \tag{4.2}$$

where P_d is the power density at a distance r from the antenna aperture. P_{da} is the power density at the antenna aperture assumed to be the input power divided by the effective aperture of the antenna. D is the diameter of the antenna in meters, and λ is the wavelength in meters. Using (4.2) we obtain an estimated power density

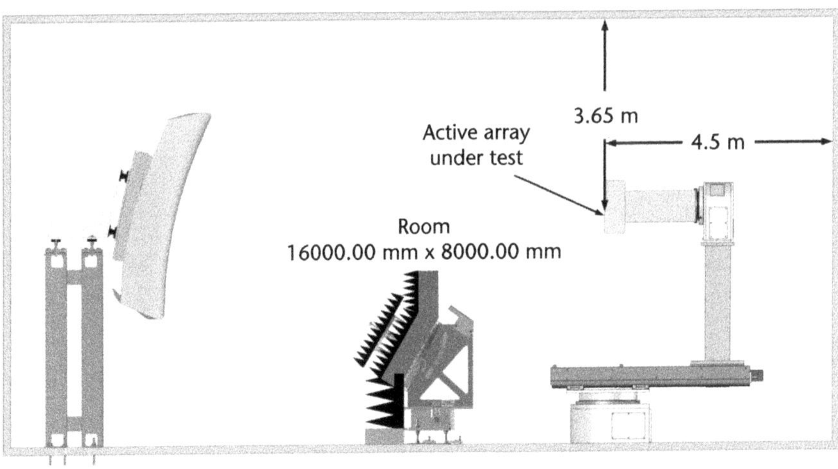

Figure 4.2 A hypothetical compact antenna test range where a hypothetical 1-m diameter array is being tested at high levels of power. The elevation is shown, but the side walls are 4m from the center of rotation of the AUT. As the antenna is being rotated to measure the pattern, the main beam will point to different walls of the range.

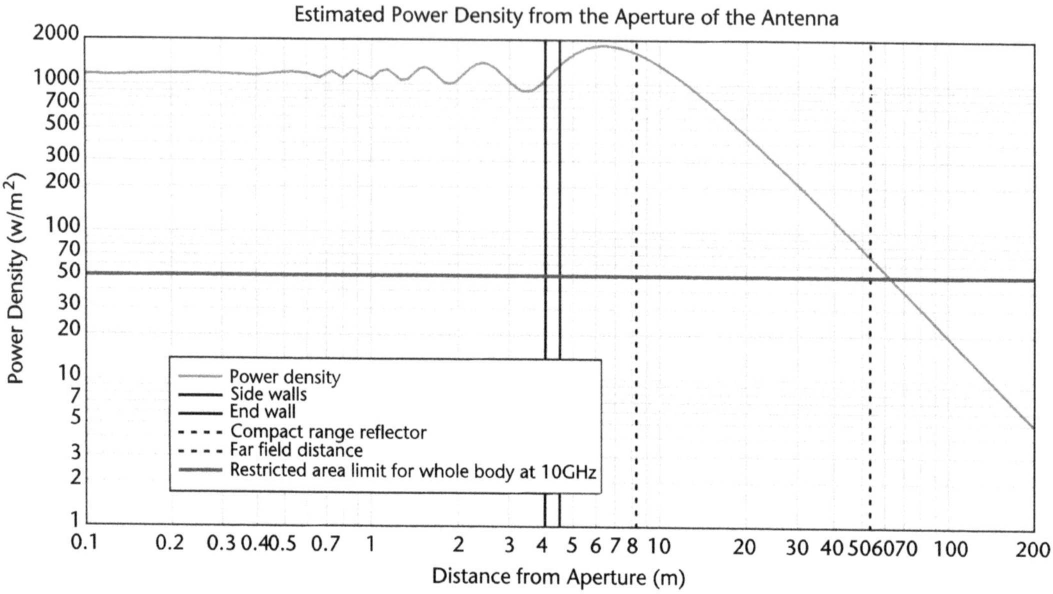

Figure 4.3 Estimated power density level for the hypothetical antenna on the range shown in Figure 4.2 The power density levels are much higher than the safety limits given on [3].

(plotted in Figure 4.3) distribution in the proximity of the antenna along the main beam direction. The results show that in the proximity of the antenna, power density levels up to 1842 W/m^2 (32.65 dB$_{W/m^2}$) are possible. The levels given by the IEEE Standard are 50 W/m^2 (17 dB$_{W/m^2}$); thus a 40-dB level of shielding effectiveness will bring the levels to −7.35 dB$_{W/m^2}$ or 0.184 W/m^2, which is 0.37% of the exposure limit. This level of 0.2W/m^2 is also lower than the exposure limit for unrestricted areas per the IEEE C95.1 standard, which is 10W/m^2 at 10 GHz.

Achieving 40 dB of shielding effectiveness (SE) is not an extremely difficult task, and the absorber on the range will add to the attenuation of any metal surfaces on the wall. The purpose of this example is to show that in many cases 100 dB of shielding over a 150-kHz to 10-GHz or 18-GHz band is overkill.

4.2 Shield

This section discusses shielding methods used to achieve high shielding levels, like those necessary for EMC measurements. For those applications, levels of 100 dB of shielding effectiveness are required. For most antenna measurement applications, the RF absorber and simple sheet metal or foil taped at the joints can provide sufficient shielding.

For high-level shielding there are different shielding materials available on the market. The most common nowadays are so-called modular shielding panels. These can be of two types. The first type, most commonly found in the United States, is panel plywood or particle board laminated on both sides with galvanized steel. The panels are clamped together with long steel clamps fastened at regular intervals with screws to a prescribed torque. While it may appear a simple approach, it requires a trained group of installers, especially to treat the connections between clamps and the corner joints. When properly installed, this method can achieve levels of 100 dB up to frequencies in the 18-GHz range and higher.

Another type of modular shielding consists of steel panels bent at the edges and bolted to each other with a metal mesh gasket between the panels. Figure 4.4 shows a sample of this type of shielding, also called "pan-type." This approach is more common in Europe, and it is preferred in certain geographical areas with wood-eating insects.

The other type of shielding is nonmodular, which typically consists of welded steel sheets. While these used to be the most expensive type of shielding technology due to the highly skilled labor required to perform the welds, the use of welding machines has reduced their cost. Welded shielding can easily achieve shielding effectiveness levels of 120 dB or higher over wide frequency ranges.

The shielding levels of the walls of the enclosure are important, but they are not the most critical. At some point it is desirable to enter the enclosure to set up measurements; it is also desirable to maintain a constant temperature through the use of ventilation and air conditioning and to bring A/C power and different electrical and optical signals into and out of the enclosure. Entering the range requires doors for the personnel to come in and out of the chamber. Bringing electrical signals and power into the enclosure requires RF filters to allow only the desirable frequency bands of the electrical signals. Finally, allowing conditioned air into the enclosure requires a special penetration that will shield the RF signals while allowing air to flow through.

Figure 4.4 Example of a pan-type shielding (so called because the panels are folded at the edges, resembling a pan). The black rubber is to protect people from hurting themselves when handling this sample. Notice the mesh gasket and the bolts. (*Source:* Author's private collection.)

4.3 Doors

Doors are the most critical part of a shielded enclosure. They are the elements of the shielding system that will endure the most wear and tear. Indeed, most manufacturers recommend checking the contact surfaces periodically to ensure that the metal-to-metal contact is not compromised.

The most common approach to door contact is the knife edge door where an edge of the door is introduced into a set of beryllium copper (BeCu) contacts to ensure metal-to-metal contact between the door leaf and the frame of the door. Figure 4.5 shows the typical BeCu finger contacts for a shielded door frame.

Hemming provides a good discussion of doors [1] including pneumatic bladder doors where air pressure is used to provide the metal-to-metal contact between the door and its frame. An inflated bladder pushes the door into the door frame to create the metal-to-metal contact. I am not aware of any bladder doors being installed in recent times, and this may be related to the maintenance required for the bladders. Pneumatics are still used in shielded doors, but the compressed air is used mainly to provide the last push to jam the door edge into the receiving contacts in the door frame.

The door leaf must make contact with the frame on all sides of the door. This means that shielded doors usually have a threshold. The threshold makes it difficult to bring large loads into the range since it can be as high as 2 inches. There are low-threshold doors, some of which are discussed in [1]. If the threshold is critical and ramps are not an option, there are sliding doors available where a pneumatic

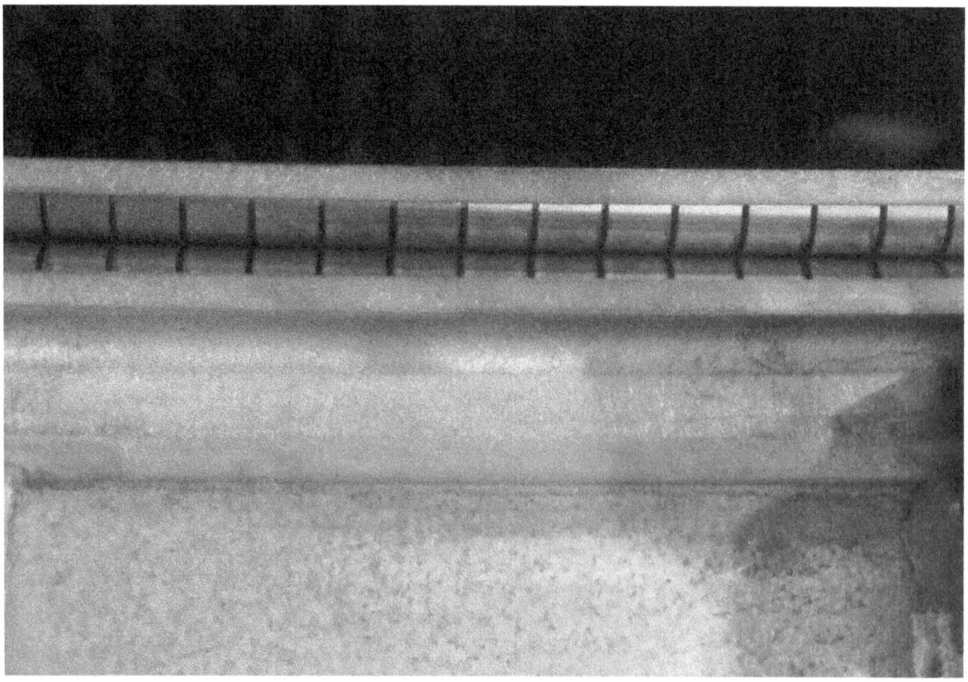

Figure 4.5 A brass frame holding the spring BeCu alloy fingers that will contact the knife edge on the door leaf. The brass frame is soldered to the door panel that is then attached to the shield panels of the enclosure.

platform is raised after the door slides out of the way. The copper finger springs for the lower edge of the door are below the floor level, and this allows for rolling carts or vehicles to be brought into the range easily. Figure 4.6 shows one of these doors while is it open. Insets in Figure 4.6 show the copper contacts and the raised level threshold or platform to enter the chamber.

Doors should be sized appropriately. It is the most critical area of the shield, and it is the way to access the range components and to bring items in and out. While doors like the one shown in Figure 4.6 may not be in everybody's budget, there are larger swing-door options, typically with double leaves and small thresholds. In addition to the cost, doors like the one in Figure 4.6 require some concrete work to cut out a recess for the platform to be lowered when the door closes. That type of civil engineering work may not be possible in some parent buildings.

Figure 4.7 shows some dual-leaf shielded doors. These doors allow for a large opening to bring in the AUT or other large equipment. One of the leaves has copper contacts, and the other leaf has a knife edge that goes in between the copper contacts (see arrows in Figure 4.7); hence, there is a primary leaf that must be opened before the secondary leaf opens. These doors can be extremely large and supported by casters or wheels to reduce the stress on the hinges.

Opening a large door means that there will be a large heat exchange between the interior of the range and the exterior. This may not be an issue if both the interior of the range and the exterior are kept at the same temperature. However, if they are not both equally climatized, it is not desirable to open the main large access door, given that for some measurements and some measurement systems like compact ranges a constant temperature is desirable. Changes in temperature can affect phase measurements and distort the planarity of near-field scanners. Hence, it is a good

Figure 4.6 A large sliding door with zero threshold. Notice the copper finger contracts on the periphery of the frame and the raised level threshold for access to the room. (Source: Author's private collection).

Figure 4.7 Large dual-leaf doors. Notice the copper contacts on the closed leaf and the knife on the open leaf. The door on the left is an 8-ft by 8-ft door (2.4-m by 2.4-m), and the door on the right is a 12-ft by 12-ft door (3.66-m by 3.66-m). (*Source:* Authors' private collection.)

idea to have an additional smaller personnel door available in the range. Opening the smaller door should minimize the heat exchange between the inside and outside and, at a minimum, reduce the time needed for the range to reach a certain operating temperature. Thus, if personnel is required to access the range to change antennas or mixers or to make adjustments or checks, the smaller personnel door is preferred to the large equipment door.

4.4 Filters

If high levels of shielding are required, as in the case of EMC measurements and for some secured facilities, the electrical signals going in and out of the enclosure must be filtered such that only the desirable signals go through the shield line. Even if measurements are being done in the X-band, it is possible for these signals to couple to the A/C network and propagate through that network to the outside of the shielded enclosure. Hence, it is desirable to have a low-pass filter to let through the 60-Hz A/C power while cutting off any other signals. Similarly, there may be Ethernet connections used to communicate and control the equipment in the measurement system, and Ethernet filters are available in the market that will cut off any other frequencies outside of the Ethernet communication band.

4.5 Penetrations

In addition to doors, there is a need to bring signals, fluids, and conditioned cooled air into the chamber. Indeed, some of the AESA antennas require liquid cooling for their amplifiers and circuits; these hoses must penetrate the shielding, since for both secrecy and safety the range must be shielded. Additionally, although not as common due to the proliferation of gas-based fire-extinguishing systems, water-filled pipes may have to penetrate the shield to feed sprinkler heads inside the shielded room.

Electrical signals and power can be fed into the shield via an RF filter, as described in Section 4.4. Optical signals can be fed through long pipes where the cutoff frequency of the pipe is higher than the highest frequency of interest.

For example, a quarter-inch inner-diameter brass pipe has its cutoff frequency at 27.67 GHz. Any signals below that frequency will propagate as evanescent modes, and thus will be attenuated as they travel down the pipe. The pipe must be long enough to get an adequate attenuation of the signals traveling through it to maintain a desired shielding effectiveness.

These pipes can be used for fiber optics, but also for liquids or gasses. However, it should be taken into account that liquids in the pipe will load the waveguide and lower the cutoff. Thus, knowledge of the dielectric properties of the material going through the pipe is important to ensure that the shielding is not compromised.

For A/C vents or just ventilation, the so-called honeycomb panels are the typical approach. The honeycomb panels are metal sheets folded and stacked and then brazed. The panel is surrounded by a frame that gets attached to the shielded wall. Figure 4.8 shows a honeycomb panel attached to a shield. The honeycomb approach can be used inside larger brass pipes to preserve the shielding while allowing fluids to flow.

Figure 4.8 A honeycomb penetration on a shield. (*Source:* Author's private collection.)

4.6 Testing the Shielding Effectiveness

The most popular method for testing shielding effectiveness are the methods described in IEEE Standard 299-2006 (R2012) [6], which is intended for room-size enclosures where the smallest linear dimension is equal to or more than 2m. The standard states that the test is to be performed prior to the installation of the RF absorber. The methods cover from 9 kHz up to 18 GHz, but it is known that the methodology is also used at frequencies above 18 GHz, and the document itself mentions a extended range up to 100 GHz [6].

The methods are divided into three different bands: the low range, covering from 9 kHz to 16 MHz, where magnetic fields are tested using electrostatically shielded loops; the resonant range covering from 20 to 300 MHz, where biconical antennas and dipoles are employed; and the high range from 300 MHz up to 18 GHz where dipoles and horns are the antenna used in the test.

All the tests call for a reference measurement where the antennas being used are located at a distance $r + t$, where t is the thickness of the shielding, and when performing the test the antennas are located at $r/2$ from the shield's outer surface. As part of the reference measurement, a dynamic range check is performed; the receive has to be linear for the entire dynamic range, and the dynamic range must be 6 dB higher than the shielding level to be measured.

The standard provides guidance on where to test the enclosure. It is no surprise that the test points are along the seams, door edges, and other penetrations. At each location, both polarizations are tested.

References

[1] Hemming, L. H., *Architectural Electromagnetic Shielding Handbook: A Design and Specification Guide*, New York, NY: IEEE Press, 1992.

[2] Federal Communications Commission, Part 15 Radio Frequency Devices, Subpart B–unintentional radiators 15.109, pp. 774–775.

[3] ANSI/IEEE C95.1-2005 IEEE Standard for Safety Levels with Respect to Human Exposure to Radio Frequency Electromagnetic Fields, 3kHz to 300 GHz.

[4] NSA, Specification NSA NO. 94-106, National Security Agency Specification for Shielded Enclosures.

[5] Baggett, M., and D. Hess, "Power Handling Considerations in a Compact Range," *Proceedings of 34th Annual Symposium of The Antenna Measurement Techniques Association AMTA 2013*, Columbus, Ohio, Oct. 6–11, 2013.

[6] IEEE Std. 299-2006 (R2012), IEEE Standard Method for Measuring the Effectiveness of Electromagnetic Shielding Enclosures.

CHAPTER 5

Anechoic Ranges for Far-Field Measurements

5.1 Introduction

As described in Chapter 2, one of the simplest approaches to measure antenna radiation parameters is to place the AUT in the range and locate the range antenna in the far field of said AUT. Although this can be done indoors and outdoors, this book concentrates on indoor measurement ranges. The most basic indoor range is the so-called rectangular chamber, which means simply that the footprint of the anechoic chamber is a rectangular shape. Rectangular footprint chambers are also commonly used in near-field measurements and in ranges where a parabolic reflector is used to measure the far field as discussed in Chapter 7. The main reason for using a rectangular footprint is that it is an easy construction versus constructing a sphere or an ellipsoid. A floor that supports four walls that support a ceiling is, after all, the simplest room that can be constructed. As we analyze the rectangular anechoic range, we determine that it has certain limitations. Some of those limitations are in part solved by the tapered anechoic chamber (also frequently referred to as simply a tapered chamber), a modification of the rectangular chamber that allows for far-field measurements at lower frequencies. As with everything in engineering, there is no "one size fits all," and tapered chambers have limitations that must be understood when operating them. This chapter describes how to size a far-field range, either rectangular or tapered, and how to size the absorber for a given desired operation at a given frequency range.

5.2 The Rectangular Anechoic Range

The rectangular footprint anechoic chamber, also known as rectangular chamber, is the most common type of range. It is based on the outdoor elevated range. The range antenna is located at a certain height over the ground, and the AUT is placed at the same elevation a distance away. The distance should be the far-field distance of the AUT. In an outdoor far-field range, certain approaches may be used to reduce the effects of the ground reflections, like selecting the height and range antenna so that the ground is not illuminated or by using fences that redirect the reflected energy away from the AUT. RF absorber is rarely used because of the cost of outdoor absorbers.

The rectangular indoor anechoic range substitutes the open space with lateral walls and a ceiling. In the rectangular range, the lateral surfaces, the side walls, and the end walls, as well as the ceiling and floor, are treated with RF absorber. The goal is to reduce the reflections from these surfaces to a level that is acceptable for the types of measurements that are to be conducted in the range. When designing a range, or analyzing a range for possible use in a set of measurements, it is first necessary to determine the acceptable error and uncertainty for those measurements.

This question of the acceptable error is related to what levels of reflected energy entering the quiet zone (QZ) are suitable to meet that error level. To answer this, let us reduce the problem to a two-dimensional chamber. This makes the physical optics approach much simpler. Let us assume a QZ of diameter D and a range antenna located at distance R inside a rectangular room with a length A and width B. Let us further assume that the range is located along the long axis of the room and that the center of the QZ is located a distance C from the closest end wall. We can then come up with a series of rays that bounce of the walls and enter the QZ as reflected energy. Figure 5.1 shows a series of these rays that bounce off the walls of the range and enter the QZ. For clarity, we limit the number of rays shown. Let us start with the rays that bounce off the end wall opposite the range antenna. These are rays labeled ξ and ζ. The angle of incidence of these rays onto the end wall is given by

$$\varphi_\xi = \varphi_\zeta = \tan^{-1}\left(\frac{\frac{D}{2}}{R + 2C}\right) \tag{5.1}$$

where φ_ζ is the angle of incidence of ray ζ, and φ_ξ is the angle of incidence of ray ξ. Equation (5.1) can be modified further by introducing the far-field distance equation as given by

$$R = \frac{2D^2}{\lambda} \tag{5.2}$$

Then substituting (5.2) in to (5.1) yields

$$\varphi_\xi = \tan^{-1}\left(\frac{\frac{D}{2}}{\frac{2D^2}{\lambda} + 2C}\right) \tag{5.3}$$

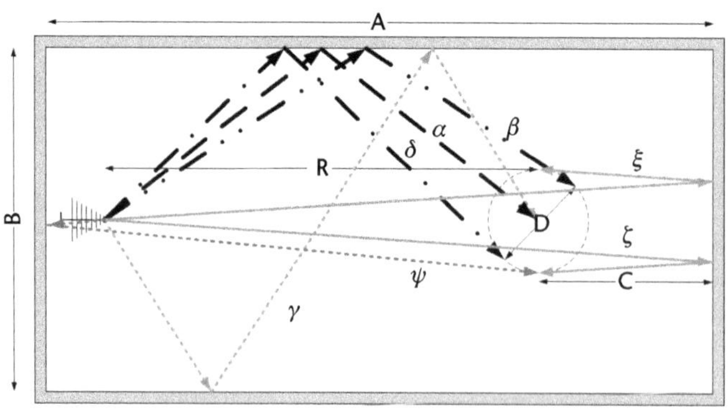

Figure 5.1 A sample of the reflected paths in a rectangular anechoic range.

Let us assume that $R \gg C$; hence (5.3) can be approximated as

$$\varphi_\xi = \tan^{-1}\left(\frac{\lambda}{4D}\right) \tag{5.4}$$

Now let us write this in terms of electrical size, such that $D = n\lambda$; then (5.4) reduces to

$$\varphi_\xi = \tan^{-1}\left(\frac{1}{4n}\right) \tag{5.5}$$

This is an approximation, but if we plot (5.5), it can be seen that as soon as the AUT or QZ is more than two wavelengths in size (that is $n \geq 2$), then φ_ξ is smaller than 10°. Figure 5.2 shows the plot of (5.5).

As shown in Chapter 3, the reflectivity of absorber changes very little for angles of incidence smaller than 10°. This means that we can assume that the reflected energy from the back wall entering the QZ will be on the level of the normal incident reflectivity of the absorber, which is commonly reported by the absorber manufacturers. It would appear that for electrically very small antennas that the back end-wall reflection could be higher. For example, for a $1/4\lambda$ antenna □ the angle of incidence on the back wall is 45°. However, recall that two assumptions have been made: (5.2) and that $R \gg C$. Nothing stops us from testing at a distance such that

$$R > \frac{2D^2}{\lambda} \tag{5.6}$$

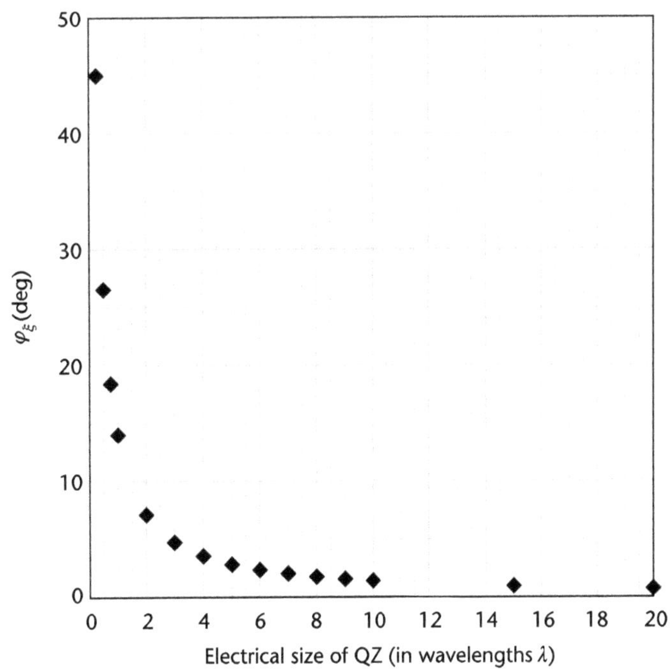

Figure 5.2 Angle of incidence on the end wall versus electrical QZ size.

Additionally, $R \gg C$ usually holds since C may be at least 2λ as shown below. In that case, (5.3) can be modified to be

$$\varphi_\xi = \tan^{-1}\left(\frac{\frac{\left(\frac{\lambda}{4}\right)}{2}}{\frac{4\left(\frac{\lambda}{4}\right)^2}{\lambda} + 2\lambda}\right) \tag{5.7}$$

Equation (5.7) reduces to $\tan^{-1}(1/18) = 3.18°$. Thus, φ_ξ is smaller than 10° so we can easily go back to having the end-wall reflectivity being equal to the normal incidence reflectivity.

Now that the end wall closer to the AUT has been dealt with, we can look at the opposite end wall of the range. Ray ψ is the reflected ray from the range antenna end wall that reflects back toward the QZ. In general, the range antenna is placed close to the end wall. There are some cases, however, where the antenna is mounted onto that end wall [1]. The angle of incidence for ray ψ is given by

$$\varphi_\psi = \tan^{-1}\left(\frac{\frac{D}{2}}{R}\right) \tag{5.8}$$

Notice that (5.8) is very similar to (5.1), and once we apply (5.2) and $R \gg C$ to (5.1), it will yield the same equation as (5.8). Hence, we can also declare that φ_ψ is smaller than 10° and that the range antenna end-wall contribution to the reflectivity is that of the normal incidence of the absorber treating said wall.

The remainder rays in Figure 5.1 are labeled α, β, δ, and γ. Ray γ is the only ray shown in the picture that has more than one bounce before entering the QZ. Both ray α and γ end up crossing the center of the QZ. The angle of incidence for the multiple bounce ray is given by

$$\varphi_\gamma = \tan^{-1}\left(\frac{R}{2B}\right) \tag{5.9}$$

Ray α has an angle of incidence onto the absorber given by

$$\varphi_\alpha = \tan^{-1}\left(\frac{R}{B}\right) \tag{5.10}$$

Equations (5.9) and (5.10) show that the tangent of φ_γ is half of the tangent of φ_α. Expressed in a different way, $\varphi_\gamma < \varphi_\alpha$. Typically, that means that the reflectivity of the absorber for φ_γ is smaller than the reflectivity for φ_α. We can state this as $\Gamma_\alpha \geq \Gamma_\gamma$. In addition, ray γ goes through two bounces so the magnitudes of the rays are $|\Gamma_\alpha|$ for the single-bounce ray and $|\Gamma_\gamma^2|$ for the two-bounce ray. Since for the absorber typically $|\Gamma| < 1$, it follows that the magnitude of ray α is larger than the magnitude of ray γ. Since the QZ level is defined as the highest level of reflected energy entering the QZ, it is a more than acceptable approximation to ignore the second-order bounces.

The other two rays that limit the reflected "beam" illuminating the QZ are the ray labeled β and the ray labeled δ. The angle of incidence for ray δ is given by

$$\varphi_{\delta} = \tan^{-1}\left(\frac{R - \left(\frac{\sqrt{2}}{2}\right)D}{B + \sqrt{2}D}\right) \tag{5.11}$$

The angle of incidence for ray β is given by

$$\varphi_{\beta} = \tan^{-1}\left(\frac{R + \left(\frac{\sqrt{2}}{2}\right)D}{B - \sqrt{2}D}\right) \tag{5.12}$$

It is obvious that $\varphi_{\delta} > \varphi_{\alpha} > \varphi_{\beta}$. In general, $R \gg 0.707D$ and also $B > 1.414D$ so using φ_{α} as the angle of incidence onto the QZ for far-field rectangular chambers is a fair approximation to get the expected QZ level.

From this discussion on the angles of incidence for the reflected rays, the most important conclusion is that for a rectangular far-field chamber, the widest angle of incidence for a single-bounce ray is the specular bounce from the lateral surfaces. Hence, the lateral surfaces of the rectangular range are the critical ones. In Section 5.2.1, the angles of incidence described here are used to size the chamber for a specific performance level. The rules and procedures can also be used for the analysis of existing chambers. This should be very useful to readers that use existing ranges where there is little information about their performance.

5.2.1 Sizing the Chamber

Sizing a rectangular chamber, or for that matter, any indoor range, starts with understanding the largest electrical and physical size of the AUT. As discussed in Chapter 2, indoor far-field chambers are usually limited to antennas that are five wavelengths in size, and these correspond to the physically largest antenna at the lowest frequency of operation.

There are some rules of thumb around in the industry, including the famous one that the width and height of the range should be three times the diameter of the QZ. I have been guilty of using this rule in the past. However, simple trigonometry, as used above, casts doubt over this rule's validity. Looking at the critical angle of incidence, the angle is given by (5.10). If we insert the far-field rule (5.2) and the rule of thumb $B = 3D$, then we arrive at

$$\varphi_{\alpha} = \tan^{-1}\left(\frac{\frac{2D^2}{\lambda}}{3D}\right) \tag{5.13}$$

This can be reduced to

$$\varphi_{\alpha} = \tan^{-1}\left(\frac{2D}{3\lambda}\right) \tag{5.14}$$

Now expressing the QZ size in terms of wavelengths, this reduces to

$$\varphi_{\alpha} = \tan^{-1}\left(\frac{2n}{3}\right) \tag{5.15}$$

For a one-wavelength quiet zone, the angle of incidence is 33°. However, the angle of incidence for a two-wavelength antenna is already 63°. As shown in Chapter 3, this requires absorber that is at about four wavelengths in size to get an adequate absorption. For a four-wavelength antenna, the angle of incidence is already in the 70° range, requiring absorber of about seven wavelengths to get close to 40 dB of absorption. I do not know the origins of this rule of thumb and what the original assumptions were. Some people claim that the rule is actually three-times the QZ diameter between the tips of absorber, not wall to wall. One explanation may come from the rule being applied primarily to the X-band range where 1-ft absorber is already 8.3λ at the lower end of the band. Whatever the origin of the rule, it should be used with care and with understanding of its potential limitations.

For the size of the chamber, we can develop an equation based on the far-field distance, which is one of the limiting factors. Following the parameters shown in Figure 5.3, the following equation can be written

$$A = R + C + L_a + S_a \tag{5.16}$$

We can rewrite C as

$$C = \frac{D}{2} + t_{e1}\lambda + B_s \tag{5.17}$$

At a minimum, B_s should be two wavelengths at the lowest frequency (λ_1). So using this lowest frequency of operation wavelength, we can rewrite the size in

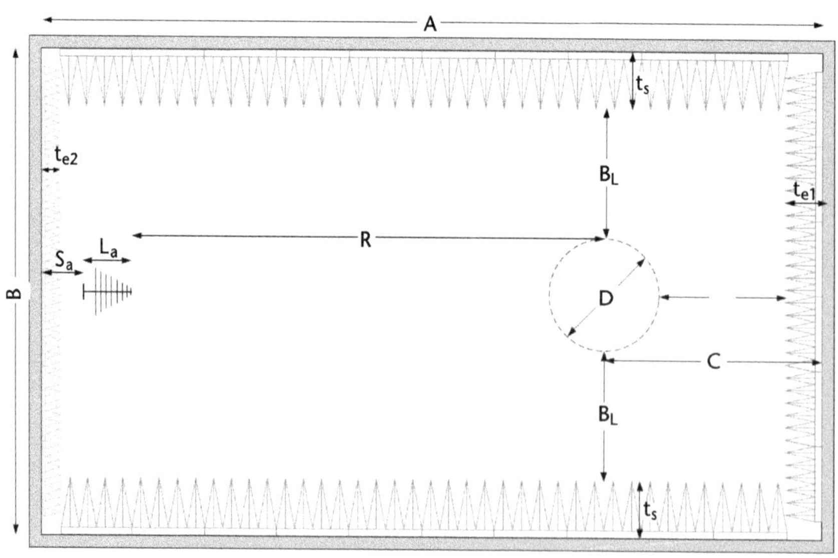

Figure 5.3 A rectangular chamber.

terms of that wavelength using also the far-field criteria (5.6). Notice that we are choosing the path length based on the lowest frequency and that we are assuming that as the frequency increases, the electrical size of the antenna remains constant or gets smaller and that the physical size of the antenna is reduced. That was one of the limitations of far-field rectangular chambers: the electrical size of the antenna is limited.

Hence, the length in terms of the lowest frequency of operation is given by

$$A = \left(2(n)^2 + \frac{n}{2} + 2\right)\lambda + t_{e1}\lambda + L_a + S_a \tag{5.18}$$

Let us stop there for the length and look at the width (and height). The width is clearly given by

$$B = 2\left(t_s\lambda\right) + n\lambda + 2B_L \tag{5.19}$$

Traditionally we have been told that $B_L = B_S = 2\lambda$. Let us at this moment start looking at the angle of incidence from (5.10) and determine an adequate angle of incidence to provide an adequate absorption. We are going to assume that the angle of incidence is from the tip; hence (5.10) is rewritten as

$$\varphi_\alpha = \tan^{-1}\left(\frac{R}{D + 2B_L}\right) \tag{5.20}$$

Let us now insert the far-field condition, and the electrical size of the QZ. Then the equation becomes

$$\varphi_\alpha = \tan^{-1}\left(\frac{2n^2\lambda}{n\lambda + 4\lambda}\right) \tag{5.21}$$

Hence, we can write the size of the chamber based on the desired angle of incidence and the QZ level we will need, and since using (5.2) and $D = n\lambda$, it follows that

$$B_L = \left(\frac{2n^2 - n\tan\varphi_\alpha}{\tan\varphi_\alpha}\right)\lambda \tag{5.22}$$

The electrical size of the space between the QZ and the tips of the absorber is a function of the angle of incidence and the electrical size of the antenna under test. Plotting (5.22) for those two parameters we obtain an interesting plot presented in Figure 5.4.

Figure 5.4 shows that an indoor rectangular range appears to be quite a useless piece of real estate. Indeed, I limit their use to 5λ, where the test distance will be about 50λ if we use the $R = 2D^2/\lambda$. Figure 5.4 illustrates that a 20-λ spacing between the QZ and the absorber is required to keep an angle of incidence better than 65°. That seems to be a lot, but consider that at 5.8 GHz, that means that the test distance is 2.60m and the height of the chamber from tip to tip of absorber is 2.34m (more than adequate for a human to walk into the range and set up the AUT onto

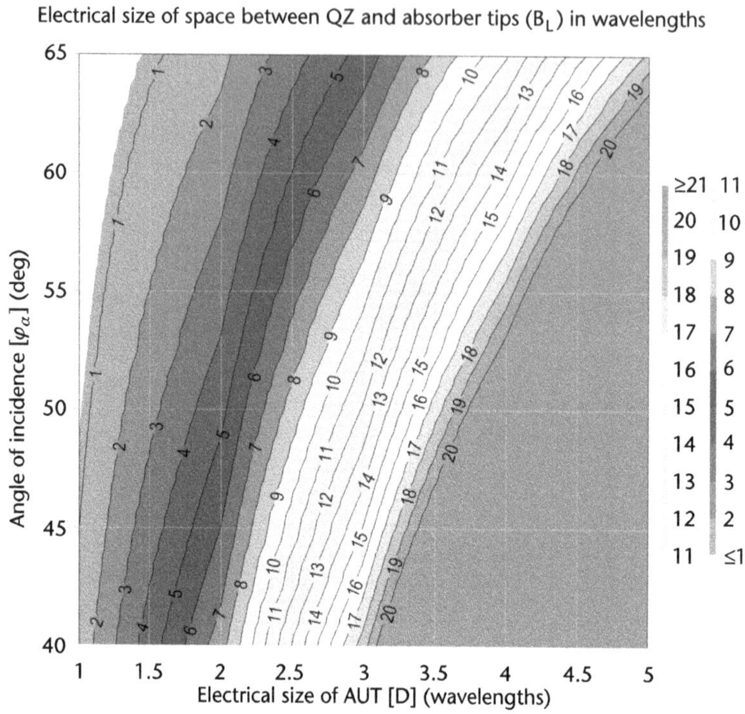

Figure 5.4 Spacing between QZ and absorber tips for different AUT electrical sizes and angles of incidence onto the absorber.

its positioner). At that angle of incidence, the 12-inch absorber is about 6λ and, as shown in Chapter 3, that provides better than 40 dB of reflectivity level. The limitation on the size of antennas that can be tested economically in a rectangular indoor range is the reason that compact ranges and near-field techniques were developed.

5.2.2 Absorber Layout

Now that we have a rough size for the rectangular range, the RF absorber required to treat it must be designed. In (5.18) and (5.19) we have not specified the required thickness of the absorber. Specifying that will start with what is the desired QZ level. We have defined the QZ levels as the highest level of reflected energy entering the QZ compared to the direct path between the source antenna and the QZ. When measuring an antenna, we are interested in measuring the features of the antenna radiation pattern. Those features include the beamwidth, the sidelobe level, and the front-to-back ratio, directivity, and gain. Let us consider Figure 5.5.

In Figure 5.5 the range antenna is measuring the pattern of the AUT. As the AUT is rotated, its main beam radiates toward the walls of the range. The sidelobes on the AUT are at about –12 dB. The range antenna is radiating about –4.5 dB in the 36° direction. That energy is absorbed by the side wall absorber, but a portion of it is reflected to the QZ. Hence, the direct path is measuring –12, but there is a –4.5 dB plus the reflected level that is being measured as well. If the absorber that lines the wall has an absorption of –25 dB at 54° of incidence (which is typical for absorber that is 2λ in thickness), then the reflected path is –29.5 dB. An interference

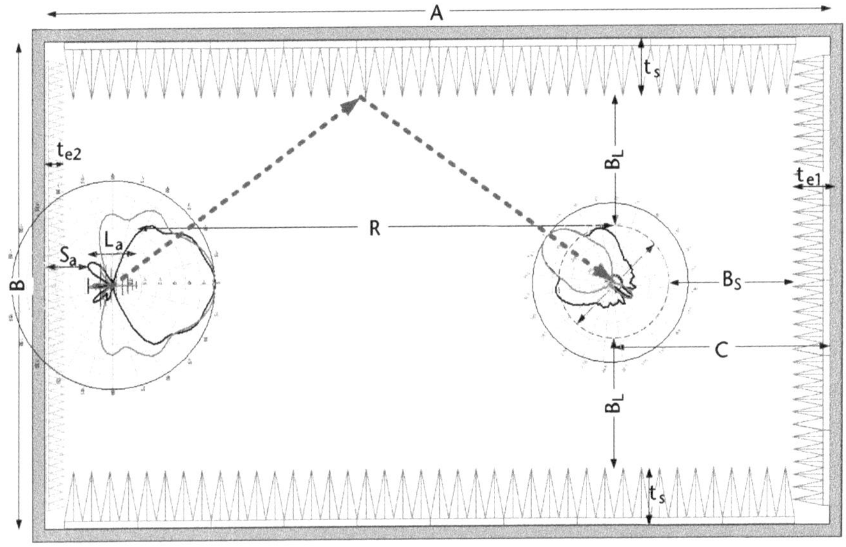

Figure 5.5 The rectangular anechoic chamber with a range antenna pattern and an AUT pattern.

pattern will be created by the two waves, the reflected and the direct. The maximum possible variation in amplitude as these two waves with different amplitude add and subtract is given by

$$\pm E_{\text{ripple}}(\text{dB}) = 20 \cdot \log_{10}\left(\left|1 \pm 10^{\left(\frac{E_{\text{reflected}}(\text{dB}) - E_{\text{direct}}(\text{dB})}{20}\right)}\right|\right) \tag{5.23}$$

The reflected minus the direct in decibels is −17.5 dB so the ripple in amplitude is −1.24 dB to +1.08 dB. That is, the sidelobe measurement will be between −10.91 and −13.24 dB, depending on the reflection being on phase or completely out of phase.

Another source of loss necessary to consider is the difference in path loss. The direct path has a length of R; hence the path loss is given by

$$PL_{\text{direct}}(\text{dB}) = 20 \cdot \log_{10}\left(\frac{4\pi R}{\lambda}\right) \tag{5.24}$$

The reflected path has a loss that is given by

$$PL_{\text{reflected}}(\text{dB}) = 20 \cdot \log_{10}\left(\frac{4\pi\sqrt{R^2 + \left(D + 2B_L\right)^2}}{\lambda}\right) \tag{5.25}$$

So the difference between the two paths is the reflected minus the direct or

$$\Delta PL(\text{dB}) = 20 \cdot \log_{10}\left(\frac{\sqrt{R^2 + \left(D + 2B_L\right)^2}}{R}\right) \tag{5.26}$$

Notice that the path loss difference is constant with frequency. Also notice that making the chamber wider and taller will increase the additional loss. Substituting $D = n\lambda$, (5.22), and (5.2) into (5.26) it yields

$$\Delta PL(\text{dB}) = 20 \cdot \log_{10}\left(\frac{\sqrt{4n^4 + \left(n + \left(\frac{4n^2 - 2n\tan\varphi_\alpha}{\tan\varphi_\alpha}\right)\right)^2}}{2n^2}\right) \tag{5.27}$$

This is an equation that provides the additional path loss of the reflected specular ray as a function of the angle of incidence and the electrical size of the antenna under test. Equation (5.27) is plotted in Figure 5.6.

It should be understood that the desired QZ level should be such that the desired features of the pattern of the antenna can be measured within some acceptable error. That error can be estimated from (5.23), which we can use to calculate the level of the QZ reflectivity that we desire to obtain a given error in our measurements of the pattern features. The reader should understand that I am concentrating on the error contribution from the chamber. There are other sources of error and uncertainty such as alignment, calibration uncertainty of gain standards, repeatability of the connections, and mismatch.

Although I like to limit the use of rectangular ranges to 5-λ antennas, this is only a recommendation. It does not mean that antennas that are 6λ cannot be measured

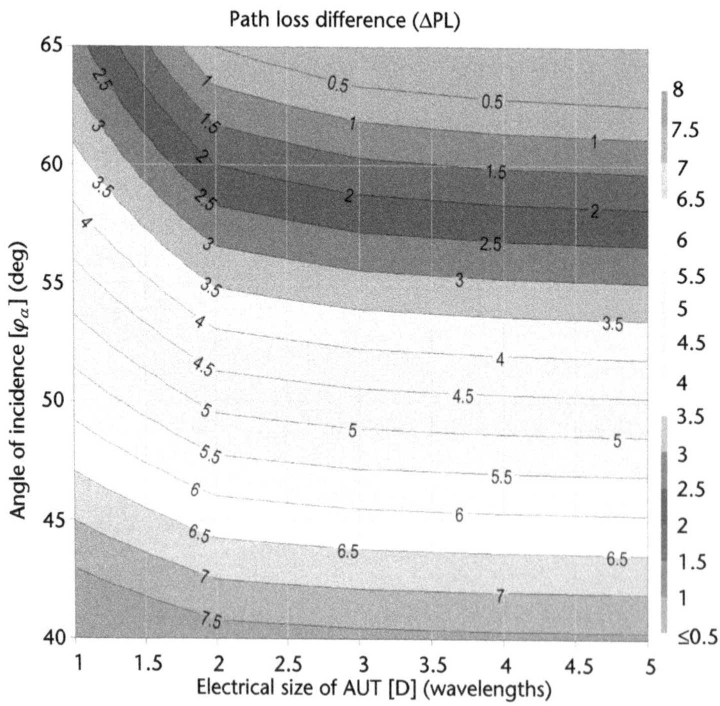

Figure 5.6 Additional path loss for reflected specular ray as a function of AUT electrical size and angle of incidence.

in a rectangular range. However, a 6-λ antenna requires a path length of 72λ to be in the far field; that could be a rather long chamber. Let us look at a typical pattern for 5-λ antennas. Let us assume a 2-, 3-, 4-, and 5-λ lines of uniform distribution. As shown in [2] the patterns for such antennas will be

$$F(\theta) = \frac{\sin(m\pi \sin\theta)}{m\pi \sin\theta}$$

where m is the number of wavelengths. If we plot such an equation in decibels, we get the radiation patterns shown in Figure 5.7.

If we look at the 5-λ line, it is the pattern that is more directive (narrower beam) and has the lower sidelobe levels, as expected. The radiation pattern shows that the first sidelobe is −13.3 dB at about 17°. The second sidelobe is about 17.9 dB below the peak at 29°. The next two sidelobes are −20.7 dB and −23 dB. A −30-dB QZ reflectivity level could potentially cause an error of −1.37–+1.18 dB for the first sidelobe. The second sidelobe could potentially have an error of −2.5–+1.9 dB. The third and fourth sidelobes could potentially exhibit errors of −3.64–2.56 dB, and −5.14–+3.2 dB, respectively. Are these errors acceptable? Let us assume that they are as we select the absorber for the different surfaces of the rectangular range.

5.2.2.1 End Wall Absorber

In Figure 5.1, the end wall behind the QZ has the critical reflection at angles closer to normal incidence. There is also some additional path loss given by the spacing

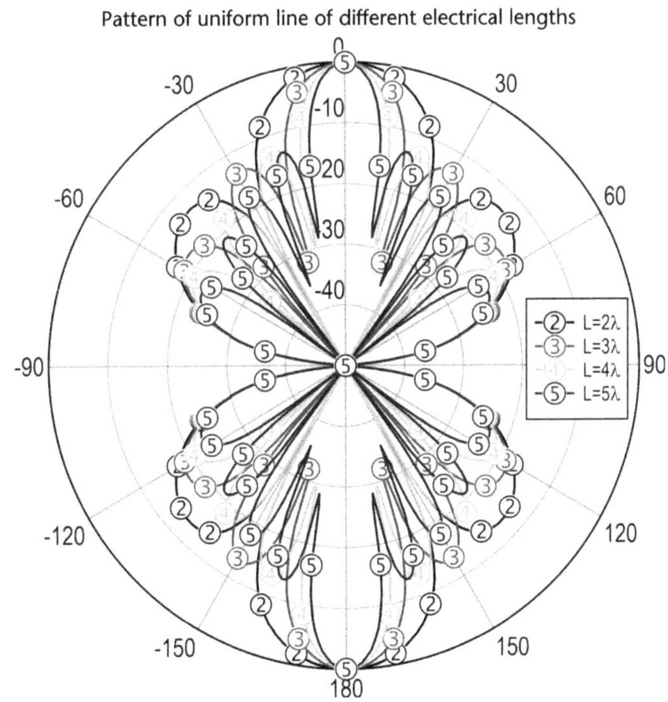

Figure 5.7 The radiation pattern of a 5-λ-long source line.

B_S as shown in Figure 5.3. Thus, for the end wall behind the QZ we choose an absorber where the normal incidence exceeds the QZ reflectivity that we desire, which, in the present example, is −30 dB.

The parameter t_{e1} in (5.18) can be set typically to 1, making the absorber 1λ. Figure 3.9 shows that such absorber typically provides 30 dB or better of reflectivity. It must be remembered that the wavelength is at the lowest frequency of interest, and that as frequency increases, the reflectivity will improve as the absorber becomes electrically longer.

The opposite end wall also exhibits angles of incidence that are very close to normal incidence as illustrated in Figure 5.1. The other factor is the pattern of the range antenna. Let us assume that the range antenna is the one whose pattern is shown in Figure 5.5. The range antenna has a front-to-back ratio of about 15 dB. That number should be added to the normal incidence reflectivity of the absorber lining the end wall behind the range antenna. Many chambers are designed following this approach where the front-to-back ratio of the range antenna is used to reduce the required thickness of the pyramidal treatment of the end wall. This should be kept in mind if the antenna is changed for one with a different front-to-back ratio. As a guide, the t_{e2} parameters in Figure 5.5 and Figure 5.3 can be chosen to be half the thickness of t_{e1}; hence $2t_{e2} = t_{e1}$. Figure 3.9 shows that a 0.5-λ absorber provides about −17 dB; hence a 15-dB front-to-back ratio will equalize the reflection from the two end walls. Because the absorber layout is dependent on the pattern of the range antenna, care should be taken when switching antennas.

The additional path loss must be taken into account. For the range antenna end wall, this is a potentially very small number since S_a is usually small. For the AUT side end wall, the difference in path length is at a minimum $2B_s$. As we have noted, we aim to have $B_s = 2\lambda$. Hence the additional path length for a n-λ antenna at the lowest frequency has a loss given by

$$\Delta PL(\text{dB}) = 20 \cdot \log_{10}\left(\frac{2B_s + R}{R}\right) = 20 \cdot \log_{10}\left(\frac{2}{n^2} + 1\right) \tag{5.28}$$

Notice that as the antenna gets larger electrically (that is, n increases) the additional path loss approaches 0 dB. For the typical largest antenna recommended for test in a rectangular range, $n = 5$, the additional path loss is 0.67 dB with the assumption that the separation to the absorber is the minimum 2λ. Increasing the separation to 4λ changes (5.28) to

$$\Delta PL(\text{dB}) = 20 \cdot \log_{10}\left(\frac{4}{n^2} + 1\right) \tag{5.29}$$

and for that separation a 5-λ-sized AUT will have an additional path loss increase of 1.3 dB.

5.2.2.2 Lateral Surface Absorber

The absorber that lines the lateral surfaces is critical in a rectangular far-field chamber. The angle of incidence onto those surfaces is not at normal incidence; hence, the absorber must be the thickest one in the range. Choosing the lateral surface absorber is a chicken and egg problem. If we are designing a chamber, then the approach is to start with the desired QZ reflectivity. If we try to stay with the −30-dB that was chosen for the end wall absorber in the previous section, then Figure 3.9 shows that a 3-λ-thick absorber will provide that level at about $\varphi_\alpha = 57.5°$. Looking at (5.22), it can be shown that for a 5-λ antenna and an angle of incidence of 57.5°, B_L is roughly 27λ. Thus, the overall width (and height) of the chamber will be 65λ. Figure 5.6 shows that, for that geometry, we have an additional 9 dB of path loss. The same 3λ has a reflectivity of −20 at 65°. B_L can be reduced to 17λ. Figure 5.6 shows for that angle and a 5-λ antenna size, the additional path loss is 6 dB. This places the reflected path at −26 dB. This is not ideal, but the pattern of the range antenna provides an additional −2 dB given its gain in the 24° direction, which will set the QZ highest reflectivity to −28 dB. The increase by 2 dB of the reflectivity of the QZ, from −30 to −28 dB, has increased the error on the hypothetical sidelobe levels from −1.37 dB to +1.18 dB to −1.77 dB to +1.47 dB for the first sidelobe. The second sidelobe could have an error of −3.25–2.35 dB instead of −2.5–+1.9 dB. The third and fourth sidelobes could potentially exhibit errors as large as −4.98–+3.15 dB for the third sidelobe and as high as −7.2–+3.9 dB for the fourth sidelobe. If this increase in potential error is not acceptable, the approach is to either increase the size of the lateral wall absorber or increase the size of the chamber to obtain a better angle of incidence.

Using (5.18) and (5.19), we see that the hypothetical indoor range will be a 45-λ width by a 45-λ height by about 60-λ-long. This range will provide −28 dB of reflectivity to measure antennas that are 5λ in size. At 1 GHz this will be 13.7m by 13.7m by 18.3m to measure an antenna that is about 1.5m. This seems to be a lot of real estate to measure such an antenna. This requirement for such a large room to measure a fairly small antenna is one of the reasons for the near-field and compact-range approaches to antenna measurements.

5.2.2.3 Implementation of the Absorber Layout

Section 5.2.2.2 provides guidance on selecting the absorber for the end wall and the lateral surfaces. Whichever are the absorber types chosen for the different surfaces of the range there are some recommended practices for using them. There is a series of questions that we must ask regarding a range layout. For example, do we need to cover the entire lateral surface with the required absorber? How do we treat the edges and corners? What is the best place to locate the door? What about vents, air conditioning, and lighting? Where do we locate the walk-on absorber? This section answers these questions. Please note that the information provided here is applicable to other ranges, such as near-field, far-field, and compact ranges.

5.2.2.3.1 Lateral Surface Treatment

In the rectangular chamber the lateral surfaces will have the longer or thicker absorber treatment. However, it is not necessary to have that treatment through the entire wall or floor. Going back to Figure 5.1, while rays α, β, and δ clearly require the thicker absorber, ray γ does not require the thickest absorber treatment for its first bounce. The so-called specular treatment is the area of the lateral surface that is covered by the critical absorber, that is, the thickest one. The rest of the areas can have shorter material as a treatment.

There is a caveat. Having a specular treatment in a section of the lateral surfaces works better on larger rooms, where the separation between the QZ and the absorber tips (B_L) is large. Experience has shown that for chambers where the separation between the QZ edge and the tips is the 2-λ minimum, the edge diffractions from the end of one type of absorber treatment and the beginning of the other type of treatment can cause additional ripple on the QZ field.

The size of the specular treatment is determined based on the size of the QZ and the section of the lateral surface where single-bounce rays exist. Hence, looking at Figure 5.1, we can look at the angles formed by rays α, β, and δ. The angles of incidence are given by (5.10) to (5.12). Notice that we are assuming the angle of incidence from the back of the absorber, as opposed to (5.20), where we assume the reflection from the tips of the absorber. This is done for simplicity, but a more rigorous analysis can be done assuming the reflection from the tips for all the rays. Assuming the reflection from the back of the absorber is a good approximation for absorbers that are up to 2.7λ in thickness, if the absorbers are larger than 2.7λ, using the tips as the point of reflection is a better approximation. From (5.10) to (5.12) an equation can be written to find the center of the bounce from the range antenna; hence, for a given φ_χ where χ is either α, β, or δ, we get

$$R_\chi = \frac{B}{2}\tan\varphi_\chi \tag{5.30}$$

The length of the specular region is $\Delta R_{\text{specular}} = (R_\beta - R_\delta) + 2\lambda_{\text{LF}}$ centered at R_α from the range antenna. The width is given by $\Delta W_{\text{specular}} = D + 2\lambda_{\text{LF}}$, where D is the diameter of the QZ.

Another approach, discussed by Hemming in the appendices to [3] is to calculate a number of Fresnel regions to estimate the size of the specular region. Hemming's approach comes from the treatment of outdoor ranges and it was originally presented in Chapter 14 of [4], in which Lyon et al. describe the use of Fresnel zones to estimate the area to be treated on the ground of an elevation range.

Starting from the geometry in Figure 5.8, the length of the specular ray path can be calculated using $R_{R0} = \sqrt{Ro^2 + (h_t + h_r)^2}$ where the grazing angle is $\psi = \tan^{-1}[(h_t + h_r)/Ro]$. The grazing angle is also 90° minus the angle of incidence. Lyon et al. go on to define the path via any other arbitrary point on the surface arriving at the equation for the length of the dashed reflected path shown in Figure 5.8.

That other reflected path is given by $R_R = \sqrt{h_t^2 + y^2 + z^2} + \sqrt{h_r^2 + y^2 + (Ro - z)^2}$. There is a phase difference between these two paths; when we set the phase difference to be an integer of π, we define the inner and outer boundaries of a given Fresnel region. Hence, the Nth Fresnel zone is limited by the outer boundary $R_r - R_{R0} = N\lambda/2$ and by the inner boundary $R_r - R_{R0} = (N - 1)\lambda/2$.

Lyon et al. [4] proceed to define the width, length, and center of the outer boundary of a given Fresnel region. To facilitate the calculation, they define three parameters using the following equations:

$$F_1 = \left(\frac{N\lambda}{2Ro} + \sec\psi\right) \tag{5.31}$$

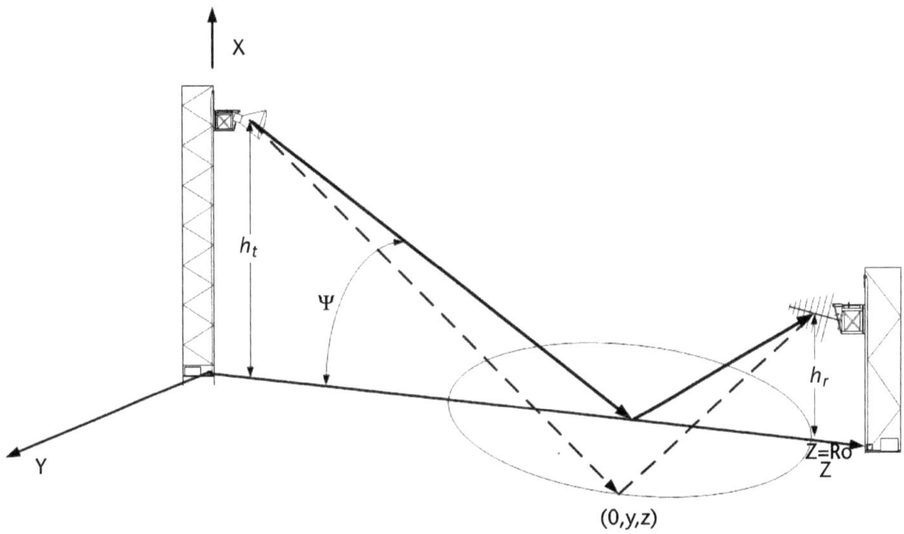

Figure 5.8 Geometry for the Fresnel zone derivation for an outdoor range.

$$F_2 = \left(\frac{h_t^2 - h_r^2}{\left(F_1^2 - 1\right) Ro^2} \right) \tag{5.32}$$

$$F_3 = \left(\frac{h_t^2 + h_r^2}{\left(F_1^2 - 1\right) Ro^2} \right) \tag{5.33}$$

Using (5.31) to (5.32), the center, as measured from the range antenna, the width, and length of the Nth Fresnel region, can be computed using

$$C_N = Ro \frac{\left(1 - F_2\right)}{2} \tag{5.34}$$

$$L_N = Ro F_1 \sqrt{1 + F_2^2 - 2F_3} \tag{5.35}$$

$$W_N = Ro \sqrt{\left(F_1^2 - 1\right)\left(1 + F_2^2 - 2F_3\right)} \tag{5.36}$$

where (5.34) is used to compute the center of the Nth Fresnel region, and (5.35) and (5.36) give the length and width of the Nth. Hemming claims that $N = 6$ is sufficient for a specular treatment on a chamber [3]. The basis of this seems to be empirical, but it may be arbitrary. I have successfully used $N = 5$ and even $N = 4$ in some cases.

How do we treat that specular area of a lateral surface of a rectangular range? One of the proven approaches is the diamond shape (Figure 5.9(a)). This treatment is more challenging to install, but it has the advantage of possibly improved performance at higher angles of incidence because no flat surfaces are presented to the incoming wave. I am not sure of its benefits. For one, as frequency increases, the material properties of typical absorber approach those of air, so the flat surfaces will have a smaller effect. On the other side, the reflections from those flat surfaces will travel toward the source of the wave (either the range antenna or the AUT) so not in the direction of the receiving antenna. The issue of backscattering is more of a RCS problem so the diamond mount does not seem to provide a huge improvement for antenna measurement ranges.

The staggering approach looks to increase the potential backscattering (Figure 5.9(b)). The idea is that increased backscatter will reduce the bistatic (which is what needs to be reduced in antenna ranges). While easier than the diamond shape to install, it is still a more labor-intensive approach. The third approach is to just cover the necessary specular region with the thicker absorber (Figure 5.9(c)). Since absorber, regardless of the pyramid height, typically comes in a standard base footprint, this is the easiest approach, and there is no apparent diminishing performance from using it.

5.2.2.3.2 Treating the Edges and Corners

At any edge of the rectangular room we face the problem of meeting the two absorber treatments that cover each of the adjacent walls. One approach is to do miter cuts. Miter cuts are a very elegant approach; clearly this type of edge treatment looks good and performs well. While the miter cut approach may be a better choice in smaller ranges where the specular treatment runs into the side walls, it is in most cases not necessary. It is a more expensive implementation, and it only works when the adjacent walls have the same type of absorber treatment. Figure 5.10 illustrates the miter cut approach. The absorber pieces are cut at an angle to treat the corner. The idea is good, but the implementation is very time-consuming. While some manufacturers advertise miter cuts available in their catalogs, these pieces often do not meet exactly at the corners during installation. Since most absorber is a foam,

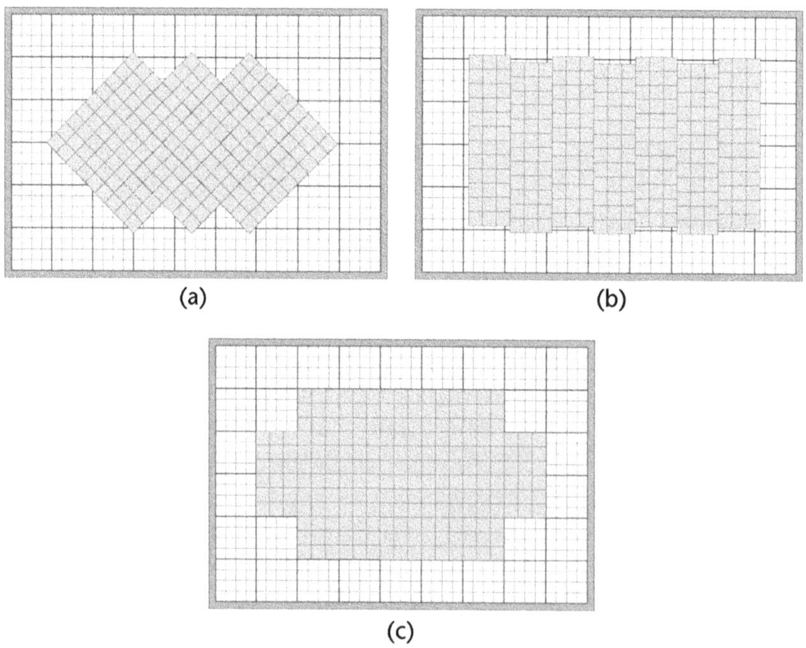

Figure 5.9 Specular treatment of lateral surfaces: (a) diamond, (b) staggered, and (c) standard.

the tolerance of the cuts can be as large as ±0.25 cm; as pieces are being placed next to each other over a large wall there may be large deviations between two adjacent rows. The other treatment for edges is also shown in Figure 5.10. The use of LFB to treat the edge is popular, and it is easier to install. Additionally, it allows for two different treatments on adjacent walls. The key to this approach is to avoid long edges visible from either antenna, as the long edge can give a large back scatter in one polarization.

5.2.2.3.3 Doors

The range must be accessible, since at some point we must enter it, mount the antenna, and then exit; therefore, doors are a necessity. Do not be shy with the size of the door. If your QZ diameter is D, do not install a door that is smaller than D in opening. Two types of doors are recommended for ranges: the equipment door and the personnel door.

The equipment door is the one used to bring the AUT into the range. It is also used during upgrades and maintenance to bring equipment into the range. It should be close to the QZ and be large enough to allow for equipment to be brought into the range.

The equipment door can also be used as a personnel door; however, I recommend a smaller door for personnel access. The reason is that the equipment door is usually large. If the range has its own climate control that is separate from the rest of the lab, having a large door open while an AUT is installed could increase the temperature of the range. Once the door is closed, there is a period of time during which the range reaches its optimum temperature for testing. (This is extremely important in near-field ranges because of the phase measurement required.) While in far-field ranges, phase measurements are not that critical, phase on cables can be affected by temperature. In the far-field range, LNAs and mixers may be part of the measurement system, and the performance of these may be affected by temperature. Temperature may not be a problem in a lot of ranges, but it is something to keep in mind.

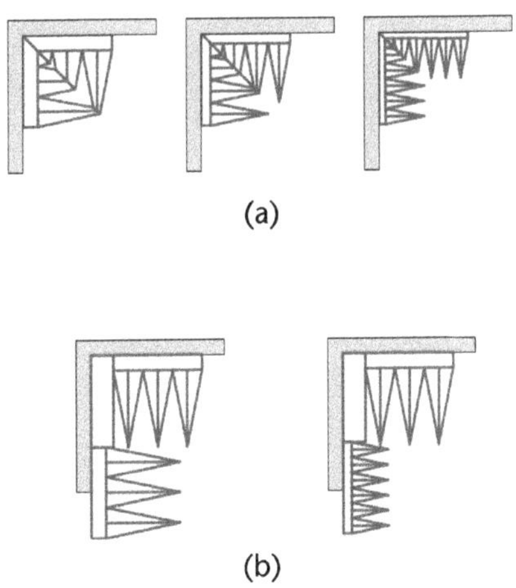

Figure 5.10 Edge treatments: (a) miter cuts for different absorbers, and (b) LFB edge treatment.

Location of the door is critical. The end wall behind the AUT is not ideal. If the door must be placed in that wall, care should be taken to avoid gaps or long flat LFB treatments toward the center of the wall. Ideally, the doors should be on the side walls, but not in the specular region; instead, they should be a bit behind or in line with the QZ. The personnel door can be in line with the range antenna location (Figure 5.11(d)). If the door must be placed on the end wall behind the QZ, if possible it should be a sliding door or a single-leaf door (Figure 5.11(b, c)). However, there is a limit to the opening size of the single-leaf door, and sliding doors are expensive. If a double door, such as the one in Figure 5.11(a) must be used, the treatment on the door is critical. One option for these doors is to use an "absorber plug." These are large false walls mounted on casters that are placed in front of the wall. Absorber plugs work, but I find them to be inconvenient.

5.2.2.3.4 Walk-on Absorber

As mentioned in Chapter 3, a walk-on is not a very good absorber. The Styrofoam insert and the hard layer that covers it reduce significantly the level of absorption of this material. Its use should be minimized to areas outside of the specular treatment. It is acceptable next to the walls and behind or on the sides of the QZ. If a walk-on is required, use the next smaller size to the surrounding absorber. Because a walk-on has a Styrofoam insert and a hard layer, its height is larger than the equivalent standard absorber. The reason for using the next size down is to hide the edge of this material under the surrounding absorber, since having the edges tower over the surrounding material will cause edge diffraction that could increase the ripple on the QZ and increase the uncertainty and error of the measurement.

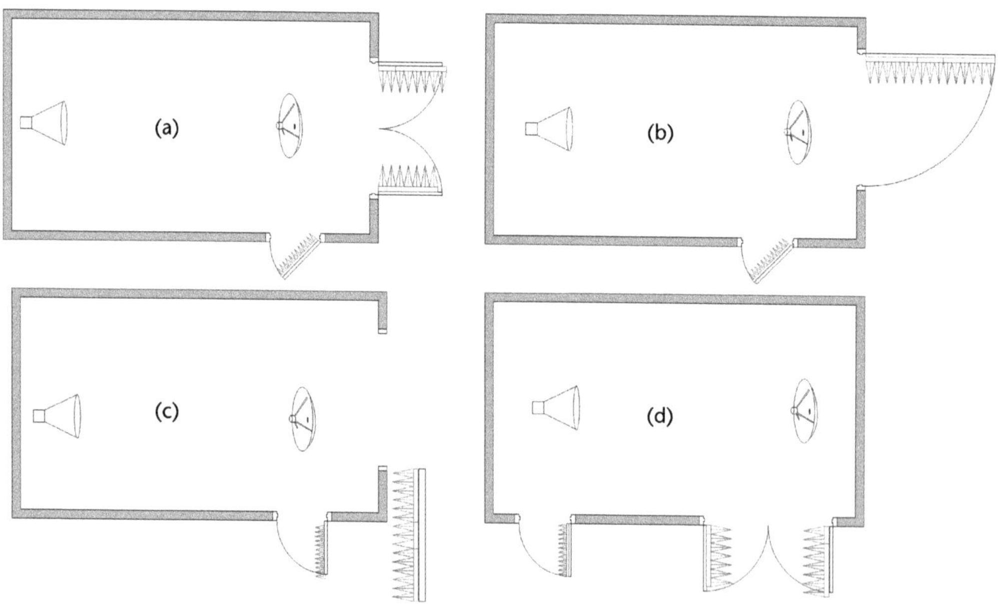

Figure 5.11 Some configurations of doors for a rectangular range: (a) equipment double door on end wall and personnel door on side, (b) single-leaf equipment door on end wall and personnel door on side, (c) sliding equipment door on end wall and personnel door on side, and (d) equipment and personnel door on side.

5.2.2.3.5 Vents, Air Conditioning, and Lighting

Temperature and humidity should be kept constant in the range. Nowadays, there are more and more manufacturers that are changing the fire retardants of the absorber, so the materials are more impervious to humidity. Temperature in the range affects the power handling of the absorber as discussed in Chapter 3. However, the importance of maintaining a constant temperature in the range is related to the measurement systems. Cables, mixers, and LNAs may change their behavior as the temperature changes.

More important, during normal range operation, when personnel access the range and install new range antennas or the AUT, this will change the temperature of the range if the outside area is not climatized.

Vents to allow cool air in and to exhaust hot air are necessary. These should be located in noncritical areas of the range [e.g., close to (but not on the) the end walls on in noncritical areas of the lateral surfaces]. If necessary, these vents can be covered with filter foam, but there is a significant reduction in pressure due to their presence, and the air flow is significantly reduced. It is desirable to work in a comfortable environment, and there is nothing worse than working inside a hot, poorly illuminated range. (I speak from personal experience.)

Regarding illumination, it is desirable to have light inside the range—at least enough to make it pleasant to work there. However, lights mean that areas of the absorber layout must be disrupted to locate lamps. Much like vents, lights should be avoided in the specular areas. Some light may be required above the range antenna location, as it makes it easier to exchange antennas if needed. The QZ should also be illuminated. When illuminating the QZ, it is better to be to the sides and a bit behind the center of the QZ location.

There are some new interesting methods for lighting up a range. For one, the LED revolution has arrived in the range business. LEDs seem to produce a light spectrum that makes ranges appear brighter than with traditional lights. Another interesting innovation is the use of dielectric light guides (basically thick fiber-optic cables) to bring light from a source outside the range to areas inside the range. These dielectric lights are less intrusive and minimize the disruption of the absorber treatments.

5.2.3 Positioners and Range Antennas

5.2.3.1 Range Antennas

Rectangular chambers are fairly forgiving when it comes to the choice of range antennas. In general, the desire is to have an antenna that reduces the illumination of the lateral surfaces, or a highly directive antenna. Horn antennas, of course, are a good choice as it is easy to achieve directivities higher than10 dBi. At lower frequencies however, horn antennas, with high gains can be extremely large and heavy. So other types of antennas such as log-periodic dipole arrays (LPDAs), are typical in the UHF range.

Another desirable quality in range antennas is bandwidth. It is ideal to minimize the setup time in an antenna range. A wideband antenna with relatively high directivity is an advantage, as there is no need to switch range antennas as different

AUTs are evaluated in the range. Log-periodic antennas, Vivaldi antennas, and ridged horns are some examples of broadband antennas that are ideal for use in ranges.

Depending on the measurements being performed, other characteristics of range antennas, such as cross-polarization performance, are more important. Hence, it may be that, depending on the testing being performed, different range antennas are used.

5.2.3.2 Positioning Equipment

The radiation pattern is a map of the radiation intensity, either power density or electric field, versus direction. Hence, it should be clear that the AUT should be rotated or positioned within the QZ to get the performance as a function of direction. The positioning equipment can go from a very simple single-axis rotator, typically an azimuth rotator or turntable, to an extremely complex multiple-axis rotator used to precisely position the antenna in the range. Similarly, the accuracy of the positioning equipment can go from half a degree to 0.03° or even tighter. The needs of the positioning equipment are related to the antenna being measured, the frequency range, and the parameter of interest, among other factors. Figure 5.12 shows different types of positioning equipment stack-ups.

The azimuth (AZ) positioner is the simplest positioning system. This rotates the antenna around one single axis, which allows the measurement of single cuts. Ideally, the antenna is placed on a foam column away from the rotator, which is usually nested in the floor absorber. To obtain other cuts, the antenna needs to be repositioned on the support. This approach is ideal for production settings where a single-plane cut is sufficient to provide a pass/fail on the AUT.

It is more common to rotate the antenna on two orthogonal axes. These positioning configurations, AZ/EL or EL-over-AZ, provide an elevation positioner in addition to the azimuth positioner and are common for large antennas. Typically, the elevation positioner is limited in range of angles due to mechanical constraints and potentially because of the RF cables, power, and feedback cables from the positioner encoders of other feedback mechanism.

The most common multiple-axis positioner is the roll-over azimuth (ROL/AZ), which is technically a horizontal axis rotary positioner over a mast over an offset linear positioner over a vertical axis azimuth positioner.

The positioners can be made out of dielectric, but for heavy loads, metallic positioners are more common. Even dielectric positioners need to be covered in absorber to minimize their effect on the antenna being supported.

5.3 The Tapered Anechoic Chamber

Section 5.2 describes the rectangular range, noting that the critical issue with this range is the reflection from the lateral surfaces, or the side walls, ceiling, and floor. In rectangular ranges, a challenge arises at frequencies where the wavelength is 1m (about 300 MHz), since at those frequencies absorbers in excess of 3m (three wavelengths) need to be used. The longest pyramidal absorbers manufactured are 144 inches (2.95m). These long polyurethane pyramids require internal reinforcement for mechanical support. They are also rare, and 96 inches (2.43m) seems to be the longest available in the market today. In the case of polyurethane, it is necessary to glue different parts during manufacturing (given available size), which increases the cost due to the added labor. In the case of other substrates, there are limitations on the machine size to manufacture such parts. More importantly, it is expensive to ship these materials.

What can be done as frequencies get closer to the 750-MHz range and lower? The answer came in 1965 when Emerson published his classic paper [5]. Prior to that paper, Emerson had filed a patent for the concept on August of the previous year (1964) [6], which was awarded in 1967. In this interesting paper [5], Emerson mentions the side walls, floor, and ceiling as the predominant error sources in antenna measurements. He then introduces his design as a chamber in the shape of a pyramidal horn that tapers from a small source end to a large rectangular test region. Emerson goes on to state that there is not yet a formulated theory for this range. He provides three possible explanations: one based on geometrical optics (GO), which is shown in Figure 4 of [7] (reproduced in numerous books without attribution and recreated here in Figure 5.13); an antenna theory explanation that considers the tapered chamber as a large horn with lossy surfaces that attenuate higher order modes; and as four orthogonally combined ground reflection ranges. Of the given theoretical explanations for the tapered chambers, the most common is the first one using GO or ray tracing. This is also the explanation that has caused a lot of the misconceptions about this type of chamber. Accordingly, Section 5.3.1 introduces a new approach, which can be combined with the GO explanation to clarify the operation of these chambers and shed light on some of their limitations.

5.3.1 Theory of the Tapered Chamber

5.3.1.1 The GO Theory

Although Emerson stated in [5] that an explanation of his operation based on EM theory was not available, his patent filed the previous year [6] favors the ray-tracing or GO theory of operation. The ray-tracing approach is definitely a way of explaining

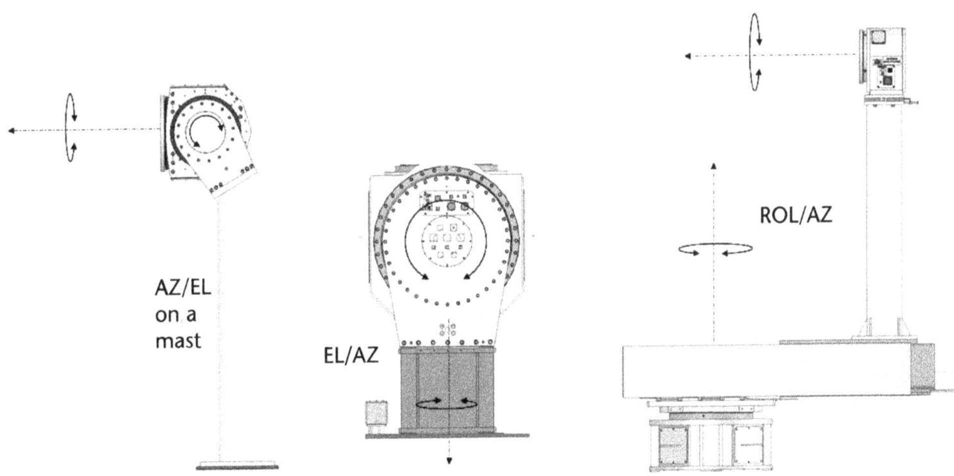

Figure 5.12 Different types of dual-axis positioners. The ROL/AZ is one of the most common types.

the operation of these chambers and some of its critical areas. The first thing that the GO model shows is that in the tapered anechoic chamber, the specular region is moved and the reflected rays are used to illuminate the chamber

Figure 5.13 shows that, for a traditional rectangular antenna chamber, the reflections from the side walls must be minimized as they combine with the direct illumination to create a ripple across the QZ. In the taper chamber, the range has been shaped, and the specular region has been moved closer to the range antenna. If properly shaped, the reflected rays and the direct ray are parallel and they help in the illumination of the QZ.

The theory goes on to claim that since the paths of the reflected and the direct ray are very close, the phase is very close. This is a total misunderstanding of basic EM, as the tapered section is always treated with absorber, and the reflected signal also has a phase shift related to the properties of the absorber. Even if not treated, for some polarizations the reflection off a metal surface will have a 180° phase shift. Additionally, if we superimpose a pattern of a source antenna onto a tapered geometry as shown in Figure 5.14, there are some interesting phenomena that seem to occur. Ray **a** is higher in magnitude than ray **b**. Also, the angle of incidence onto the absorber by ray **a** is larger than that of ray **b**. Hence, the reflection coefficient for ray **a** is possibly larger than that of ray **b**. Based on that ray **a′** has a higher magnitude than ray **b′**. Hence, the amplitude distribution across the QZ may have a ripple associated with the amplitude. This theoretical explanation of the tapered chamber is an oversimplification of the problem.

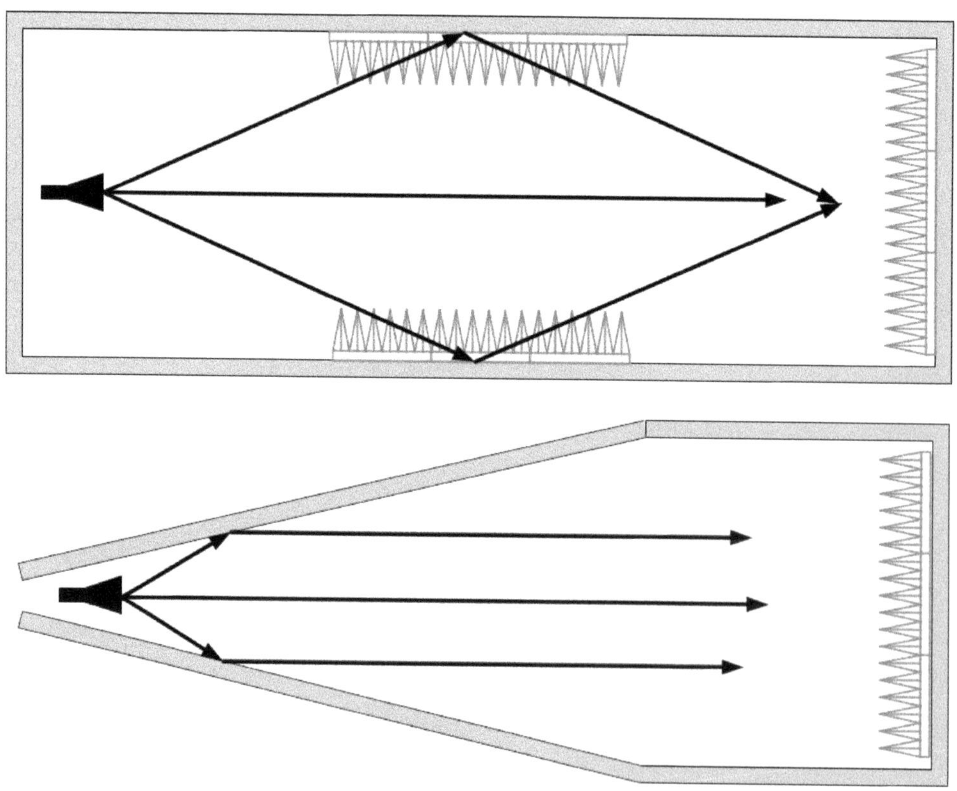

Figure 5.13 Basic Idea behind the tapered chamber. The specular reflection is not reduced or eliminated, but it is used to illuminate the quiet zone.

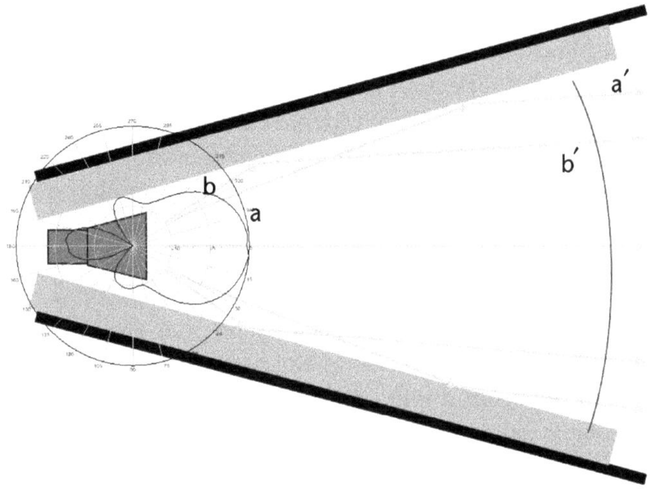

Figure 5.14 Criticism of the common misconception that the tapered chamber produces a collimated plane wave at the AUT.

Ray tracing does help to explain some areas of tapered chamber design, and GO does show the importance of maintaining a good constant angle on the tapered section. Figure 5.15 shows that changing the angle of the absorber treatment in the tapered section can caused undesired reflections that affect the quality of the quiet zone.

Readers should understand that the goal is not to discredit the GO explanation of the tapered chamber operation, but rather to show its limitations. Basically, the GO approach provides some insight on how the tapered chamber operates, but it does not clearly explain all the issues and it may be misleading in some cases (as with the assumption that we have a constant phase across the quiet zone). Nevertheless, without GO, it is not possible to explain the importance of a constant angle for the tapered absorber treatment.

5.3.1.2 The Ground Reflection Range Theory

In [5, 8], Emerson credits Hollis of Scientific Atlanta as the proponent of this theory. Hollis was one of the editors of the famous "Blue Book" of antenna measurements

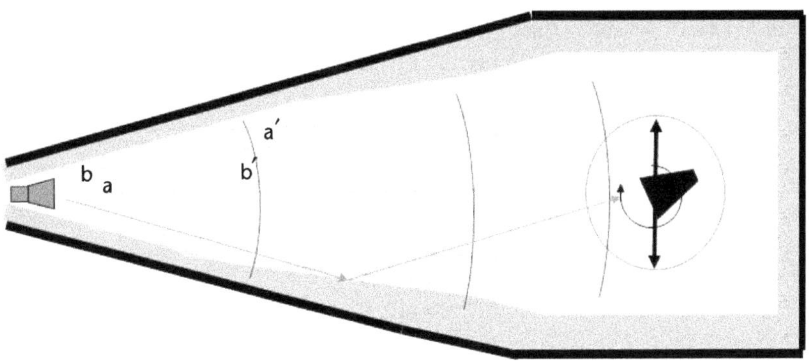

Figure 5.15 Importance of the tapered treatment and of maintaining a constant angle on the tapered section.

[4]. Additionally, in other papers written in the 1960s and 1970s, such as the ones by Kummer and Gillespie [7] and Hemming and Heaton [9], the tapered range is treated as a ground reflection range of sorts. Indeed in Chapter 14 of [4], Lyon, Hollis, and Hickman mention the tapered chamber as a reflection range. They divide the reflection ranges into two major types, the outdoor ground reflection range and the tapered anechoic chamber. Looking at [7, 9], which discuss the ground reflection range and the similarity between these ground reflection ranges and a tapered anechoic chamber, is that both use reflected energy to illuminate the antenna under test. In [3], Hemming mentions the theory of the tapered chamber in terms of the ground reflection range; however, he states that the similarity is in the approach to the analysis. That similarity in the approach is where the tapered chamber is analogous to the ground reflection range. For the ground reflection range in [7, 9], image theory is used to make the analysis. Hence, the approach is to look at the behavior of the tapered chamber by looking at the source or range antenna and its reflections or images. Hemming [3] proceeds to prove the validity of this methodology by looking at a tapered geometry for a 7.63-m-wide tapered chamber with a 3.05m × 3.05m QZ located centered in inside the 7.92-m-long rectangular section of the QZ. The tapered section of the chamber is 17.68m. The length to the apex of the tapered section is 21.33m, and hence the angle of the tapered chamber is 20.3°. Figure 5.16 shows the tapered geometry provided by Hemming.

Hemming's description is taken from [4, 10]. It is better to go to that original document because the discussion in [3] is missing information and has some errors (possibly typographical) that make the discussion difficult to follow. Even then, the discussion in [10] fails to provide some important information, and it states that the results shown are an approximation. For example, the analysis does not provide the pattern of the range antenna but it states that its pattern, and that of the probe antenna, are accounted in the data shown. Appendix 5A provides a Matlab script with the equations derived in [10] to analyze the QZ performance. Using the script, the results shown in Figure 5.17 are obtained. These are not the same results as given by Hickman and Lyon [10] and repeated by Hemming [3]; however, in reading [10] it is clear that some information is missing to properly replicate the results. The method is useful in showing that as the distance from the range antenna to the walls of the tapered increases, a large ripple can be inserted into the QZ field distribution.

The theoretical approach used by Hickman to explain the operation of tapered chambers takes us to the next possible theoretical approach. This last approach, the array of sources theory, also uses image theory, but it treats the images as elements in an array of sources.

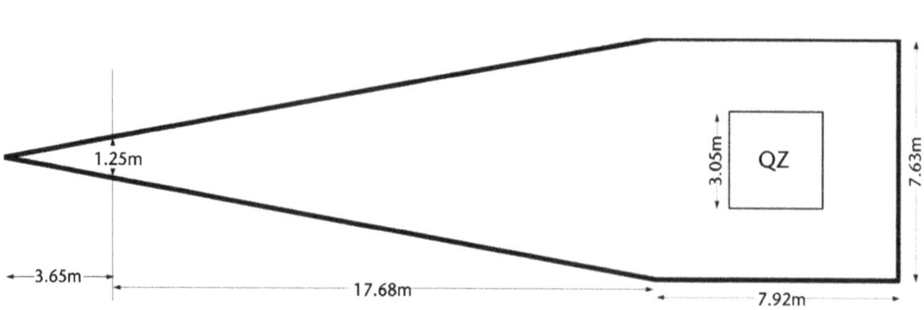

Figure 5.16 The geometry analyzed by Hemming in his explanation of tapered chambers.

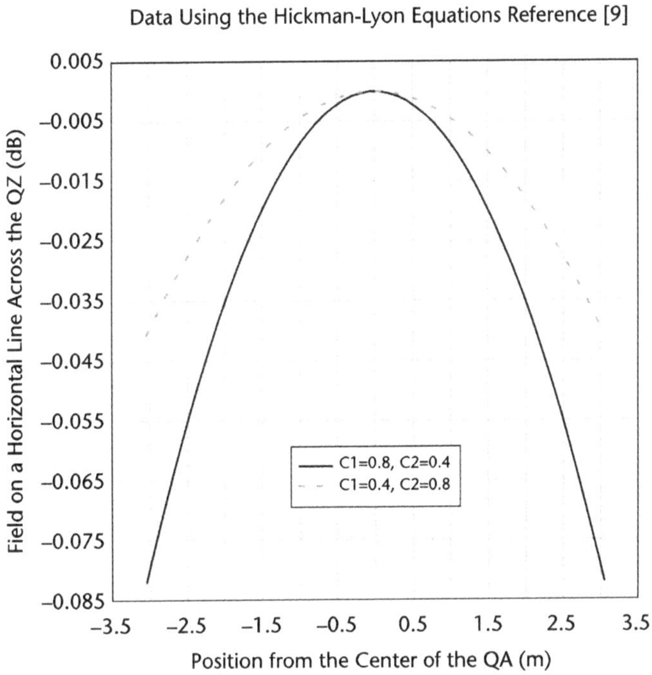

Figure 5.17 Results from the Hickman and Lyon analysis.

5.3.1.3 The Array of Sources Theory

Hemming does mention the idea of source array in [3] when describing his theory of operation based on the ground reflection range theory. Continuing with this idea, one can think of the source as an array of the source and its images. We can then define the array factor of the source. For defining the array factor we can assume these sources to be isotropic sources.

If we observe the feed from the QZ on a typical tapered chamber, we see the range antenna surrounded by absorber. This thin layer of absorber typically under 0.2λ has a limited absorption, usually between −4 and −7 dB for that thickness. Figure 5.18 shows a typical geometry of a tapered chamber showing the feed and its attenuated images from the absorber treatment.

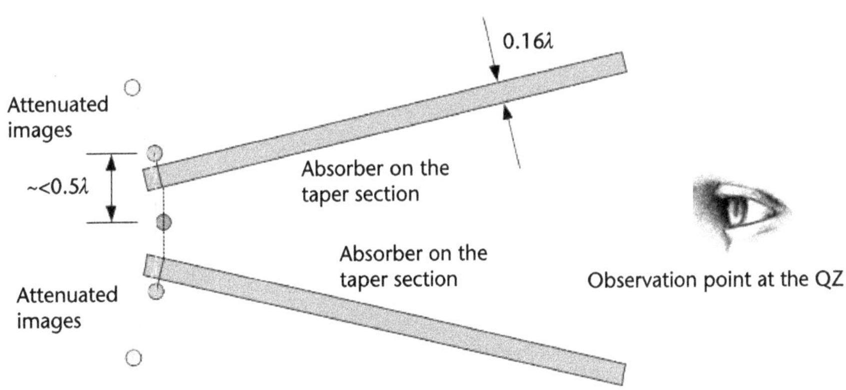

Figure 5.18 The array of sources theoretical model for tapered chambers.

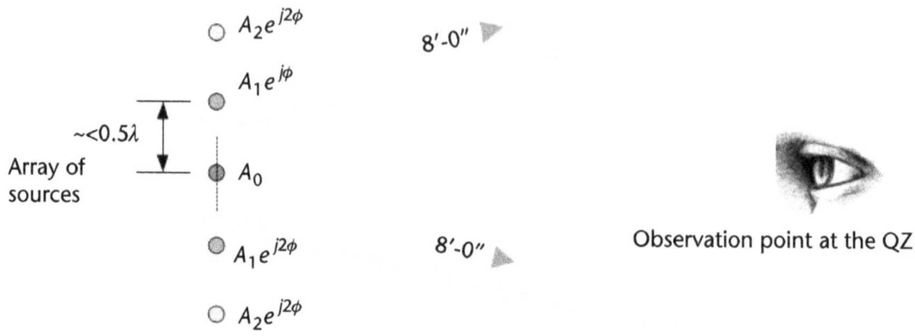

Figure 5.19 Approximation of the array of source.

Continuing with the approach used in the previous theoretical explanations we will reduce the problem to a two-dimensional case. Hence, we can further approximate these as a linear array with equally spaced elements and nonuniform amplitudes. If we assume that the reflected wave from the absorber has been attenuated by 7 dB and has a phase shift of φ, then the third image, created by the third reflection, will have an amplitude that is –21 dB or (0.8% of the original feed amplitude). If we ignore these third order bounces, we can reduce the tapered section to a five-element array, like the one shown in Figure 5.19.

We can write the array factor of this array as

$$\begin{aligned} AF = A_0 + A_1 e^{j\beta d\cos(\theta)+j\varphi} + A_1 e^{j\beta(-d)\cos(\theta)+j\varphi} \\ + A_2 e^{j\beta 2d\cos(\theta)+j2\varphi} + A_2 e^{j\beta(-2d)\cos(\theta)+j2\varphi} \end{aligned} \tag{5.37}$$

where d is the distance between the central element and the first image, and φ is the phase of the reflected wave from the absorber. The angle θ is the angle of observation from the quiet zone. The amplitude values are given by the reflectivity of the absorber, so

$$A_1 \approx \Gamma A_0 \text{ and } A_2 \approx 2\Gamma A_0 \tag{5.38}$$

Using these values and computing the phase for the reflection from the tapered absorber we can calculate the array factor (AF). The phase can be calculated by looking at the problem as a slab of absorber over a PEC and looking at the normal reflectivity. Solutions for this slab of dielectric over PEC problem can be found in Chapter 3. Appendix 5A provides a MATLAB script to determine the array factor of the tapered chamber illumination, assuming up to four images.

Solving for the AF, it can be shown that if the images are electrically close and they are sufficiently attenuated in magnitude, for a range of observation angles the AF is close to 0 dB. Once we insert the actual feed antenna, its pattern is multiplied by the AF. Since for these angles of observation the AF is 0 dB, the feed pattern is not affected by the array in that set of angles of observation. Hence, for those angles of observation the feed appears to be by itself in free space. This is a quasi-free-space condition that is created in tapered chambers.

It should be noted that this is all very frequency-dependent and that as frequency changes it may be necessary to reposition the feed antenna to maintain an adequate AF that will provide this quasi-free-space condition.

Let us assume a LFB, of 1 ft. in thickness with a dielectric property of $2.4 + j3.48$. This material will provide a normal incidence reflectivity of −5 to −13 dB depending on the frequency for the 100–500-MHz range. At 200 MHz, the reflectivity is −13.13 dB with a phase of −165 degrees. If we use these values for estimating the phase difference and magnitude for the elements of the array and assume that the reflected signals are at ¼ wavelength, we get the array factor shown in Figure 5.20.

It can be seen that the array factor is almost 0 dB in magnitude for the angles of interest. This means that the amplitude taper will be the one given by the pattern of the range antenna. Notice also that the phase of the array factor is also close to zero for the angles of interest; hence, for the aspect angles that are observed from the quiet zone, the phase is close to that of the far field.

5.3.2 Sizing the Chamber

Like many other aspects of the tapered chamber, sizing it is based on empirical results. To size the tapered range let us start from the QZ size. The QZ must be large enough to encompass the largest antenna being tested. It should be noted that like the rectangular range and the outdoor reflection range, which it mimics, the

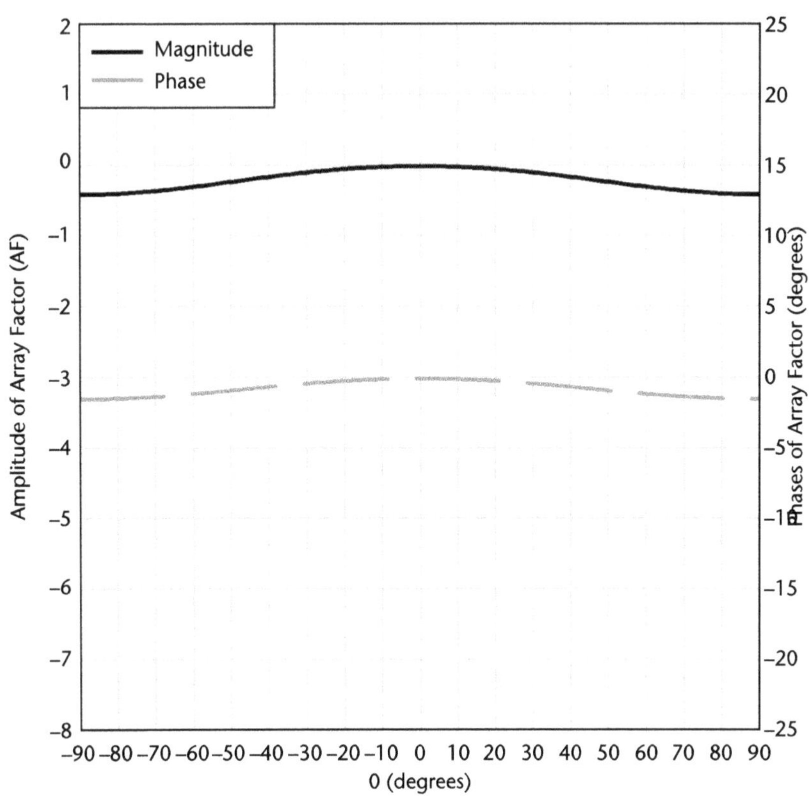

Figure 5.20 The array factor of the sample described. Notice the array factor of 0 dB for the −20 to 20 degree angular range.

tapered range is a far-field range. Like the rectangular range, the electrical size for the largest antenna should be limited to 5λ. How large should be the width, height, and length of the chamber? To answer this question let us split the tapered range into a rectangular section, where the QZ is located, and a tapered section that we use to illuminate the taper.

5.3.2.1 The Rectangular Section: A Starting Point

The rectangular section of the tapered range houses the QZ. There are lateral surfaces and one end wall. The end wall is the main one determining the reflectivity of the QZ. In general, the width and height of the rectangular section is determined by

$$W_{\mathrm{RS}} = H_{\mathrm{RS}} = D + 4\lambda + 2t_l\lambda \tag{5.39}$$

where W_{RS} is the width of the rectangular section, and H_{RS} is the height of the rectangular section. The parameter t_l is the pyramid height of the absorber treatment on the lateral surfaces in terms of wavelengths. D is the diameter of the QZ. Equation (5.39) uses the typical two-wavelength spacing between the QZ and the tips of the absorber.

The length of the rectangular section is usually given by a similar equation where $L_{\mathrm{RS}} = D + 4\lambda + 2t_e\lambda$ (where t_e is the electrical size of the end wall pyramidal treatment) and the two-wavelength spacing is maintained. It is common as well to see the old approach of using the width and height as three times the QZ. So

$$W_{\mathrm{RS}} = H_{\mathrm{RS}} = 3D \tag{5.40}$$

5.3.2.2 The Tapered Section Angle

The tapered section is the critical part of the taper chamber. Hemming states that for operation to high frequencies, the angle of the taper should not exceed 30° [3]. In my career, I tried to never exceed 28°. I was told that this was developed empirically. As a young engineer starting to do chamber design, this concept seemed acceptable. However, with time I started questioning the origin of the 30° number. Empirically means that somebody built a tapered chamber, proceeded to change the angle of the tapered section, and made measurements to arrive to a desirable angle. That would have been an extremely expensive approach.

From the theory of the array or sources, it is understood that we want certain array factor. We want the array to be close to a planar array. To achieve this "quasi-planarity" the smaller the angle of the taper the better. However, the smaller the angle, the longer the taper. Was the 30° a compromise? The problem with this theory is that the array of source theory itself has not been discussed in any of the original papers. Hence the reasoning for the 30° angle is not based on a desire for planarity. Reference [10] provides an important clue; it is a technical report written as part of the commissioning of a tapered anechoic chamber in 1968, three years after Emerson's original paper [5]. The report provides a section on a theoretical analysis of the tapered chamber. The analysis is based on the ground reflection range. Thus, the ground reflection range theory must hold a clue to the taper angle choice.

In the ground reflection range, the direct path and the reflected path are set to add in phase at the test aperture, producing a smoothly varying interference pattern [4, 10]. There must lay the key to the tapered chamber angle. Figure 5.21 illustrates the typical ground reflection range geometry with a test zone D a distance R from the range antenna as defined in Chapter 14 of the "Blue Book" [4]. The range antenna is located a separation h_t from the reflecting surface. The derivation shown by Lyon et al. in Chapter 14 of [4] shows that the relation between the height of the test zone and the height of the transmitter for a smooth varying interference pattern is given by

$$h_t = \frac{\lambda R}{4h_r} \tag{5.41}$$

If we assume that we have a perfect reflection from the wall (which is not really the case here), let us assume that our wall absorber is short enough that $h'_t = 0$. I am also going to use the infamous (5.40), and I am going to substitute the following values based on (5.40) into (5.41). Hence $D = n\lambda$, and $R = 2D^2/\lambda$, thus

$$h_t = \frac{\lambda \frac{2D^2}{\lambda}}{4(1.5D)} = \frac{2n^2\lambda^2}{6n\lambda} = \frac{\lambda}{3} \tag{5.42}$$

Incidentally, this shows an array spacing of less than a wavelength. Let us now continue with the derivation. The angle of the range to the reflecting surface is given by

$$\alpha = \tan^{-1}\left(\frac{1.5n + 1/3}{2n^2}\right) \tag{5.43}$$

Let us plot α as a function of the electrical size of the QZ, that is n. Figure 5.22 shows that as the electrical size of the QZ gets larger, the angle of the range gets smaller. So theoretically, the taper angle of less than 30° is to be able to have electrically large QZ. This answers one of the statements in Hemming's book [3] and one that I heard in my early years in the anechoic range business. Hemming states, "Because we wish to operate up to 18 GHz, set the cone angle at no more than 30

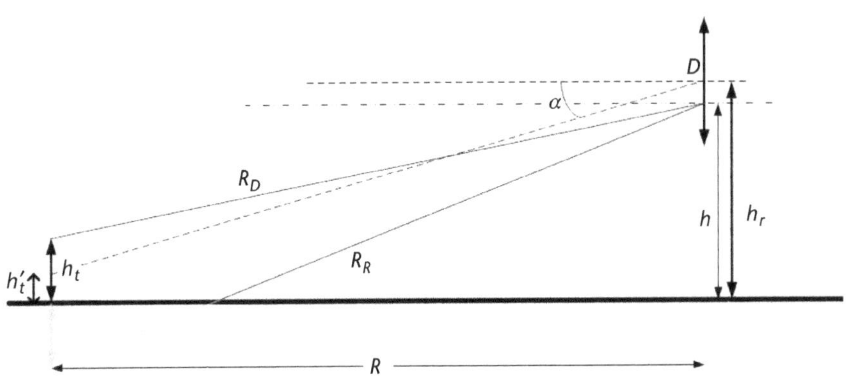

Figure 5.21 Geometry of a ground reflection range.

degrees" [3]. There are no reasons given as to why, but it seems to be related to being able to have an electrically larger antenna in the QZ at higher frequencies. Indeed, Hemming recommends a cone angle of less than 36 degrees for 6-GHz operation and less than 30 degrees for 18-GHz operation.

In general, the taper angle should be 30 degrees or smaller. This means that the length of the tapered section is approximately given by

$$L_t \geq 2W_{RS} \tag{5.44}$$

using the 28 degrees that I prefer as the maximum angle. It should be noted that the entire derivation for the angles of the taper is based on [4], which uses some approximations to arrive at some of its equations. One of the approximations is that the range length is much larger than the test zone elevation h_r. Also $h_r = 1.5D$ was used in our derivation. (For a good amplitude taper of 0.25 dB across the test zone, Lyon et al. [4] recommend $h_r \geq 4D$.). $R \gg h_r$ means that $R \gg 1.5D$, or written in terms of wavelengths, $2n^2 \gg 1.5n$, which seems to hold better for electrically larger QZ.

5.3.3 Absorber Layout

Even now, over 50 years since their introduction, tapered chambers' inner workings and theory are still fuzzy. Yet these chambers are used, and they have been implemented around the world. As a matter of fact, as described in [11], they are more cost-effective at lower frequencies. How is their implementation? How is the absorber layout for the tapered anechoic chamber? This section answers these questions, starting with the rectangular section in Section 5.3.3.1, and then moving to the critical part of the taper range, the taper section, Section 5.3.3.2.

5.3.3.1 The Rectangular Section

The reflectivity of the QZ in a tapered chamber is mainly limited by the reflectivity of the end wall. This is because, unlike the rectangular chamber, there are no

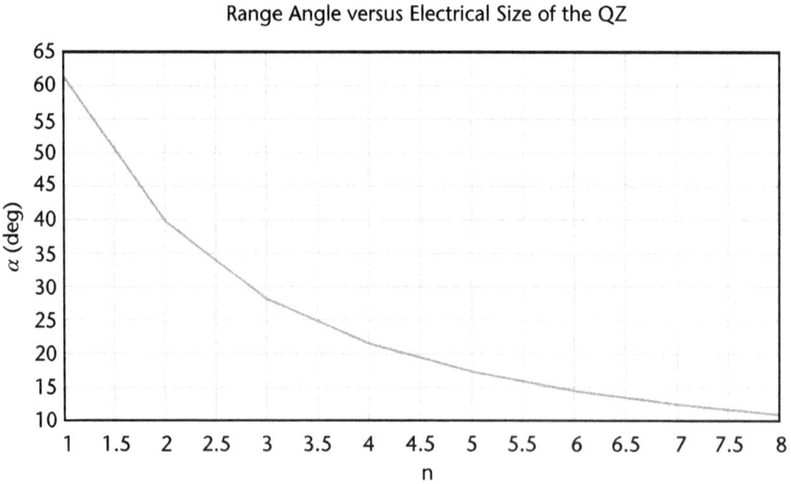

Figure 5.22 Range angle versus the electrical size of the QZ.

specular reflections from the lateral surfaces (see Figure 5.13). Hence the back-wall absorber must be such that the reflectivity at normal incidence exceeds the level of reflectivity required in the QZ.

This raises an interesting conundrum. On one side, tapered chambers came about to provide better QZ reflectivity at lower frequencies (under 750 MHz), but the longest absorber on the market is about 96-inches-long, and per the manufacturer's specifications it only provides −32 dB at 100 MHz; indeed, $\lambda_{100\text{MHz}} = 118$ inches so the 96-inch absorber is only 0.8λ, which per Figure 3.9 will give −25 dB at normal incidence. That seems to limit how low in frequency tapered chambers are expected to operate. Typically, the longest type of absorber is 72 inches, and that one will limit the use of tapered chambers at 100 MHz with QZ reflectivity levels in the range −25 to −20 dB.

There are ways to try to increase the absorption; see, for example, the discussion of curvilinear absorbers in Chapter 3. Chebyshev arrangements are other ways of increasing the absorption by means of phase cancelation of the reflected energy (see Chapter 10). Other ideas, such as tilting the reflective wall, do not seem to provide any improvement for spherical wavefronts. A deeper discussion on tilting the wall can be found in Chapter 8 and in [12]. In general, 100 MHz is the lower limit for tapered chambers and for indoor anechoic ranges in general. The end wall absorber is such that

$$t_e\lambda = \begin{cases} 2\lambda \\ \text{or} \\ \text{largest available size} \end{cases} \tag{5.45}$$

The lateral surfaces of the rectangular range are not as critical, yet should still be treated. A rule of thumb to follow is for the lateral surface absorber to be such that $t_l = t_e/2$. Figure 5.23 shows the end wall of the rectangular section of a tapered chamber. The treatment is not different than that of a standard rectangular chamber. The absorber is pyramidal, and it extends to the lateral surfaces (side walls, floor, and ceiling) or to an absorber "frame" build out of LFB. On the side view in Figure 5.23 we see that the frame extends the length of the end wall absorber.

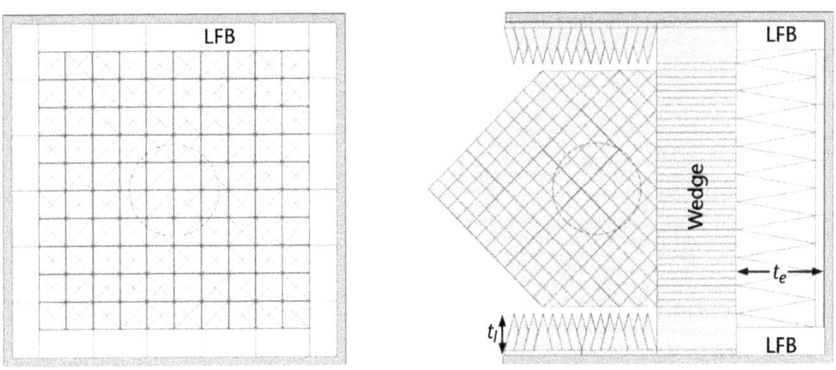

Figure 5.23 End wall and side view of the absorber treatment on the rectangular section of a taper range.

From the end wall absorber to the main lateral treatment, a wedge-type absorber is typically used. The reason for the wedge is that at higher frequencies, it is desirable to minimize any backscattering that could increase the ripple in the QZ.

The following case study shows one such case where wedge was not used behind the QZ treatment. The reason for avoiding the wedge was that for this particular tapered chamber, there was an additional spherical near-field range orthogonal to the main taper range. Hence the lateral surface was used as the end wall for the additional range [13].

Figure 5.24 shows a picture of the back corner of that range. Since the side wall for the taper range becomes the end wall for the spherical near-field range, wedge could not be used as it will have different response depending on the polarization. When the QZ was probed at 800 MHz, the chamber was out of specifications by several decibels as shown on the left polar plot in Figure 5.25. The fix was to add additional tips on the treatment to reduce the spacing between the tips and increase the backscattering frequency. The additional tips can be seen in Figure 5.24. The right polar plot in Figure 5.25 shows the improvement toward the corner after the additional tips were inserted. This particular case shows the importance of the proper treatment in the rectangular section of the tapered chamber as well as an in-situ technique to correct a problem.

Figure 5.23 shows the lateral treatment of the tapered chamber, where the absorber has been rotated by 45 degrees on the lateral wall similar to the treatment done on some rectangular chambers. The rotated absorber is also seen on the left side of Figure 5.24. I find this treatment approach better in smaller tapers.

Very large tapers have sometimes smaller pyramidal absorber treatments on the lateral surfaces surrounding the QZ and wedge treating the balance of the lateral surfaces. In those larger tapers, the pyramidal treatment is not rotated (see Figure 5.26).

Figure 5.24 Side wall treatment (to the left of the picture) and end wall treatment (to the right of the picture), showing no wedge. Additional pyramids are inserted in between the pyramids of the lateral wall pyramidal treatment to attenuate the problem to the side wall pyramidal treatment to attenuate the problem.

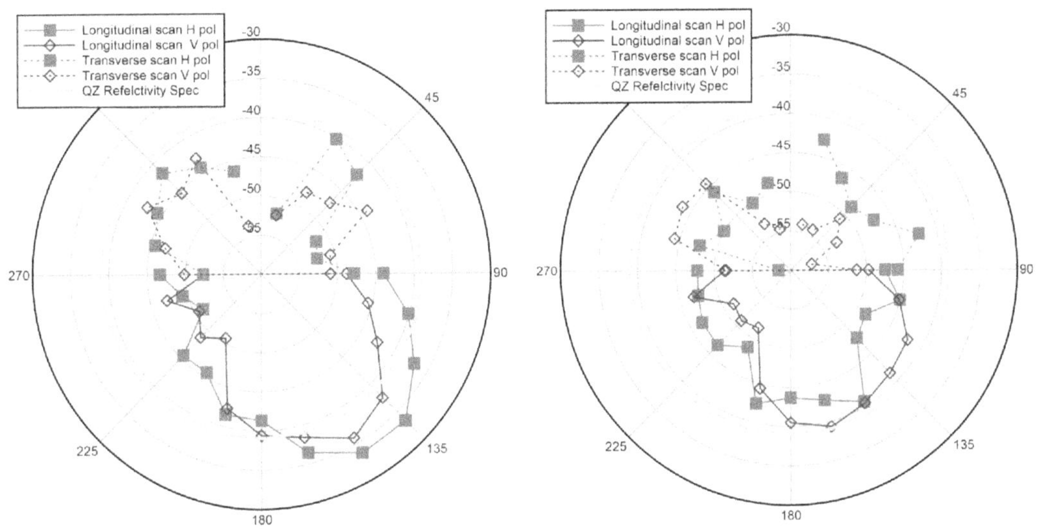

Figure 5.25 Free-space VSWR data for the chamber shown in Figure 5.24 before and after the modulation.

5.3.3.2 The Taper Treatment

The taper treatment is the most critical. The measured performance of the anechoic chamber is dependent on the proper treatment of the taper. It is important to keep a constant angle on the absorber treatment of the taper range. Figure 5.15 shows that a change in the angle of the taper can cause interfering waves to come at angles of incidence at which they will cause a significant ripple on the amplitude distribution across the QZ. It is interesting that despite that need for a constant angle, some

Figure 5.26 Lateral surface treatment on a very large tapered chamber showing the pyramidal treatment not rotated. (Photo courtesy of PPG Aerospace—Cuming Lehman Chambers.)

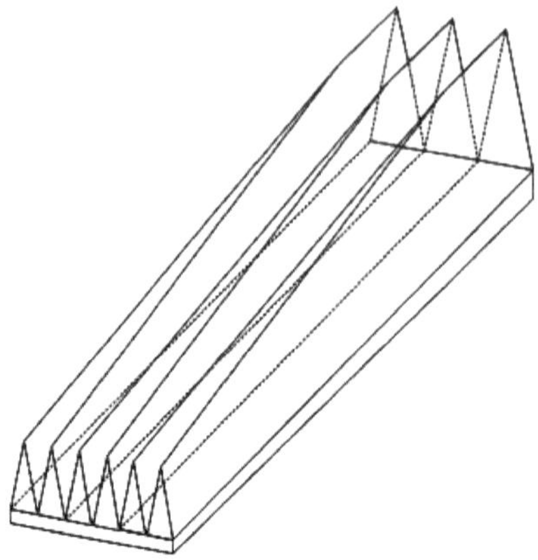

Figure 5.27 A wedge transition piece. The transition pieces are done from wedges with a base width of *x* to wedges with a base width of *x*/2.

absorber suppliers went through the effort of making some rather complex cuts to transition from shorter wedge to longer wedge pieces. One such piece is shown in Figure 5.27. These transitions are done usually from a wedge piece that has a base of a given width x to a piece that has a wedge of a given width $x/2$. Since, as it was shown in Chapter 3, it is desirable to keep a given ratio for the absorber pyramid (or wedge) height to the base width, this means that these transition pieces are done from a given size of absorber to a size that is half of that. For example, a 24-inch wedge piece is transitioned to a 12-inch wedge piece. The length of these transitions depends on the manufacturability, but is rare to see them in sections longer than 6 ft. (1.82m). These have been used successfully in several tapered chambers. But in my experience, the change in the angle of the taper causes additional reflections, and these cause high amplitude ripples on the amplitude illumination of the QZ. I do not recommend their use, and I question the performance of chambers that use them at higher frequencies (above 2 GHz in general). If such a chamber were to be used at frequencies above 2 GHz, I would suggest that the QZ is probed to ensure that there is not excessive ripple.

If possible, it is better to use the same wedge thickness across the entire range. Most of the original tapered chambers use this approach. Only in the feed section are different materials (such as LFB) used.

Another approach is to use shaved wedges. Shaved wedges are shown in Figure 5.28. The wedge goes from a thicker absorber to a thinner one. The approach has been used in several chambers with satisfactory results.

The shaved wedge approach provides a constant angle through the length of the taper section. Figure 5.29 shows an installation of shaved wedge on a tapered chamber. The wedge gets shorter as we approach the feed point, and more and more of the flat surface shows. In most cases the flat surface will match with LFB on the conical feed section. It is important in these cases to use a low carbon loading on the wedge, as it will allow for better penetration of the waves on the flat surfaces.

A final approach to designing the taper section is the conical approach described in [13]. The idea of continuing the conical feed section all the way from the feed

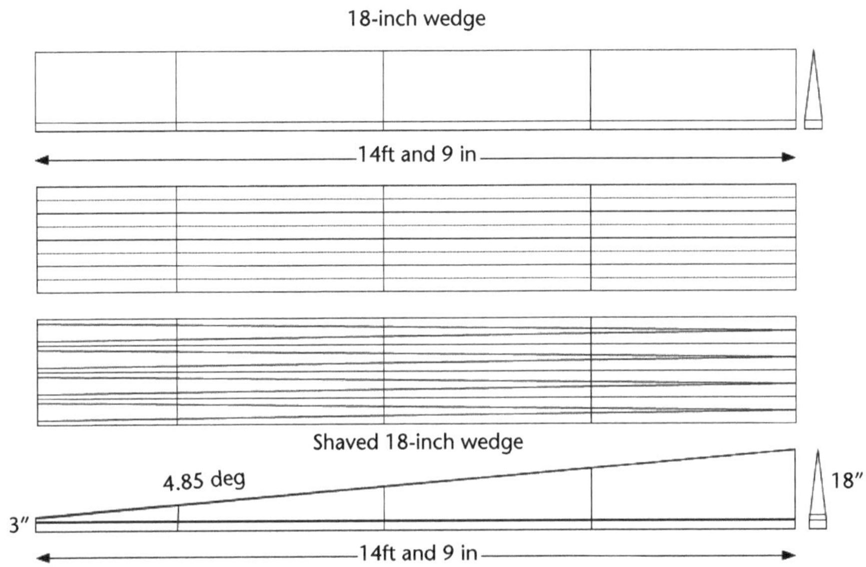

Figure 5.28 A set of 18-inch wedge shaved down to three inches in thickness.

area to the rectangular section was proposed by Randy Stack and myself (we originally discussed it during dinner in London in May 2009). We ran with the idea and developed the chamber described in [13]. The approach for the absorber is almost a "reversed-shaved" wedge. The bottom of the wedge is shaved, and this shaving narrows the footprint of the wedge. The reduced footprint does not leave voids on the coverage since the conical section has a reduced cross-section as the apex is approached. The result is a wedge that extends all the way to the apex as shown in Figure 5.30.

The conical tapered illustrates a new out-of-the-box approach to designing tapered chambers, and it is the only change to the established approach to designing these chambers since their inception.

Figure 5.29 Use of shaved wedge in a tapered chamber. The left picture shows a view from the range antenna; the picture on the right shows a view from the door toward the range antenna.

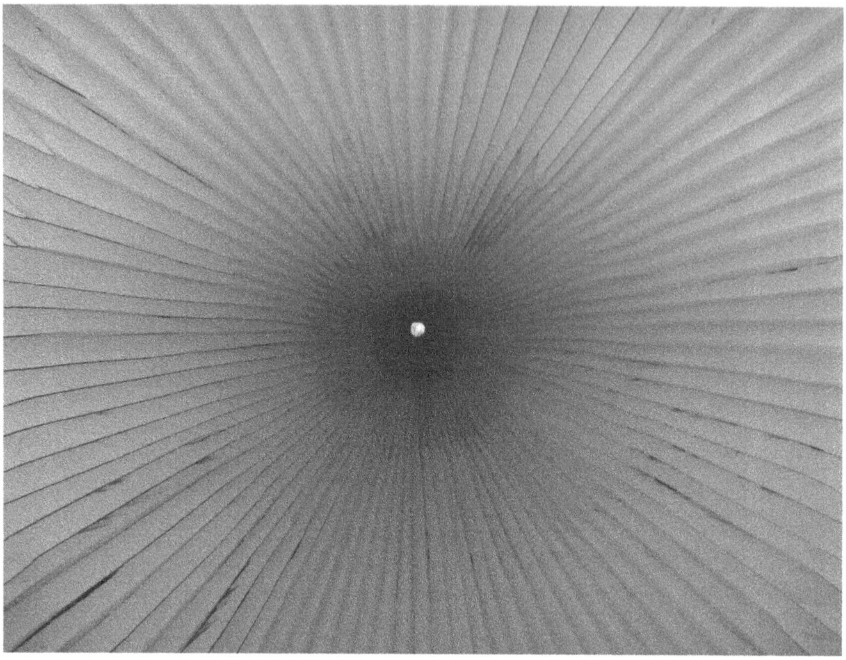

Figure 5.30 The conical tapered chamber looking from the QZ toward the range antenna location at the apex of the cone.

5.3.4 Concerns and Limitations of the Tapered Chamber

Tapered chambers are a great tool. In [11], the author shows that a tapered chamber can provide a reduction of 33% in height and width compared to a rectangular chamber to achieve the same performance with an increase of only 20% in length. More important, the increase does not increase the area to be covered by absorber (because of the tapering), and the volume of material is smaller (because of the shorter pyramids and wedges required). The range, however, is a little trickier to implement; hence the cost reduction may not be as dramatic as the absorber and height reduction will suggest.

It seems that the taper range is a win-win option. Certainly, tapers have their place, as mentioned in Chapter 2 and in [14], in the 100- to 750-MHz frequency range as the most efficient type of indoor facility for measuring antennas. They are, however, not a panacea.

There is a belief that tapered chambers generate a collimated wavefront. This is not true and can clearly be seen if we do a phase probing of the QZ of a tapered chamber. In general, the phase distribution across the QZ of a tapered chamber follows the far-field expected phase distribution. As an example, consider the measurement of two tapered chambers, with the phase recorded along a horizontal line transverse to the range crossing the spherical QZ.

Figure 5.31 shows the results of such a scan. Chamber A had a 1.2-m (4-ft) spherical QZ, and the path length changes from about 10m to 12m depending on the location of the source antenna. Chamber B was a smaller tapered chamber with a 0.9-m (3-ft) spherical QZ and a path length that varied from 6.9m to 7.8m. The plots are for vertical polarization in both cases. The plots show how as the frequency

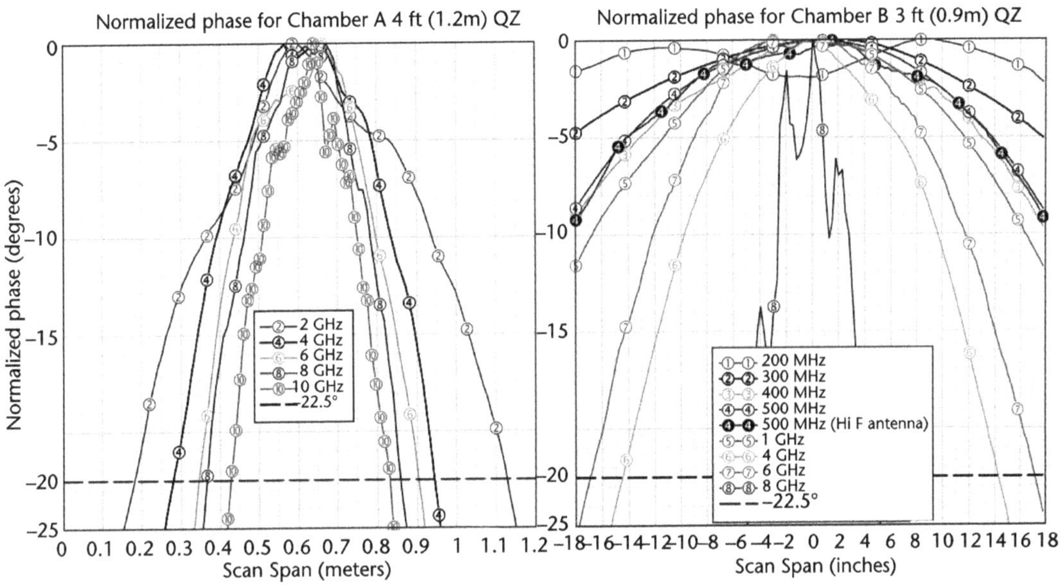

Figure 5.31 Normalized phase across the QZ of two different tapered chambers. Chamber A, on the left, has a roughly 10- to 12-m path length and a 1.2-m diameter QZ. The plot on the right is for Chamber B, which has a 0.91-m QZ and a path length that changes from 6.9m to 7.8m

increases and the QZ becomes larger electrically, the usable QZ for the given range of test distances gets smaller. Further analysis of Chamber B is shown in Figure 5.32, and it demonstrates that with a couple of exceptions, 3 GHz being the most notable, the far-field illuminated QZ size follows the far-field equations' expected size.

From the results presented in Figure 5.31 and Figure 5.32 it should be clear that the tapered chamber is like the rectangular range, a far-field range. There are

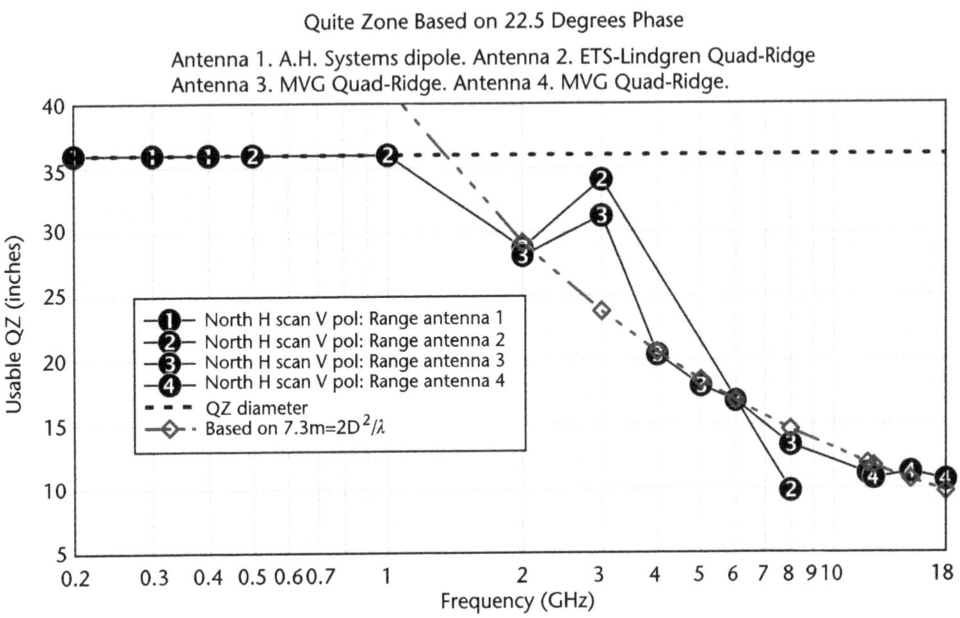

Figure 5.32 Measured far-field QZ versus frequency.

other peculiarities to the tapered chamber that should be kept in mind. These are discussed in Sections 5.3.4.1–5.3.4.4.

5.3.4.1 Frequency Dependency

The tapered chamber is very frequency-dependent. The illumination depends on the location of the range antenna and on the type of array factor that can be generated. The two Matlab scripts provided in Appendix 5A can be used to see how the illumination of the QZ can change as the electrical location of the source versus the lateral surfaces of the taper is changed. Figure 5.33 illustrates the potential problems that can arise in a tapered chamber if care is not taken. The chamber is illuminated with a broadband antenna that covers the entire Ku-band. The antenna was located at the feed section, and its position was optimized for the center of the Ku-band. Without changing the location of the antenna, scans were performed at 12.4 GHz, the lower end of the band, and at 18 GHz, the upper end of the band. The plot in Figure 5.33 shows that at the lower end, there is a skew on the amplitude and that the edges of the QZ are higher in magnitude than the center. Additionally, a slight ripple can be observed. The ripple is more pronounced at 18 GHz. The user of a tapered chamber should make sure that the QZ is scanned at the frequencies of use and that a proper illumination is achieved for a given range antenna at a given frequency.

5.3.4.2 Absorber Dependency

The illumination is dependent on the RF absorber. Recall that while in the rectangular range we are trying to minimize the reflectivity from the lateral surfaces; in the taper range we are trying to use the reflected energy to illuminate the AUT. More important, the reflectivity of some absorber cuts is dependent on the polarization of the incoming wave. Even for LFB there is a different reflectivity depending on the polarization of the incoming wave, as shown in Chapter 1. Hence the illumination changes with polarization. This is illustrated in Figure 5.34. The difference

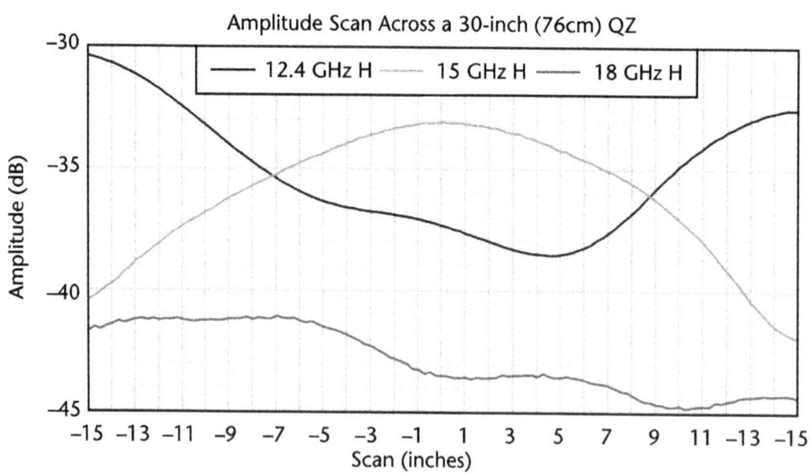

Figure 5.33 Amplitude scans for a 30-inch QZ of a tapered chamber at three frequencies.

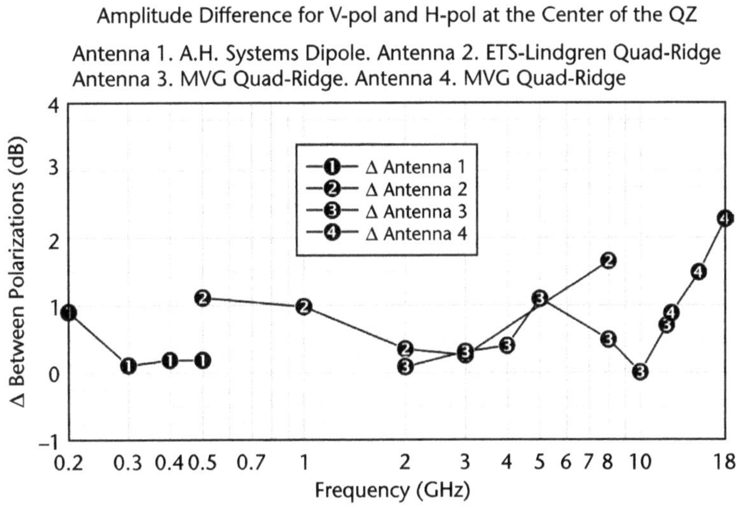

Figure 5.34 Absolute value of the difference between vertical and horizontal polarization on a tapered chamber.

between polarizations in a tapered range can be as high as 1 dB, even at higher frequencies where the absorber has more absorption and better penetration due to lower permittivity. This dependency on the absorber should be taken into account when performing polarization measurements, and the difference should be taken into account to reduce the potential error of the measurement.

The other effect from the absorber is that even when perfectly mechanically aligned, small variations in the absorber covering the different sides of the feed section will affect the amplitude profile of the illuminating wave. The effect will be seen as a skew or asymmetry of the illuminating wavefront.

The effect is illustrated in Figure 5.35. Notice that the phase peaks at −134.5 degrees at 7.7 inches from the center of the QZ. The amplitude skew is not as

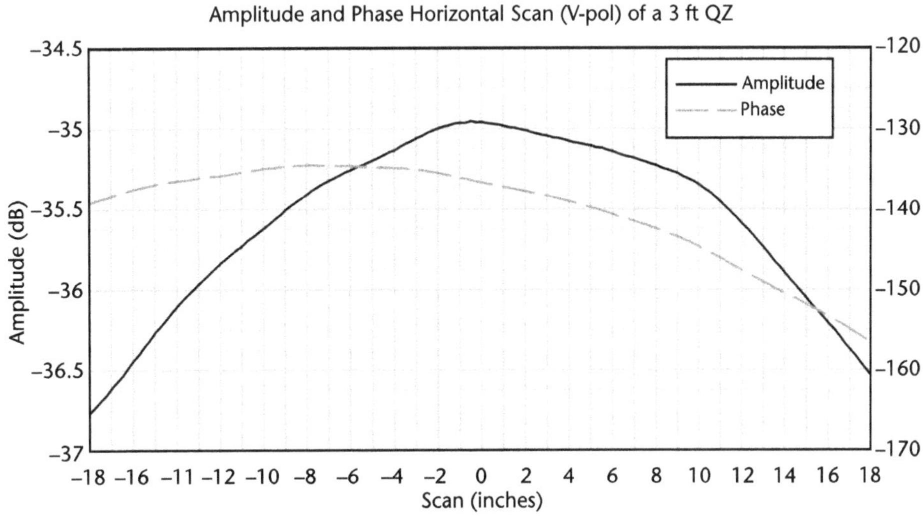

Figure 5.35 Measured amplitude and phase across a 3-ft QZ after mechanical alignment of the range antenna and the probing scanner.

noticeable, but the peak is at −34.95 at about a half-inch from the center of the QZ. Furthermore, there is a noticeable asymmetry at the edges of the measured QZ. The phase at −18 inches is about 20 degrees higher than the phase at the 18 inches. The amplitude also shows an asymmetry. At the −18-inch mark the amplitude is 0.23 dB lower than the amplitude at 18 inches. While the asymmetry may not be an issue in most cases, it shows the potential effect from the illumination of the QZ using the reflections from the absorber.

The fact that the illumination depends on the absorber will constrain the range antenna being used. The range antenna for a taper is discussed in Section 5.3.4.3.

5.3.4.3 The Range Antenna

The range antenna of a tapered chamber is critical. It must work well with the chamber geometry to properly achieve the illumination that is desired. Finding a feed that will properly illuminate a tapered chamber across wide bands of frequencies is a complex problem. There are several such designs described in the literature [15–17]. Horns are the most common antenna used in tapered chambers, because they tend to fit well within the geometry of the chamber. Additionally, broad band horns such as ridge horns have their phase center shift toward the feed point as frequency increases. This shift of the phase center also follows the geometry of the taper. However, typically the frequency range of the ridge horn, when used in the tapered chamber, does not match the frequency range of the antenna itself. At some frequency, the array factor that it forms with its images does not provide the right illumination of the QZ. At low frequencies, where horns may be physically too large to position at the apex of the chamber, dipole antennas can be used successfully as range antennas. LPDAs, however, are not recommended. While they may work at one frequency (typically the lowest), as frequency increases the higher frequency elements toward the front of the LPDA will radiate, and these are electrically and physically much further away from their images, causing a poor illumination of the QZ.

5.3.4.4 The Automotive Half-Taper

One interesting modification of the tapered chamber is the half-taper geometry. This type of range, as far as I have been able to find, originated in Japan. It seems to have been developed as an approach to do both antenna measurements and automotive EMC in the same anechoic range. Since a lot of the automotive antennas operate as low as 87 MHz (i.e., FM radio), the tapered range appeared to be a good approach. However, EMC applications usually require a metallic ground plane. If we look at the geometry of the typical half-space tapered chamber (see Figure 5.36) it is really obvious that we are looking at a ground reflection range. However, the ground reflection range rules [4] are not followed. In the half-space taper, both antennas, the range antenna and the AUT, are typically close to the ground. In the best case the AUT may be mounted on top of the vehicle, but for a typical sedan that may be 1.52m (5 ft.) above ground. Using (5.41) and the geometry shown in Figure 5.36, the ideal height of the secondary antenna in an optimum ground reflection range can be calculated. The results of that experiment are shown in Figure 5.37. As we see for an antenna fixed at 1.5m, the optimum height of the other antenna in the range,

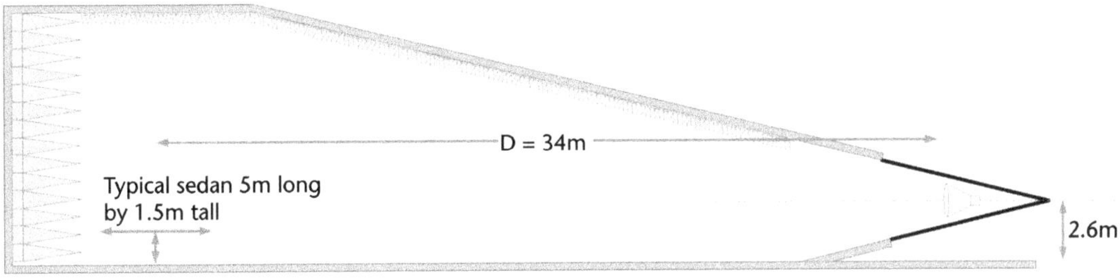

Figure 5.36 A typical geometry for an automotive half-taper range, approximated from different implementations of the concept at different locations.

for a range that is 34-m-long, varies from over 20-m-high to well below a meter. It is apparent that for most of the range of frequencies, the half-taper automotive geometry, while potentially usable, is not optimum, and care should be taken to avoid potential errors in the measurement related to the amplitude taper across the AUT. Anyone using a half-taper range is encouraged to revisit the derivations in [4].

5.4 Error and Uncertainty Analysis in a Far-Field Range

There are several sources of uncertainty and error in a far-field range. Some of them are related to the RF system, some to the positioning equipment, and some to the

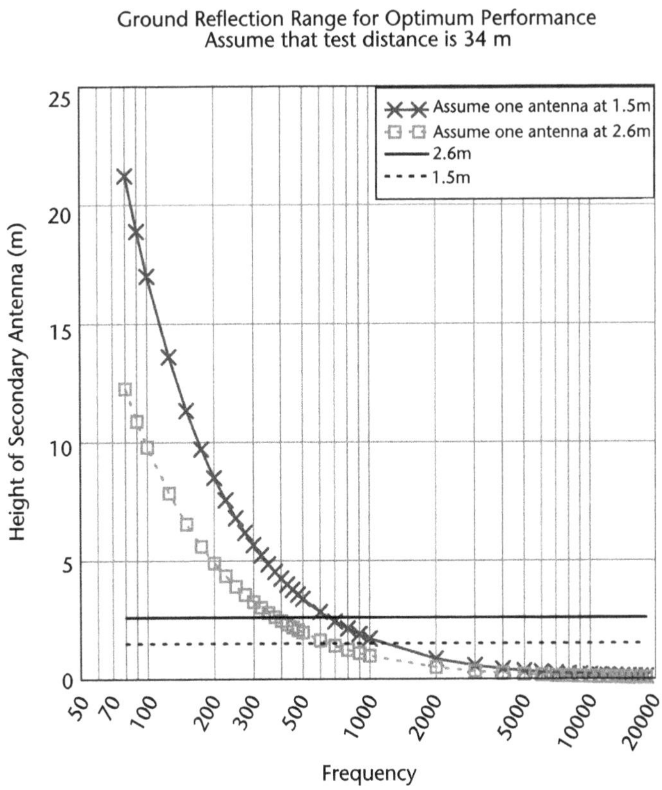

Figure 5.37 Height over ground for antennas in a ground reflection range when the geometry of the half-taper is applied

range itself. This book concentrates on the error or uncertainty contribution from the absorber treatment and from the positioning equipment.

5.4.1 Contributions from the Absorber Treatment

The main goal of the anechoic range is to reduce the reflected signals in the antenna measurement range. However, as shown in Chapter 3, it is not possible to have a perfect absorption and there is going to be a level of reflected energy in the anechoic range's QZ. This level will affect the measurement.

5.4.1.1 Potential Error

If we assume that the reflectivity level in the QZ is 30 dB, what does that translate into errors for our measurements? Figure 5.38 shows a well known set of plots that was presented in a slightly different way in [4]. Both plots show the error in the measured quantity as a function of the direct versus the potential reflected signal. The plot in Figure 5.38 (a) shows the potential error for reflected to direct-signal ratios of –55-–25 dB. The plot in Figure 5.38(b) shows the potential error for reflected to direct ratios of –25–20 dB. For ratios smaller than –25 dB, the in-phase and out-of-phase error are very close so the plot in Figure 5.38(a) shows the error in measured level as a positive and negative quantity. The plot in Figure 5.38(b) shows the error as two possible extreme cases: one case when the reflected and direct signals are in phase and another case when they are not in phase.

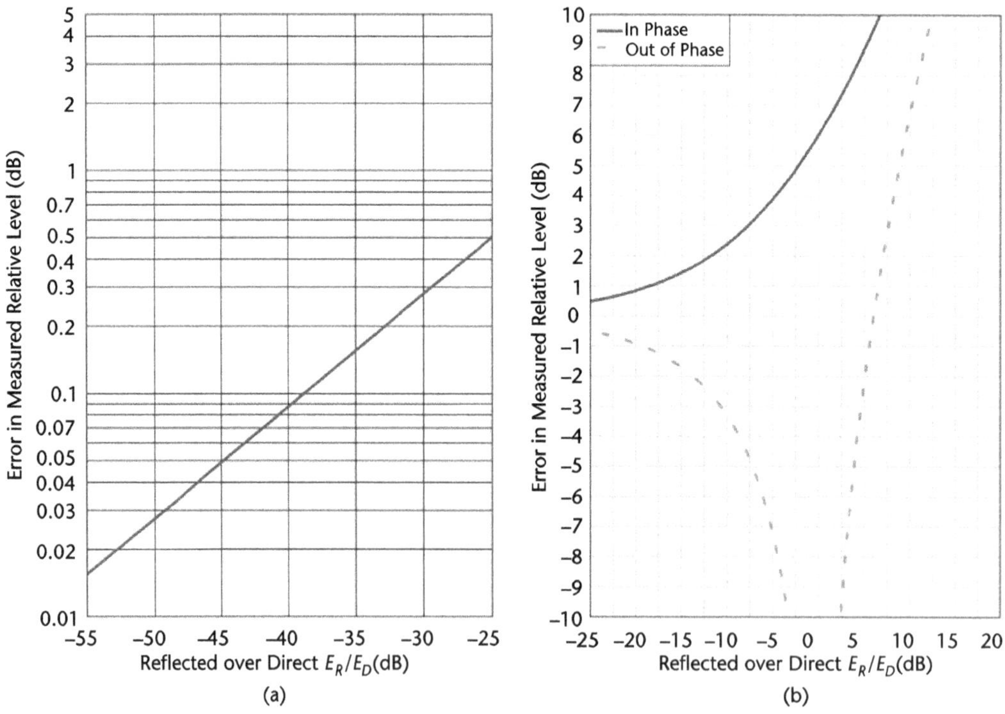

Figure 5.38 Potential error of measured signals.

5.4.1.2 Main Beam, HPBW, and SLL Error

Let us continue with the assumption that the QZ reflectivity is −30 dB. In general, ranges will provide that number without specifying from which direction the reflected signals come. Because we ignore exactly where these signals may be arriving from, rather than an error (which can be corrected), we are dealing with an uncertainty. Uncertainty means that we know with a given degree of confidence that the true value lies within a certain range from the measured value.

For the main beam, we assume the relative signal level to be 0 dB. The main beam is the reference value that we measure. Hence if the QZ reflectivity level is −30, the error may be slightly under ±0.3 dB at ±0.279 as illustrated in the left plot on Figure 5.38. If we want to be a bit more rigorous, we could use

$$\Delta_{\text{in phase}} = 20 \cdot \log_{10}\left(\left|1 - 10^{((E_R(\text{dB}) - E_D(\text{dB}))/20)}\right|\right) \tag{5.46}$$

$$\Delta_{\text{out of phase}} = 20 \cdot \log_{10}\left(\left|1 + 10^{((E_R(\text{dB}) - E_D(\text{dB}))/20)}\right|\right) \tag{5.47}$$

which show that the potential error is between +0.270 dB and −0.279. 3 dB from that level are the 3-dB points that mark the HPBW. The ratio between direct and reflected for those is −27 dB. The potential error for those features of the pattern is approximately ±0.397 dB.

How about sidelobes? Those are going to depend on the level of sidelobe that we are trying to measure. If we assume a −10-dB SLL, then the ratio between direct and reflected is potentially −20 dB, which translates to a potential error of +0.828 dB to −0.915 dB, while a −20-dB SLL will correspond to a ratio of reflected to direct of −10 dB, which means a potential error of +2.39 to −3.30 dB.

All these potential errors of the signals being measured will translate also to uncertainty regarding the angular HPBW as well as the gain. Just the measurement error of the main beam has an error of 0.28 dB; hence, the gain measurement will have that contribution from the absorber in the uncertainty. If the gain is obtained from measuring a gain standard, such as a standard gain horn (SGH), the potential uncertainty of that calibration standard must be accounted for as well. Thus, for example, let us assume that a SGH with an uncertainty of ±0.25 dB is used. Then combining the calibrations' standard error with the error from the reflected levels in the range gives us a measurement uncertainty for the gain of $\sqrt{U_{\text{SGH}}^2 + U_{\text{reflect}}^2} =$ ±0.375 dB. This two-term uncertainty estimation does not take into account other potential sources of error, such as errors from rotary joints in the positioners and uncertainties from mismatch and positioning equipment. The reader should be aware of the potential errors to the measurement that are the result of the anechoic layout of the range itself.

5.4.2 Contributions from the Positioning Equipment

Alignment is critical in an antenna range. The positioning equipment can contribute to the overall error of the measurements. If we assume a typical ROL/AZ positioner, the main sources of error in that case are the nonorthogonality of the two axes and the nonintersection of the two axes of rotation.

As an example of error contributions from the positioning equipment, we can look at the nonintersection error. The nonintersection can be measured as a distance between the two axes (the axis of rotation of the azimuth positioner and the axis of rotation of the roll positioner). The potential error of the nonintersection on the main beam value can be estimated as a 1 dB per $\lambda/10$ of deviation. Hence, a nonintersection of 0.2 inches at 1 GHz will potentially provide an error of 0.2/1.18 dB or 0.169 dB. It should be noted that the error is frequency-dependent.

As discussed in Chapter 3, the absorber performance improves with frequency. This leads to an important conclusion: At lower frequencies, the contribution from the misalignment errors to the uncertainty is smaller than the reflectivity contributions, while as the frequency increases the effects of the range absorber diminish as the effects of the positioning become dominant.

5.5 Range Validation Testing

Currently, the industry's accepted methodology for measuring the performance of a far-field indoor range is the free-space VSWR. The method dates back to 1963 to work performed by the University of Michigan for NASA [18].

The original test procedure was intended for intended for high frequency (x-band) and for rooms with square cross-sections. The test is performed on a horizontal plane, and only the side walls and the end wall are measured; the ceiling and floor reflectivity contributions are not accounted for in this horizontal plane measurement.

The test consists of a series of transverse and longitudinal scans for both principal polarizations. The interference pattern of the fields in the room is sampled. Figure 5.39 illustrates the scanning of the QZ. A reference trace is done by boresighting the range antenna to the antenna probing the QZ. The antenna probing the QZ should be a high-directivity antenna such as a horn or a LPDA. The transverse scans are

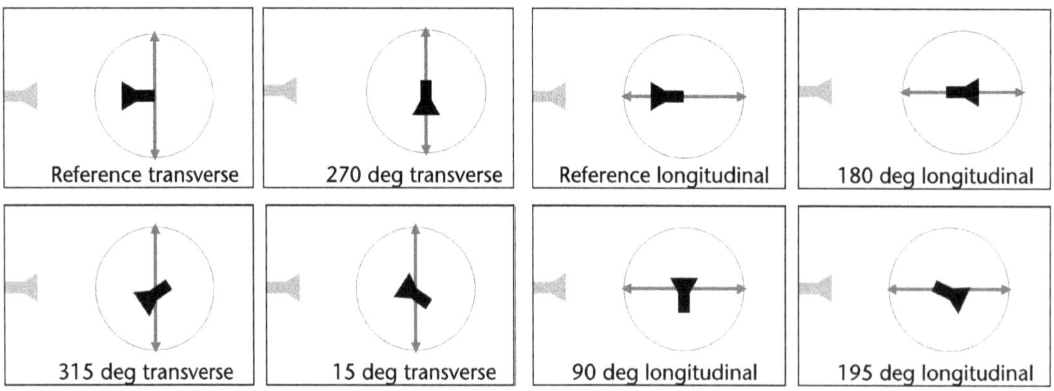

Figure 5.39 Sketches showing the scanning of the QZ for the free-space VSWR test for some angles. Sampling is typically done every 15 degrees.

repeated with the probe antenna looking at angles from –90 degrees to 90 degrees passing through 0 degrees, with 0 degrees being the boresight angle.

The longitudinal scans also have a reference trace where the probing antenna is boresighted to the range antenna. The longitudinal scans are repeated with the probing antenna pointing at different angles from 90 to –90 passing through 180. (See samples in Figure 5.39.)

Figure 5.40 shows an apparatus for measuring the free-space VSWR. Figure 5.40(a) shows the apparatus set for longitudinal scan and Figure 5.40(b) shows the apparatus set for transverse scan. The rails allow for linear motion for the scan, and the carrier on the rails has an azimuth stage to rotate the probing antenna to the desired angle. Figure 5.40 shows an LPDA and a horn antenna being used as the probing antennas for the test.

Once the data is acquired it needs to be analyzed. The procedure for analyzing the data can be tedious, and it is not easy to automate. The technician has to rely on experience to analyze the data and decide what is the ripple caused by reflections versus the ripple caused by mechanical oscillations of the positioning system or even other sources of noise. Brian Tian does an excellent job of explaining the test in his AMTA paper [19]. To evaluate the reflectivity on a given direction, the trace at a given angle is compared to the reference, and the difference in magnitude between the two traces is added to the reflectivity related to the ripple of the trace.

Let us follow an example. Figure 5.41 shows the transverse reference scan and the scans with the probe pointing at –90 degrees and –30 degrees. A polynomial second-order curve fit is done on each of the scans to compare the levels and to obtain the ripple from the scans. Let us concentrate on the –30-degree trace.

In Figure 5.42 the curve fit of the reference trace is compared to the level of the curve fit of the trace taken with the probe pointing at –30 degrees at the region where the ripple will be estimated. (Tian suggests doing this at several locations along the traces to get an average value [19].)

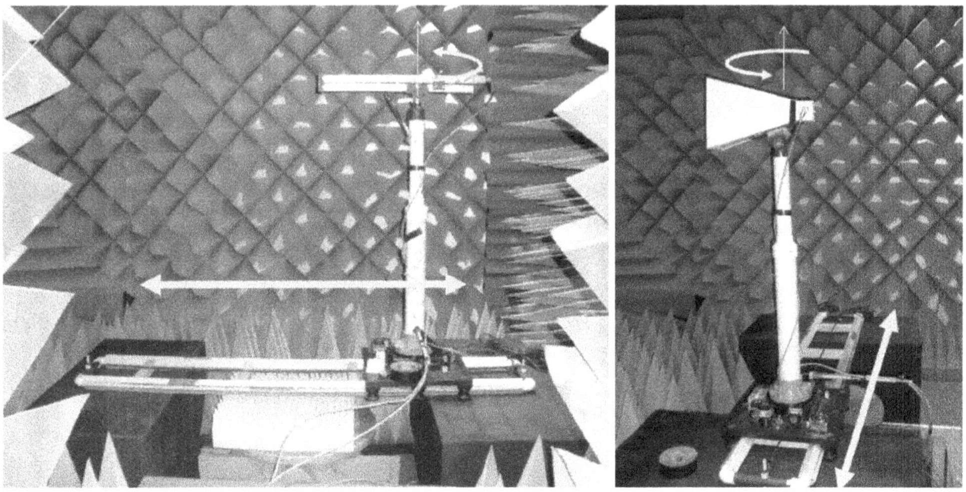

(a) (b)

Figure 5.40 Free-space VSWR setup. The rails allow for transverse or longitudinal motion of at azimuth positioner that rotates the probing antenna to the desired angle. (*Source:* Author's private collection.)

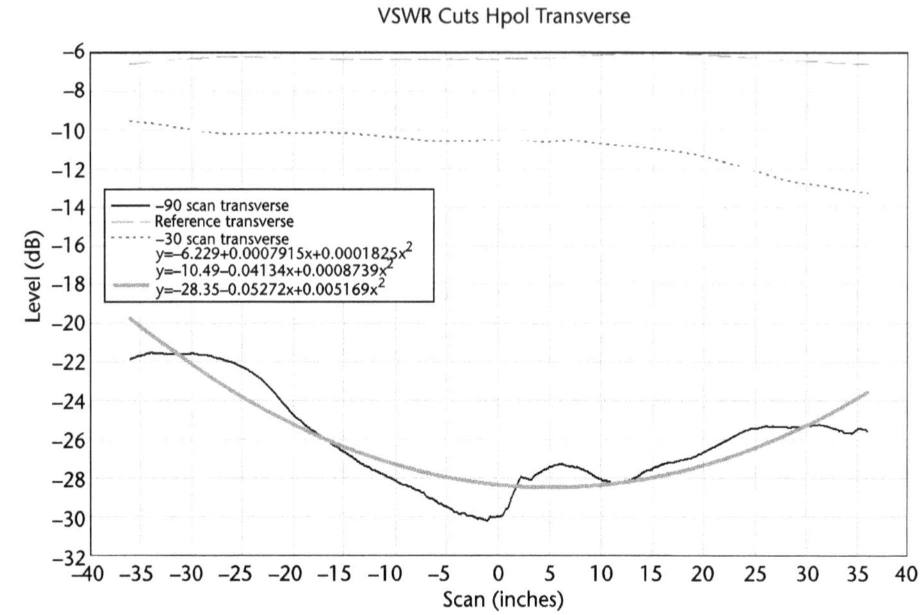

Figure 5.41 Sample data from a free-space VSWR acquisition, showing the transverse scans at −90 degrees and −30 degrees.

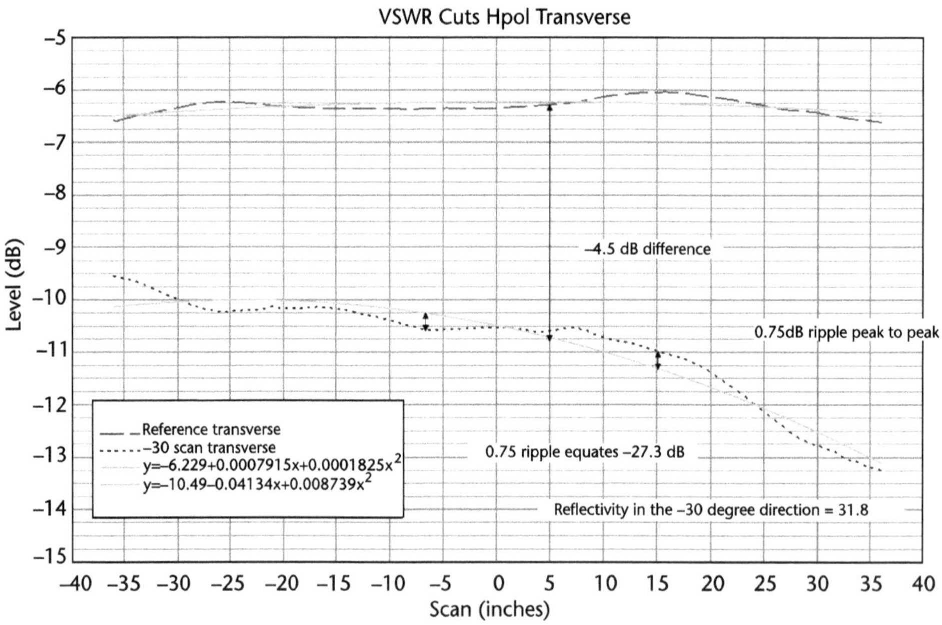

Figure 5.42 Comparing the trace at a given angle to the reference and extracting the ripple from the trace.

The equation used to evaluate the reflectivity is

$$R_{\text{at}\,\theta} = 20\log_{10}\left(\frac{10^{\left(\frac{r_{p-p}}{20}\right)} - 1}{10^{\left(\frac{r_{p-p}}{20}\right)} + 1}\right) - \Delta L \tag{5.48}$$

where ΔL is the level difference between the reference trace and the trace at the angle θ, and r_{p-p} is the peak-to-peak ripple of the trace at the angle θ. Hence, in the case shown in Figure 5.42, ΔL is 4.5 dB and r_{p-p} is 0.75 dB giving a reflectivity in the $\theta = -30$ direction of −31.8 dB.

It is important to understand that there are limitations to this test procedure. The ceiling and floor are not tested. The QZ also needs to be large enough to encompass a full cycle of the interference pattern. If these limitations are understood, the method is fairly accurate as the comparisons with QZ estimates have shown.

5.6 Conclusion

This chapter exposes readers to a typical analysis of the simplest of ranges, the rectangular range. There are a lot of constraints with these types of ranges. While some of these constraints, such as the low frequency limit of the typical rectangular range, can be circumvented by using a taper range, there are also limitations in the use of tapered anechoic ranges. The biggest constraint is that as frequency increases the electrical size of the antennas being measured is reduced. To solve that issue other types of antenna measurement approaches have been developed: the near-to-far-field ranges and the compact range. Those ranges are investigated in the following chapters. Readers, please note that some of the guidelines given in this chapter about location of doors and treatment of corners can be applied to near-field ranges and compact ranges.

References

[1] Rodriguez, V., "Open Boundary Quadridge Horn Antenna for the 80 MHz to 1 GHz Range: A Dual Polarized Solution for Testing Antennas in the VHF and UHF Ranges," *2010 Proceedings of the Fourth European Conference on Antennas and Propagation (EuCAP 2010)*, Barcelona, Spain, April 2010.

[2] Olver, A. D., et al., *Microwave Horns and Feeds*, New York, NY: IEEE Press, 1994.

[3] Hemming, L. H., *Electromagnetic Anechoic Chambers: A fundamental Design and Specification Guide*, New York: IEEE-Press/Wiley-Interscience, 2002.

[4] Hollis, J. S., T. J. Lyon, and L. Clayton, Jr., eds., *Microwave Antenna Measurements* (Second Edition), Atlanta, Georgia: Scientific-Atlanta, 1970.

[5] Emerson, W. H, and H. B. Sefton, Jr., "An Improved Design for Indoor Ranges," *Proceedings of the IEEE*, Vol. 53, No. 8, Aug. 1965, pp. 1079–1081.

[6] Emerson, W. H., "Anechoic Chamber," U.S. Patent 3,308,463, Mar. 7 1967.

[7] Kummer, W. H., and E. S. Gillespie, "Antenna Measurements—1978," *Proceedings of the IEEE*, Vol. 66, No. 4, April 1978.

[8] Emerson, W. H., "Electromagnetic Wave Absorbers and Anechoic Chambers Through the Years," *IEEE Trans. on Antennas and Propagation*, Vol. 21, No. 4, July 1973.

[9] Hemming, L. H., and R. A. Heaton, "Antenna Gain Calibration on a Ground Reflection Range," *IEEE Transactions on Antennas and Propagation*, Vol. 21, No. 4, July 1973, pp. 532–538.

[10] Hickman, T. G., and T. J. Lyon, "Experimental Evaluation of the Massachusetts Institute of Technology, Lincoln Laboratory, Anechoic Chamber," Final Report, *Scientific Atlanta*, Submitted Under Contract AF 19(628) 5167, June 1968 (http://www.dtic.mil/dtic/tr/fulltext/u2/686066.pdf).

[11] Rodriguez, V., and J. Hansen, "Evaluate Antenna Measurement Methods" *Microwave and RF*, October 2010, pp. 62–67.

[12] Rodriguez, V., "On the Disadvantages of Tilting the Receive End-Wall of a Compact Range for RCS Measurements," *39th Annual Meeting and Symposium of the Antenna Measurement Techniques Association AMTA 2017*, Atlanta, GA, October 2017.

[13] Rodriguez, et al., "A Cone Shaped Tapered Chamber for Antenna Measurements Both in Near Field and Far Field in the 200 MHz to 18 GHz Frequency Range and Extension of the Quiet Zone using an RF Lens," *Journal of the Applied Computational Electromagnetic Society*, Vol. 28, No. 12, December 2013, pp. 1162–1170.

[14] Rodriguez, V., "On Selecting the Most Suitable Range for Antenna Measurements in the VHF-UHF Range," *2018 IEEE Conference on Antenna Measurements and Applications*, Västerås, Sweden, September 2018.

[15] Burnside, W. D., et al., "An Ultra-wide bandwidth, tapered Chamber Feed," *18th Annual Meeting and Symposium of the Antenna Measurement Techniques Association AMTA 1996*, Seattle, Washington, 30 September to 3 October 1996.

[16] Lee, K.-H., et al., "Numerical Analysis of a Novel Tapered Chamber Feed Antenna Design," *24th Annual Meeting and Symposium of the Antenna Measurement Techniques Association AMTA 2002*, Cleveland, Ohio, 3–8 November 2002.

[17] Lee, K.-H., C.-C. Chen, and R. Lee, "UWB Dual Linear Polarized Feed Design for Tapered Chamber," *25th Annual Meeting and Symposium of the Antenna Measurement Techniques Association AMTA 2003*, Irvine, California, 19–24 October 2003.

[18] Hiatt, R. E., E. F. Knott, and T. B. A. Senior, "A Study of VHF Absorbers and Anehcoic Rooms," *University of Michigan Report 5391-1-F*, February 1963.

[19] Tian, B., "Free Space VSWR Method for Anechoic Chamber Electromagnetic Performance Evaluation," *30th Antenna Measurement Techniques Association Annual Symposium (AMTA 2008)*, Boston, MA.

[20] Rodriguez, V., "Comparing Predicted Performance of Anechoic Chambers to Free Space VSW Measurements," *39th Annual Meeting and Symposium of the Antenna Measurement Techniques Association AMTA 2017,* Atlanta, GA, October 2017.

Appendix 5A: MATLAB Scripts

5A.1 Hickman and Lyon Ground Reflection Range Analogy

This script follows the analysis given in [9] of this chapter. It is also the analysis presented by Hemming in [8]. Used with some of the slab absorber calculators, could be a useful tool in checking the overall geometry for a tapered chamber.

```
% Equation for the analysis of cuts of the QZ per Hickman and Lyon
% The Script provides a plot of the field across the QZ
% the reflectivity of the absorber is given by C1 and C2
% C1 is the side wall absorber while C2 is the floor and ceiling
% absorber in the Taper. the values for C1 and C2 are switched to
% model the polarization of the source antenna
% The script can be used to check how as frequency increases the
% QZ illumination gets worse
j=0.0+1.0i;
ED=1; %Direct Field Strength
lambda=3E8/230E6;% calculate lambda
QZrad=3.05; %QZ radius
Stepping=2*QZrad/100;% take 100 points in the QZ transverse
P=(-QZrad:Stepping:QZrad); %calculate the points in the QZ
PL=21.6 ; % test distance from source to center of QZ
spacing= .625; %spacing from the source to the absorber treatment
Alpha=(20.54/2)*pi/(180); %Taper angle in radians
S=2.*spacing*cos(Alpha);
beta=atan(P/PL); %angles to different points at the QZ
for ii=1:1:101;
RD(ii)=sqrt(PL^2+P(ii)^2);%Distance from source to the point across the QZ
Rr1(ii)=sqrt(RD(ii)^2+(4*spacing)^2+4*spacing*(PL*sin(Alpha)-...
 P(ii)*cos(Alpha)));
Rr2(ii)=sqrt(RD(ii)^2+(4*spacing)^2+4*spacing*(PL*sin(Alpha)+...
 P(ii)*cos(Alpha)));
Ro(ii)=sqrt(RD(ii)^2+(4*spacing)^2+4*PL*spacing*sin(Alpha));
%Distance from source to the point across the QZ
end
C1=.8*exp(j*pi); % reflectivity of absorber wall 1
C2=.4*exp(j*pi); % reflectivity of absorber wall 2
E=ED*(exp(-j*2*pi*RD/lambda)+C1*((RD/Rr1)*exp(-j*2*pi*Rr1/lambda)+...
(RD/Rr2)*exp(-j*2*pi*Rr2/lambda))+2*C2*RD/Ro*exp(-j*2*pi*Ro/lambda));
Elog=log10(E/E(51));
plot (P*3.28,Elog)
```

5A.2 Tapered Array Factor

This script allows the user to enter some values for the reflectivity of a flat piece of absorber of given thickness and adjust the distance from the center element to the absorber to see how, as the electrical distance increases (i.e., the frequency increases), the array factor moves away from unity, hence showing how frequency-dependent the tapered anechoic chambers can be.

```
% Taper chamber array factor approximation
% this Matlab routine allows you to set the magnitudes and phases for the
% reflectivity off the absorber around the main antenna to see what the
% theoretical array factor is.
%ideally the array factor should be 0dB at the critical aspect angles
% A0 is the center element so its magnitude is unity
% A1,A2...A4 are the magnitude of the additional reflections
A0=1;
j=1i;
ref=-13; % this is the reflectivity of the absorber around the antenna
amplitudefactor= 10^(ref/10);
A1=amplitudefactor*A0;
A2=amplitudefactor*A1;
A3=amplitudefactor*A2;
A4=A3;
frequency=100.E6;
c=299792458.;
lambda=c/frequency;
d=0.257*lambda; % d is the electrical spacing between the center antenna
% and its images (twice the distance from the center element to
  the absorber
dtr=(pi/180);
phi=-164*dtr; %this is the phase is degrees of the reflected energy
k=(2.*pi)/lambda;
for i= 0:1:180;
 theta(i+1)=i-90;
 AF(i+1)=0.;
 AF(i+1)= A0 + A1*exp(j*(k*d*cos(theta(i+1)*dtr)+phi))+ ...
  A1*exp(j*(k*(-1*d)*cos(theta(i+1)*dtr)+phi))+ ...
  A2*exp(j*(k*(2*d)*cos(theta(i+1)*dtr)+ 2*phi))+ ...
  A2*exp(j*(k*(-2*d)*cos(theta(i+1)*dtr)+ 2*phi))+...
  A3*exp(j*(k*(4*d)*cos(theta(i+1)*dtr)+ 2*phi))+ ...
  A4*exp(j*(k*(-4*d)*cos(theta(i+1)*dtr)+ 2*phi));

end

AFr=10*log10(abs(AF));
AFphs=angle(AF)*(1/dtr);
plot (theta,AFr)
grid on
title ('AF at 100MHz')
xlabel('angle observation from horizon (degrees)')
ylabel ('AF (dB)')
```

CHAPTER 6

Anechoic Ranges for Near-Field Measurements

6.1 A Bit of History

Near-field measurements are an accepted method for measuring the radiation pattern of antennas. Like many inventions used nowadays, the idea of near-field measurements goes back many years, in this case, to the 1960s. Hess relates the experience of one company, Scientific Atlanta, in the implementation of near-field systems in 1963 [1]. The system described in [1] already had the basic components of a near-field system: a planar scanner, a receiver to measure magnitude and phase, and (in 1963) an analog computer to compute the far field from the near-field data. Hess also relates further advances in the 1970s, with the introduction of digital computers. The 1980s saw the introduction of spherical near-field (SNF) measurements, which allowed for near-field measurements on any kind of antennas—not only the high-gain antennas that could be tested with planar near-field (PNF) measurements.

One of the assumptions for near-field antenna measurements is that the range antenna does not affect the field radiated by the AUT. The theory is that there are not multiple reflections between the AUT and the range antenna. To accomplish this, the range antenna is typically electrically small; hence it is usually referred to as a near-field probe or probe, rather than a range antenna.

For high-gain antennas, the PNF range was usually devoid of absorber. The only wall covered was the one behind the scanner. Indeed, the IEEE Standards Association issued a recommended practice for near-field antenna measurement (IEEE STD 1720-2012) that specifies that for PNF, absorber should be placed in front of the antenna to absorb the main beam of the antenna [2]. It also recommends absorber on the immediate structure to the measuring probe [2]. The standard adds that more absorber coverage is needed if the probe is not directive and describes how proper placement of the absorber can reduce multipath and leakage. However, it does not give any guidance for SNF or cylindrical near-field (CNF) ranges.

This chapter describes the main geometries of PNF and SNF systems, also touching on the CNF, which is essentially a combination of PNF and SNF techniques. For further detail on near-field to far-field theory, readers may refer to [3].

6.2 The PNF Range

As Hess mentions in [1], PNF systems are the earliest implementation of near-field measurements. However, there is an important condition PNF ranges must meet to be suitable for measuring an antenna: most of the power radiated by the antenna must be incident onto the plane where the near field is being measured [3]. It follows from this that the AUT must be highly directive. In general, a directivity higher than 15 dBi is sufficient.

This condition should be kept in mind when designing the absorber layout of a PNF range. Since the AUT is fixed in this range (not rotated around any axis), knowing where most of its radiation is pointing is critical in designing the range.

The range must be large enough to encompass the scanner, and the scanner should be large enough to provide an adequate scanning plane. In a PNF, we are not measuring the radiation all around the antenna but the radiation from the radiated energy that crosses the scan plane.

Consider Figure 6.1, where the angular range for a valid measurement is limited by the size of the scanning plane. In [2], guidelines are provided regarding the valid pattern, where the angular range of far-field pattern validity is calculated using

$$\pm\theta = \tan^{-1}\left(\frac{\pm L - \frac{a}{2}}{d}\right) \tag{6.1}$$

where a is the antenna size that can be electrically written as $n\lambda$; d is the separation between the scan plane and the AUT, which we can write in terms of wavelengths as $k\lambda$; and L is the length of the portion of the scan up and down, L_y, or left to right, L_x, in front of the antenna being tested. Equation (6.1) can be rewritten in such a way that the range angle is one of the known variables. This is a better equation, since usually we are interested in knowing the pattern out to certain angle, and we have a priori knowledge of the angular range that we are interested in measuring. The unknown becomes the scan size; hence for the horizontal scan we can rewrite (6.1) as

$$L_x = \left(n + 2k\tan\left(\theta_s\right)\right)\lambda \tag{6.2}$$

where the aperture size a is now expressed in wavelengths as $n\lambda$, and the test distance d is expressed in wavelengths as $k\lambda$, as shown in Figure 6.1. In [2], it is recommended that the minimum value for k should be between 3 and 5. In general, the separation is such that $3 \leq k \leq 10$.

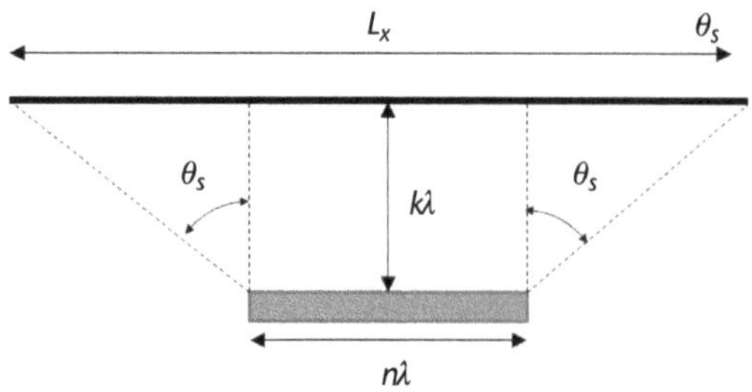

Figure 6.1 Planar field geometry scan plane, valid angle, separation, and antenna aperture.

6.2.1 Sizing the Chamber

Once we know the angle of interest and the size of the antenna (6.1), the information can be used to obtain the size of the scanner. Equation (6.1) can be used to determine the required scan plane size in the vertical and the horizontal plane. We mainly focus here on vertical plane scanners, but the equations presented can be applied to horizontal planar scanners. Horizontal scanners are common in satellite antenna testing applications where gravity may deform the antenna, so it is preferable to have the AUT facing up. The width of the range should be at least wide enough to house the scanner, hence

$$W \geq L_x + (4 + 2t_s)\lambda + \Delta_{scn} \tag{6.3}$$

It should be noted that while in (6.2) the wavelength λ is the wavelength at the lowest frequency of operation for the antenna being tested, when sizing a range we will use λ for the lowest frequency of operation of the range, which will be the lowest frequency for any antenna being tested in that particular range.

Figure 6.2 shows the different parameters that define the minimum width of a PNF range. The electrical thickness of the absorber treatment is discussed in Section 6.2.2. The biggest unknown is usually Δ_{scn}, which is dependent on the scanner construction. It is a good practice to leave enough room around the scanner to facilitate its service and, potentially, to place part of the RF or position control system racks in the range.

The height of the range follows a very similar approach to the approach used for estimating the width. The following can be used to estimate the required range height:

$$H \geq L_y + y_o + \Delta_h + (2 + t_s)\lambda \tag{6.4}$$

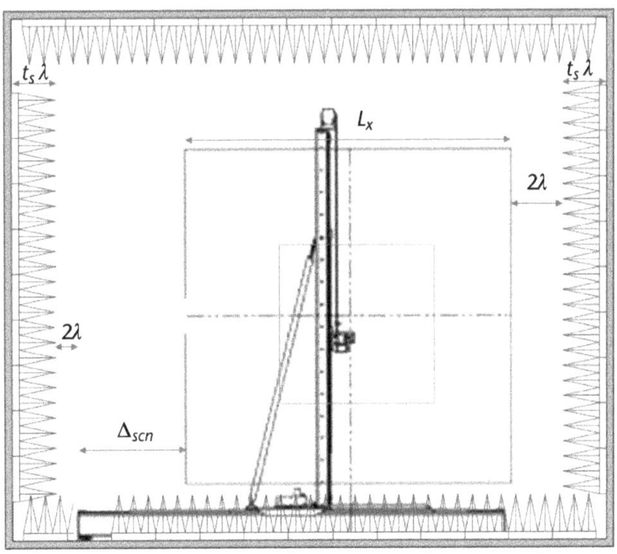

Figure 6.2 Minimum width of a PNF range.

There are two unknown parameters in (6.4). One is Δ_h. This parameter accounts for the additional height of the scanner above the scan plane. It basically accounts for the pulley used to hoist the probe carrier up and down the vertical stage of the planar scanner. The other parameter is y_o. This parameter denotes the distance from the bottom of the scan plane to the floor of the range. The value of y_o should be such that the following condition is met:

$$y_o \geq t_s\lambda + 2\lambda \tag{6.5}$$

Figure 6.3 shows the parameters that define the height of a range. Again, this is a recommended minimum. While for scanners that provide scan areas smaller than 1.8m by 1.8m this may be sufficient; for larger scanners, care should be taken to leave enough room for cranes or forklifts to assist in raising the vertical stage during installation.

The last dimension that needs to be defined is the length of the range. The length of the range is defined by the following:

$$L \geq S_{clr} + A_d + (4 + k + t)\lambda \tag{6.6}$$

Figure 6.4 shows the different parameters that appear in (6.6). From Figure 6.4, is clear that A_d is the depth of the antenna, and S_{clr} is the scanner footprint. (Recall that this is a minimum size.) For help with servicing the scanner it may be desirable to leave some space between the back wall absorber and the scanner structure. Furthermore, although I am recommending four wavelengths behind the AUT, it may be desirable to have more space for ease of making the connections to the AUT.

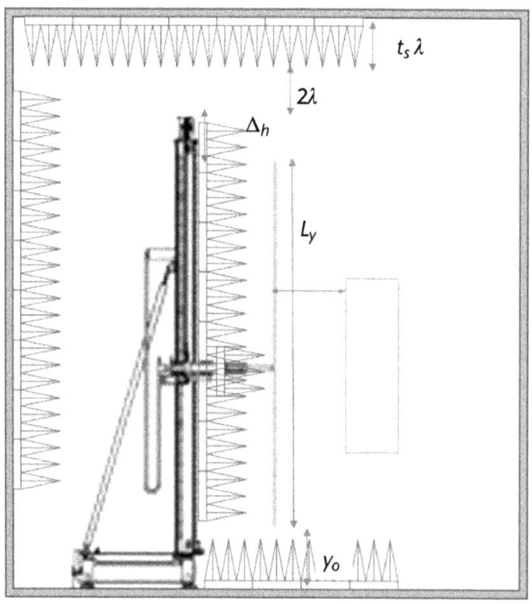

Figure 6.3 Minimum height of a PNF range.

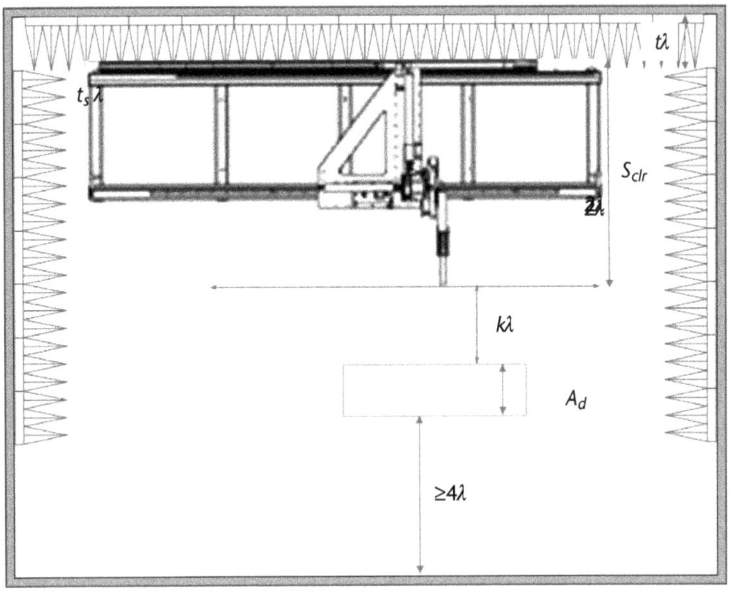

Figure 6.4 Minimum length of the planar scanner range.

6.2.2 Absorber Layout

In (6.3), (6.4), and (6.6) we have not defined the absorber treatment thickness. This section defines the typical electrical thickness for the absorber treatment of the PNF range. Let us start first with defining t. The parameter t is the electrical thickness of the absorber on the wall behind the scanner and in front of the antenna aperture. As a minimum we want $t \geq 2$. Choosing this value should provide at least 40 dB of absorption at normal incidence. That same size absorber should be applied to the scanner. This is, however, not absolutely necessary; the scanner structure in general presents a small surface to the radiating antenna under test. Hence, a small portion of the radiation will be reflected from this structure. It is not uncommon to see the scanner structure absorber be a size smaller than the back wall absorber.

How about the lateral surfaces of the range? In the IEEE recommended practice [2], there is no recommendation for the absorber on the lateral surfaces for PNF ranges. This is partly because usually very high directivity antennas are being tested; hence, the level of radiation toward the lateral surfaces is very small. This is a valid argument. However, the lowest directivity antenna that will be tested should be considered.

One of the reasons that antennas are measured is to obtain the gain. There are several methods to measure the gain as discussed in Chapter 2, but it is common to use a gain standard such as a pyramidal horn. This SGH is measured in the range, and the gain of the AUT is then related to the levels measured for the known standard as described in Chapter 2 and [4]. In many cases, for PNF ranges, the SGH used to measure gain is the lowest directivity antenna that is tested. Again, a directivity of 15 dBi is the minimum for an antenna to be successfully measured in a PNF range. A 15.5-dBi SGH has an aperture of about 2.62λ by 1.96λ; hence a $3.27\text{-}\lambda$ for the diagonal, which, at frequencies below 1 GHz, makes for a physically large antenna.

To find the typical range of angles of incidence onto the lateral surface absorber (Figure 6.5), we start by substituting (6.2) into (6.3) to obtain

$$W = \left(n + 2k\tan\left(\theta_s\right) + 4 + 2t_s\right)\lambda + \Delta_{scn} \tag{6.7}$$

If we assume that

$$\left(n + 2k\tan\left(\theta_s\right) + 4 + 2t_s\right)\lambda > \Delta_{scn} \tag{6.8}$$

which usually holds true for most scanners, then we can rewrite (6.7) as

$$W \approx \left(n + 2k\tan\left(\theta_s\right) + 4 + 2t_s\right)\lambda \tag{6.9}$$

Looking at Figures 6.2 and 6.5, we can solve for the angle of incidence. If we assume that the reflection is mainly from the tips of the absorber, then the $2t_s$ terms fall out of (6.9). The angle of incidence when the probe is at the center of the scan area is approximately given by

$$\theta = \tan^{-1}\left(\frac{k}{n + 2k\tan\left(\theta_s\right) + 4}\right) \tag{6.10}$$

Equation (6.10) shows the angle of incidence being related to the electrical size of the AUT and to the electrical separation between the AUT and the scan plane and the angular range of interest for the far-field pattern. Plotting the angle θ for different antenna sizes, we see that in general the angle of incidence is less than 25 degrees. (See Figure 6.6.) As shown in Chapter 3, those angles provide a reflectivity that is close to that of normal incidence; that is, the absorber is very efficient.

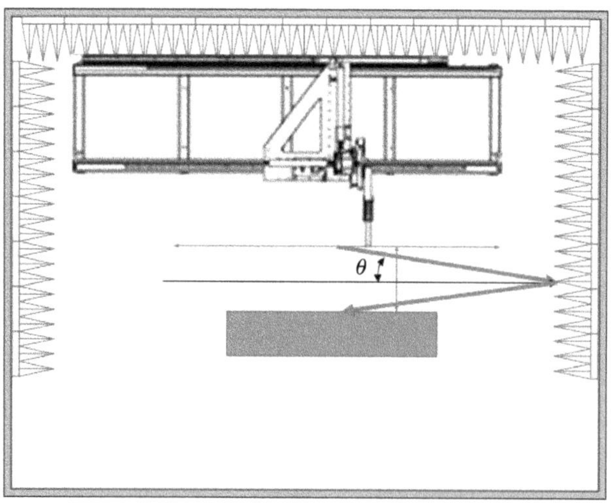

Figure 6.5 The angle of incidence for the specular bounce onto the lateral surfaces of the range.

Unlike the far-field chamber, in the PNF range, the probe is not static. The probe moves on a plane. Because of this movement the specular angle of incidence will change as the probe moves. Figure 6.7 shows the probe at the extreme of the scan angle. From Figure 6.7, we can obtain an equation for the angle of incidence when the probe is at that position. The following equation can be used to obtain the angle of incidence at the extreme of the scan plane:

$$\theta = \tan^{-1}\left(\frac{k\lambda}{\frac{L_x - n\lambda}{2} + 2\lambda}\right) \tag{6.11}$$

If we substitute (6.2) into (6.11) then the equation is reduced to

$$\theta = \tan^{-1}\left(\frac{1}{\tan(\theta_s) + \left(\frac{2}{k}\right)}\right) \tag{6.12}$$

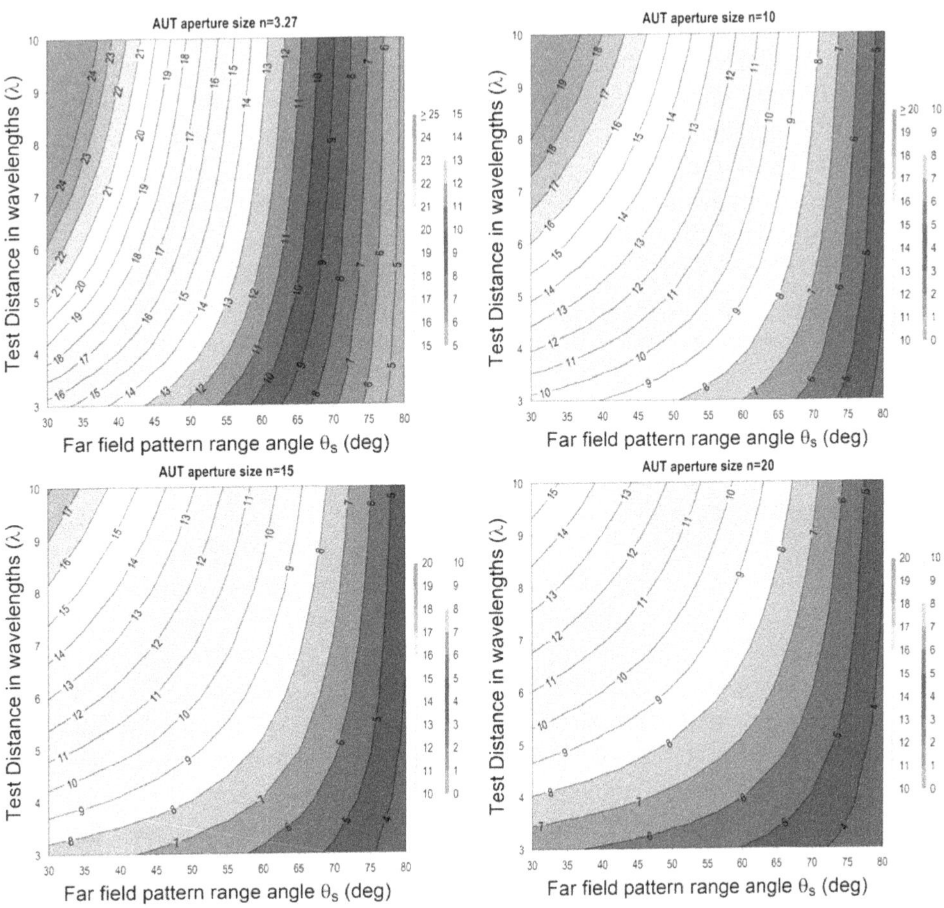

Figure 6.6 Angle of incidence onto the lateral surfaces for different AUT sizes.

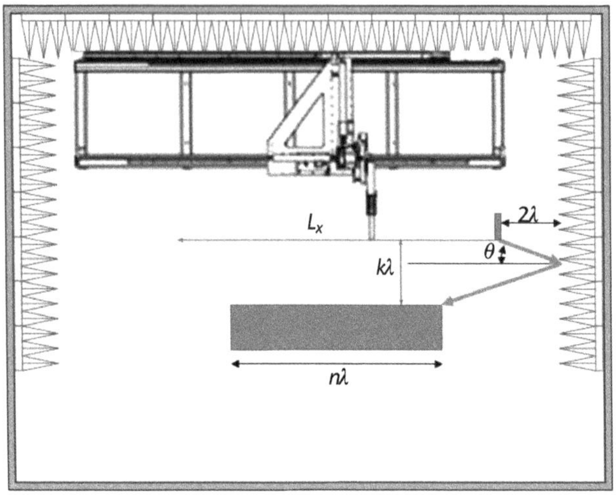

Figure 6.7 The angle of incidence on lateral surfaces when the probe is at the edge of the scan plane.

Notice that when at the extreme of the scan and looking at energy radiated from the edge of the AUT, as shown in Figure 6.7, the angle of incidence is only dependent on the far-field pattern maximum range angle and the test distance. Figure 6.8 shows that for the minimum distances given in the IEEE recommended practices [2], the angle of incidence is less than 45 degrees in the worst case, and this angle improves as the angular range for the far-field pattern increases. Even for larger separations, the angle of incidence does not exceed 55 degrees. Readers may question the use of larger separations from the radiating aperture to the measurement plane. It is because parabolic reflector antennas are one of those cases where a larger separation is required to account for the feed and focal distance of the reflector antenna.

Based on the angles of incidence onto the lateral surfaces of a PNF range, it will appear that $t_s = 2$; that is, the pyramidal absorber should have a thickness of 2λ. This is certainly a safe choice but also potentially overkill. Let us recall that the antennas that we are trying to measure in the PNF range are highly directive antennas, and in the worst case, they will exhibit directivities around 15 dBi. If we look at the typical SGH pattern for a 15.5-dBi horn (shown in Figure 6.9), it is apparent that for the worst case, the angle of incidence on to the lateral surfaces (for $k = 10$, $\theta_s = 30°$) is 52 degrees. Looking at back at the SGH pattern in Figure 6.9, it is apparent that in that direction, the antenna is radiating less than 10 dB compared to the direct path. Because of this lower level being radiated toward the lateral surfaces, the value of t_s can be as low as 1. This one-wavelength thick pyramidal absorber on the lateral surfaces is typically sufficient for treating a PNF range.

The reason for the smaller absorber on the lateral surfaces is that the AUT on a PNF is not rotated in any way. However, there is a family of antennas that are suitable for PNF measurements, phased arrays. Although not rotated, phased arrays, or specifically active electronically scanned array (AESA) antennas, point their main radiation toward the lateral surfaces. For PNF ranges designed to measure these antennas, it is safer to have $t_s = 2$. The ceiling and the floor on the PNF range can have the same absorber as the two lateral walls.

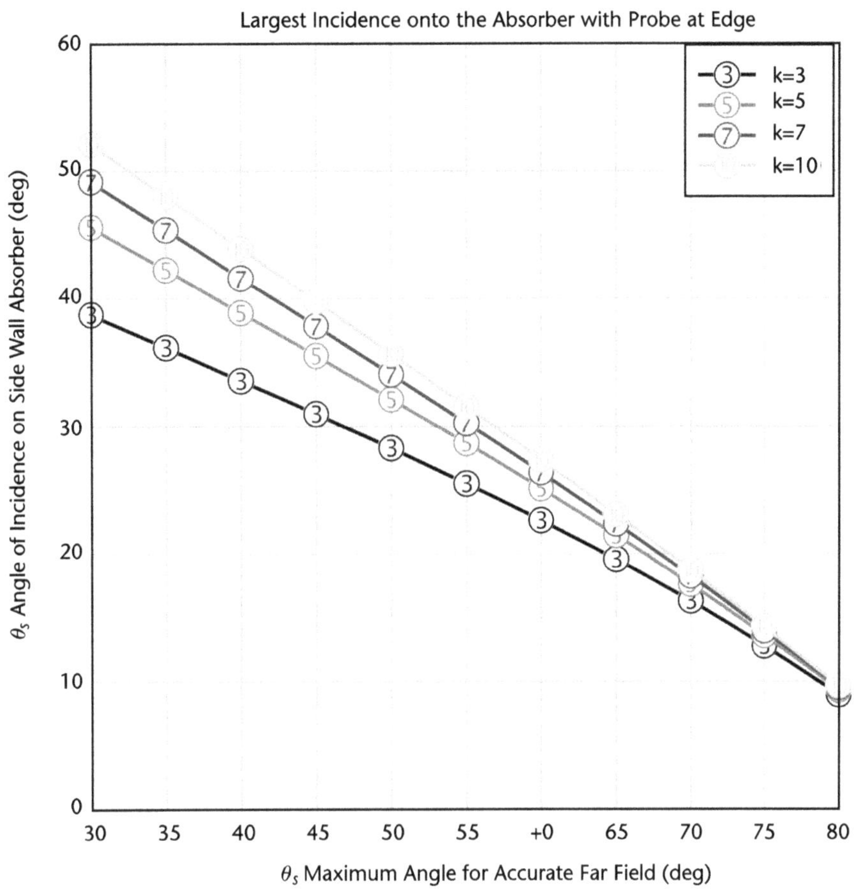

Figure 6.8 Angle of incidence for different probe-to-AUT separations.

Finally, the last surface in the range is the wall behind the AUT. It is common to leave this wall untreated. After all, the back lobe on these antennas is very low. The SGH shown in Figure 6.9 has a back lobe that is almost 30 dB lower than the main beam, and this is the antenna with the lowest directivity that should be measured in a PNF range. In most cases this back-lobe radiation will be hidden from the measurement by the AUT itself. In the case of parabolic reflectors where there may be some spillover from the feed, some absorber may be applied on some areas to reduce a potential reflection from the back wall; this absorber can be typically as short as $\lambda/4$ and still provide a reflected level that is smaller than –40 dB.

6.3 The SNF Range

Recall from Section 6.1 that SNF ranges appeared in the 1980s [1]. The basis for this method was introduced a decade earlier. In the SNF, the fields on a sphere that encloses, or almost encloses, the AUT are measured. In the SNF approach, there are no assumptions about the directivity of the AUT, so the methodology is valid for omnidirectional antennas as well as low-directivity antennas. It is important to

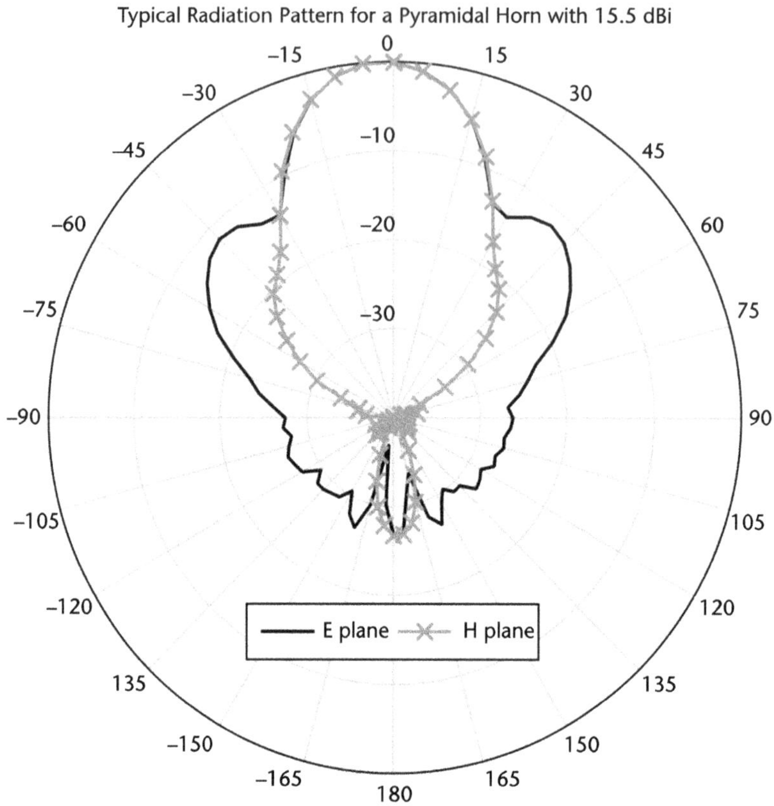

Figure 6.9 Typical pattern for a 15-dBi SGH.

understand that this provides flexibility to the system, meaning that low-directivity and high-directivity antennas can be measured. SNF systems are not inexpensive pieces of equipment, and their operation requires understanding of the transform theory, so if the only antennas that need to be measured are low-directivity, a far-field system such as those described in Chapter 5 is a more economical approach.

In the SNF, the fields are measured over a spherical surface. A spherical wave expansion is then obtained to describe the fields on (and outside) what is called the minimum sphere, minimum radiating sphere, or maximum radial extent (MRE). This is a spherical surface that encloses the radiating object and its structure. While typically the MRE is not necessarily the QZ, here, we will refer to the MRE as the QZ, understanding that for the SNF case the QZ is the largest MRE that we can have in a given range.

Because on near-field measurements we are electrically and physically close to the AUT, there is more flexibility on how to sample the data. Based on how the data is sampled, we have three different approaches to the SNF ranges:

- Fixed probe;
- Movable probe, movable AUT;
- Fixed AUT.

In the fixed-probe or phi over theta systems, the AUT is rotated on a roll-over azimuth positioner as described in Chapter 5, while the probe is fixed in location. This approach is very similar to the rectangular chamber but with a test distance that is not driven by the far-field equation.

In the second approach, also known as theta over phi, both the probe and the AUT are positioned. The AUT is placed on an azimuth positioner, while the probe describes an arch either on a fixed track or suspended from a rotating arm or gantry. This family can include systems where the probe is not physically moved, but where multiple probes located around the AUT are switched on and off to measure the field at different locations (see Figure 10.8). For range purposes the absorber layout is similar in both cases, but for the multiple probe the absorber around the probe itself is more critical.

The third approach is the most complex and least common. In this approach the AUT is fixed, and a mechanical arm, or robot, positions the probe around the AUT. This complex approach is only used in extreme cases, where in addition to RF, power, and communication lines, the AUT requires liquid cooling to chill its internal amplifiers. Managing all those cables and hoses is extremely complex if we were to rotate the AUT. Another application for the fixed AUT approach is on-chip antennas, where the chip is powered and fed via probes that prevent the rotation of the AUT. It should be noted that the last two configurations could be, and have been, used for far-field ranges, but only for cases where the far-field test distance is sufficiently small.

Figure 6.10 shows the three possible configurations of a SNF range. Figure 6.10 (a) is a typical roll-over azimuth positioner with a fixed probe a distance away from it. While the probe may rotate to measure both orthogonal polarizations of the field, the AUT is rotated in phi by the roll state at the top of the vertical mast and in theta by the azimuth positioner under the offset section.

(a) (b) (c)

Figure 6.10 Three configurations of SNF ranges: (a) a fixed-probe system, (b) a movable probe, and an AUT system, and (c) a movable probe fixed AUT system.

In Figure 6.10(b), the AUT, in this case a horn mounted at an angle with respect to the vertical, is rotated by an azimuth positioner around a vertical axis. The probe is mounted on a carrier and scans along the curved track to move the probe to different theta angles with respect to the vertical while the AUT spins in phi.

Figure 6.10(c) shows one of the possible approaches to a fixed AUT SNF. The probe moves along a curved track that will position the probe along different theta angles. The curved track or arch is mounted to a rotary positioner, similar to a roll positioner, that rotates the entire arch and positions the probe on phi.

The first two approaches are the most common ones. Sections 6.3.1, 6.3.2, and discuss the sizing of the range and selection of the absorber.

6.3.1 Fixed-Probe Implementations

Recall that fixed-probe implementations are similar to rectangular chamber ranges. In these implementations, the AUT is typically mounted on a roll-over azimuth positioner with the probe mounted a distance away, at a minimum of three wavelengths (3λ). Figure 6.11 shows a typical SNF system, which is similar to a rectangular range like the far-field ranges discussed in Chapter 5.

The main difference between the far-field chambers discussed in Chapter 5 and the fixed-probe-SNF range is the test distance and the range antenna (i.e., the probe). The probe is required to have certain characteristics to make the SNF algorithm provide the correct results [2]. The IEEE recommended practice for near-field measurements recommends circular waveguides as the probe but also suggests that rectangular open-ended waveguides can be used [2]. Typically, these probes have directivities in the 8-dBi range.

Figure 6.12 shows the typical SNF range shown in Figure 6.11 with the main features of the range labeled in terms of wavelengths. The first and most important item in sizing an SNF range is the size of the AUT. Let us assume that the electrical size of the antenna is $n\lambda$. Using the far-field equations, we know that the minimum far-field distance is $2n^2\lambda$. However, for the SNF the distance will be $k\lambda$, where k is limited on the lower bound to be $k = 3$. It should be mentioned that the SNF algorithms and transformation can be performed at any distance; the measurement can

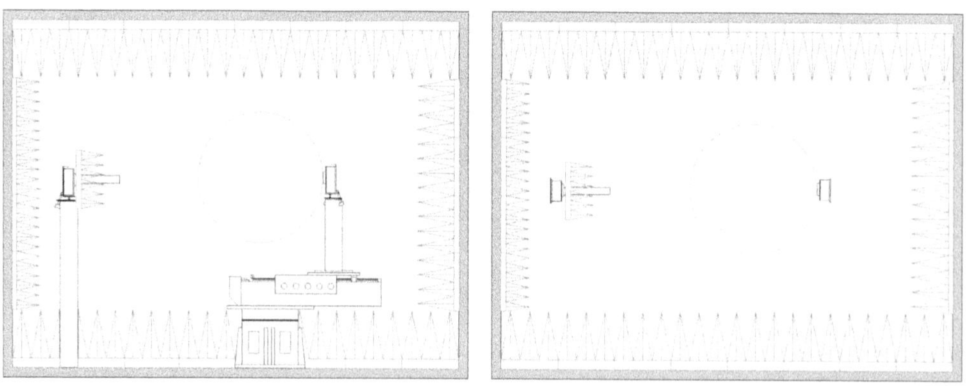

Figure 6.11 Typical SNF range with a fixed probe: (a)elevation side view, and (b) plan view.

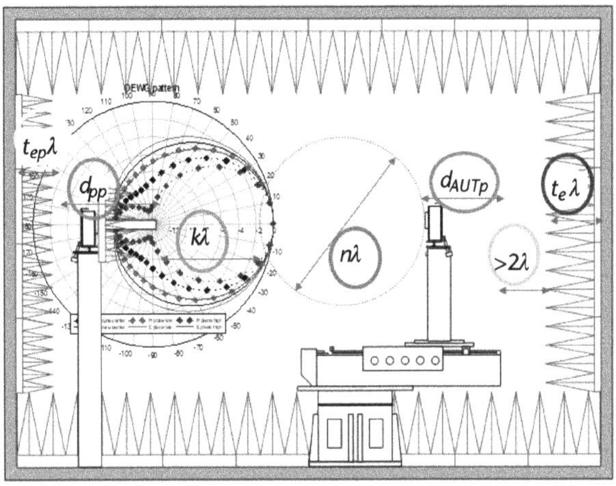

Figure 6.12 Elevation of a typical SNF range showing the length of the range and the probe pattern for an open-ended waveguide (OEWG) superimposed.

even be conducted at $k = 2n^2$, the far-field distance. The result is the true far-field pattern, $R \to \infty$, where R is the test distance. This is one of the advantages of the near-field to far-field mathematical transformations: They provide the true far-field pattern, not an approximation as the $2D^2/\lambda$ distance does.

However, real estate is valuable, and the other advantage of this approach is that we can reduce the test distance to a minimum. Indeed, experience has shown that $k = 2$ is acceptable, provided that the probe is physically very small and minimizes the effects of multiple reflections between AUT and probe.

6.3.2 Sizing the Chamber

As shown in Figure 6.12, the length of the range must be larger that $(k + n)\lambda$ to fit the test distance and the AUT. In addition, the range must allow for enough space for the probe and the positioning system as well as the RF absorber. As was the case for far-field ranges (Chapter 5), we want at least a 2-λ space between the positioner and the absorber tips. Hence, the range requires a minimum length L as given by

$$L \geq d_{pp} + d_{\text{AUT}p} + \left(n + 2 + k + t_e + t_{ep}\right)\lambda \tag{6.13}$$

Typically, a reflectivity of −40 dB from the walls will give us a potential error of less than ±0.1 dB. Chapter 3 shows that a 2-λ absorber provides that reflectivity level; hence $t_e = 2$. The absorber behind the probe benefits from the front-to-back ratio of the probe. The pattern of a typical OEWG probe superimposed in Figure 6.12 shows a front-to-back ratio of 10 dB. In actuality, the ratio may be higher given that the probes are usually nested on a backing of absorber as the pictures in Figure 6.10 show. The front-to-back ratio suggests absorber that provides −30 dB of normal incidence may be sufficient. As shown in Chapter 3, a one-wavelength pyramidal absorber provides 30 dB of normal incidence reflectivity; hence $t_{ep} = 1$.

These are recommended sizes, if a smaller error from the reflected signals is desired it can be accomplished by increasing the absorber size from the recommended values.

The only unknowns left in (6.13) are related to the size of the positioners, which is related to the physical size of the AUT as well as its weight. As a rule, the estimate could be that an additional 60 cm (about 2 ft.) is necessary. This approximation is good provided that the lowest frequency is not extremely low, making the antenna extremely heavy and physically large. For the probe and the probe support, based on the typical probe sizes the largest value for d_{pp} = 150 cm (about 5 ft.), but for frequencies above 3 GHz that estimate can be cut in half.

The next dimension is the width of the range. Figure 6.13 shows the plan view of the typical fixed-probe SNF range. It is obvious that the range should be wide enough to house the largest antenna being measured and allow for the positioner to rotate freely and maintain a minimum 2-λ distance from the absorber tips. In addition to the antenna and the positioner and spacing, the width must also fit the necessary absorber. In general, the width W of the range can be given by

$$W \geq 2 * d_{\mathrm{AUT}p} + \left(n + 4 + 2t_s\right)\lambda \tag{6.14}$$

The value for $d_{\mathrm{AUT}p}$ could be approximated from the same value given when estimating the length. The other unknown is the absorber size for the lateral walls. To estimate the size of that material, let us make the assumption that $d_{\mathrm{AUT}p}$ is much smaller than 2λ. This is a worst-case scenario, since we are looking at a "narrower" range. However, it will help in estimating the worst-case absorber required on the lateral walls. Additionally, as shown below, it will help with the ceiling absorber estimate. Readers can do a similar analysis of a range to estimate the effects of the absorber reflectivity using the approach below and the absorber approximations shown in Chapter 3. This can be used during the uncertainty analysis of the near-field range, as it will be shown in Section 6.5.1 for a PNF range.

The simplification of the width provides us with an equation for the angle of incidence:

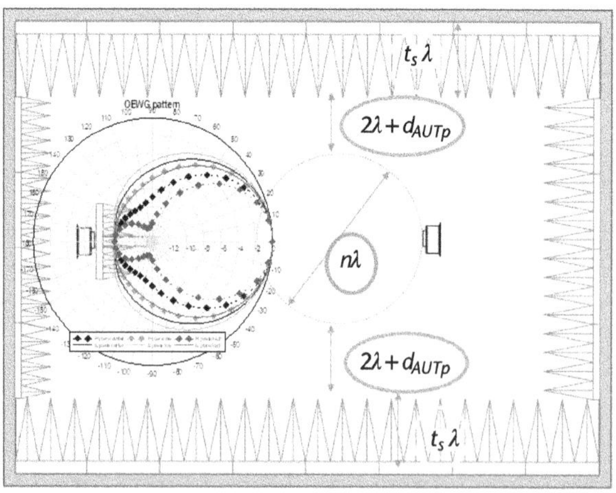

Figure 6.13 Pan view of a typical fixed-probe SNF range.

$$\theta = \tan^{-1}\left(\frac{\frac{n}{2} + k}{n + 4}\right) \tag{6.15}$$

Figure 6.14 shows how the angle for the reflected ray to the center of the QZ is calculated. This is a worst case, since generally the chamber will be wider to account for the positioner. We can look at the edges of the QZ and arrive at two other possible angles that will give us a "range" of potential angles of incidence. Figure 6.14 shows an angle α and an angle β. The angles are given by the following two equations:

$$\alpha = \tan^{-1}\left(\frac{k}{n + 4}\right) \tag{6.16}$$

$$\beta = \tan^{-1}\left(\frac{\frac{n}{2} + k}{\frac{n}{2} + 4}\right) \tag{6.17}$$

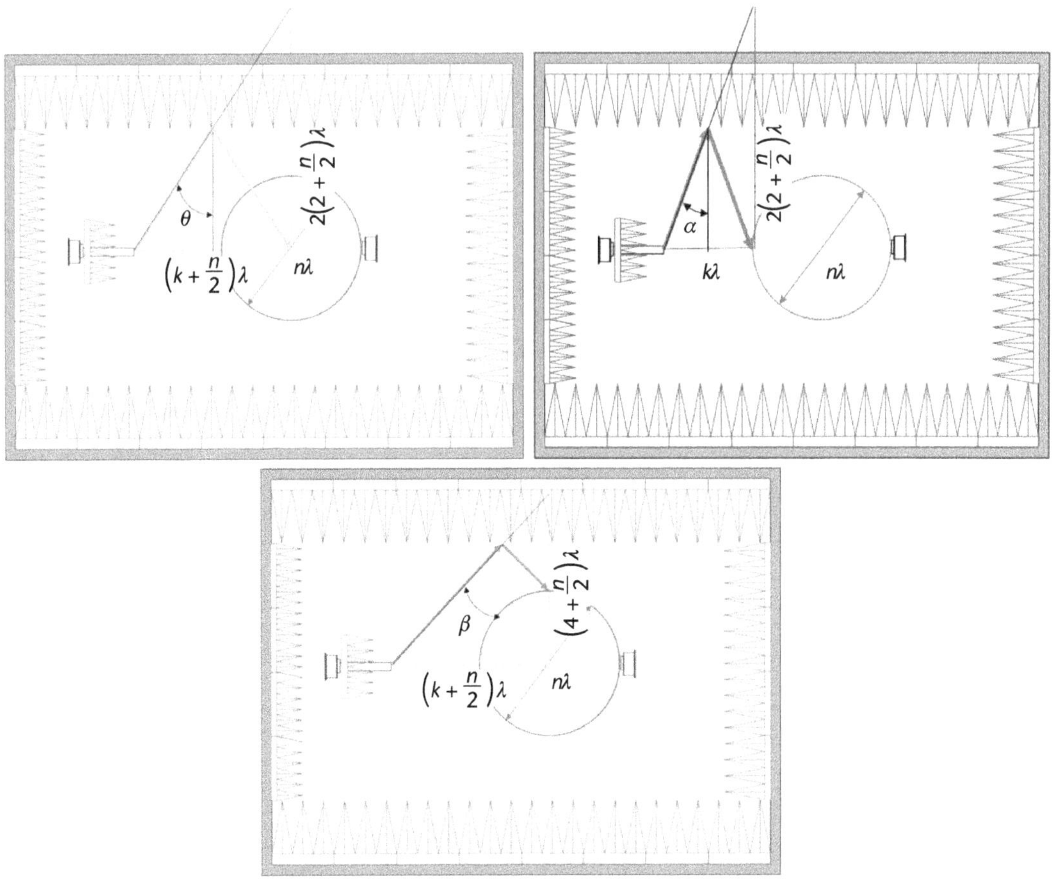

Figure 6.14 Estimating the angle of incidence for reflections to the center of the QZ.

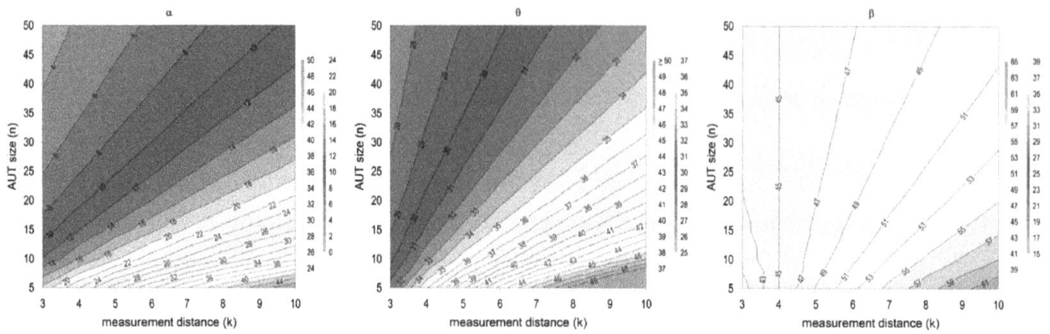

Figure 6.15 Plot of the angles of incidence for different AUT sizes and test distances.

We can analyze these angles for different electrical sizes of AUTs and different distances to the probe.

Computing α, β, and θ as a function of the test distance probe separation k and the AUT size n, the plots shown in Figure 6.15 are obtained. As expected, the worst angle of incidence is the one to the side edge of the QZ, angle β. Analyzing β in more detail shows that for smaller antennas it is better to be closer to the 3-λ test distance to the edge of the QZ. Ideally, the shorter the test distance the better.

To achieve a reflectivity of 40 dB for angles of incidence as wide as 50 degrees, an absorber of 3λ is necessary. However, in practice, 2.5λ may suffice, given the additional path losses and the probe pattern and the fact that this is a worst-case scenario where the range is narrower than it should be since we assume that the positioner size is ignored.

The range height is very similar to the width. Following the sketch in Figure 6.16, the equation below gives the height:

$$H \geq h_p + (n + 4 + t_s)\lambda \tag{6.18}$$

Notice that given the separation of 2λ between the ceiling absorber and the AUT, the same angles that were calculated for the lateral surfaces (i.e., the side walls) can

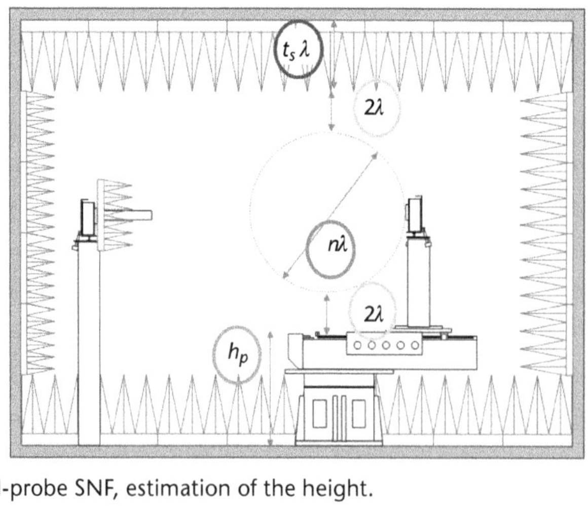

Figure 6.16 Fixed-probe SNF, estimation of the height.

be used for the ceiling and floor. The other parameter is the height of the positioner azimuth rotator and the offset member, so that these do not interfere with the floor absorber, typically $h_p > t_s\lambda$.

6.3.3 Absorber Layout

In general, it has been shown that the fixed-probe SNF range has a combination of absorber that ranges from 3 to 2.5λ for the lateral surfaces, 2λ on the end wall closer to the AUT, and 1λ on the end wall closer to the probe. The wavelength λ is again assumed to be at the lowest frequency. If the reader wants to perform an analysis of the potential reflectivity on a given range, the same approach can be followed to arrive at the reflectivity levels.

6.3.4 Movable Probe and Movable AUT SNF

The second type of common SNF ranges are the ones where both the probe and the AUT are repositioned (other than polarization rotation of the probe, which is common in the fixed-probe systems). One such configuration of the type of SNF systems is shown in the center picture of Figure 6.10. Figure 6.17 shows another configuration where a rotating arm or gantry repositions the probe (in the case of Figure 6.17, a dual-polarized LPDA) around the AUT, which is rotated around an orthogonal axis.

Regardless of arches (i.e., curved tracks that position the probe) or gantries (i.e., rotating arms that support the probe) or multiple electronically switched probes, the chamber sizing is very similar and so is the absorber treatment. These ranges are in general much larger, since the rooms are typically cubical in shape. The reason for this approach varies, but it is generally better for large platforms with antennas, such as automotive applications where the antenna is mounted on a vehicle. In those cases, rotating the vehicle and its antennas in two axes is challenging. In some cases the AUT cannot be rotated as chemicals may spill. This last case may seem

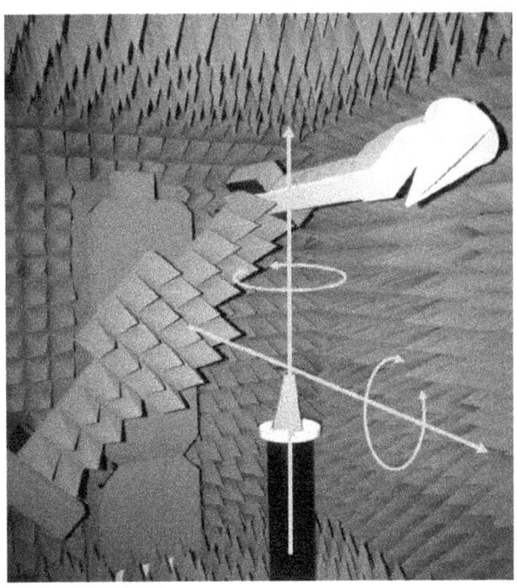

Figure 6.17 A gantry style SNF system.

odd, but consider laser printers and copiers with Wi-Fi or Bluetooth connectivity. Rotating the printer may lead to toner powder spilling everywhere.

Additionally, consider antennas where gravity may distort the antenna. In those cases it is important to maintain the gravity vector constant in a given direction.

From the range perspective, the problem is that the line of sight between the probe and the AUT changes as does the electromagnetic environment. Minimizing that change is important and will reduce the uncertainty and the contribution of the absorber treatment during the acquisition.

6.3.5 Sizing the Chamber

For the purpose of estimating a volume for an SNF range with a gantry arm or arch, we can start with the geometry shown in Figure 6.18. This geometry moves the probe from one side to the other of the AUT; that is, $-\gamma \leq \theta \leq \gamma$, while the AUT is rotated half a revolution. Another approach is to move the probe from $\theta = 0$ to $\theta \leq \gamma$ while the AUT is rotated a full revolution. Both arches and gantries are manufactured in either configuration, but typically most arches or curved tracks are built so that the probe moves from $\theta = 0$ to $\theta \leq \gamma$.

Figure 6.12 shows how the antenna range, or line of sight, changes for each measurement point. In addition, shown in Figure 6.18, Figure 6.10, and the system shown in Figure 6.17 is that the measurement does not take place over a full closed spherical surface. There is a truncation of the surface, and this truncation can cause errors on the transform. The assumption is that, as assumed in the PNF approach, the bulk of the radiation goes through the fraction of the spherical surface over which the measurement takes place and that little energy crosses the area of the sphere that is not scanned.

Thus prior to choosing a range to perform a measurement, the user needs to have a little a priori knowledge of the pattern of the AUT. Is the maximum θ angle of scan $\theta = \theta_{max}$ sufficient to measure the antenna? What is the potential truncation error from the available θ_{max}?

Figure 6.19 shows the geometry with the different dimensions. The maximum θ angle, can be defined as $\theta_{max} = 90 + \varphi$. The height of the range can be defined as

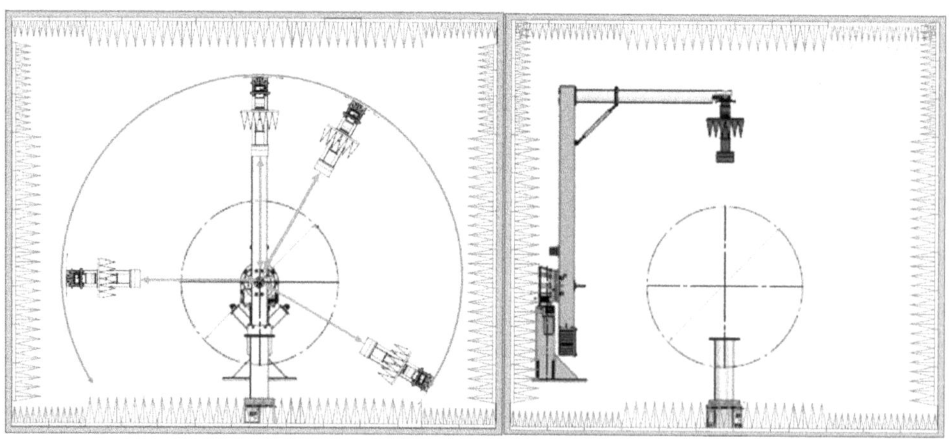

Figure 6.18 A typical SNF geometry using a Gantry arm to position the probe.

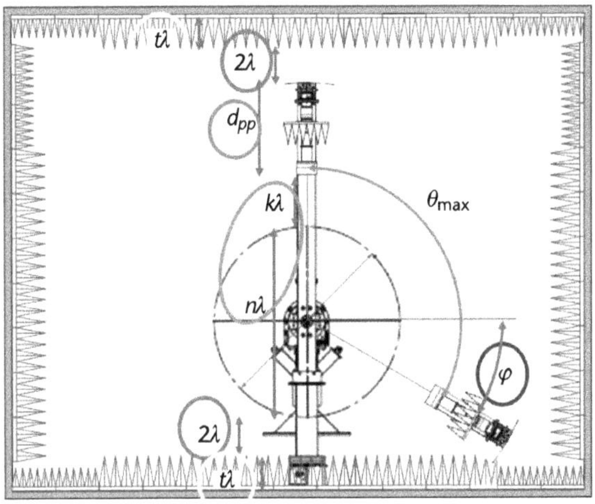

Figure 6.19 Geometry of a gantry SNF system showing the different dimensions and the maximum angle θ_{max}.

$$H \geq d_{pp}(1+\sin(\varphi)) + \left(\left(\frac{n}{2}+k\right)+4+\left(\frac{n}{2}+k\right)\sin(\varphi)+t\right)\lambda \tag{6.19}$$

If the angle φ is small enough, then (6.19) can be rewritten as

$$H \geq d_{pp} + (2t+4+n+k)\lambda \tag{6.20}$$

Now that we have a set of equations for the height of the range, let us consider the other two dimensions of the range.

Figure 6.20 shows a sketch of the width and length of a gantry SNF range. The width can be made smaller by defining the range of movement of the probe as 0 to θ_{max}. The reduction in the width of the range will be the gantry and probe structure d_{pp} (or the arched track and probe) and the test distance $k\lambda$: but be aware that

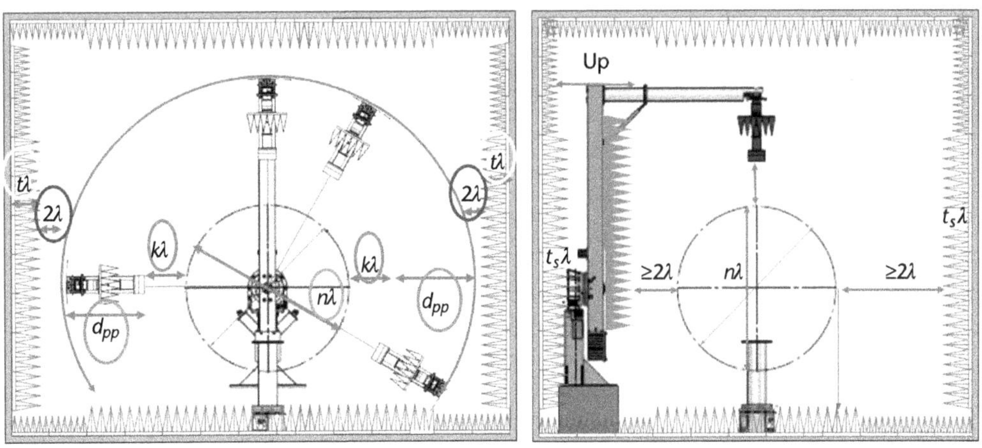

Figure 6.20 Width and length of a gantry SNF system.

mechanically there may be a need for a counterweight for the gantry movement. For the cases shown in Figure 6.20, the width and length of the range are given by the following:

$$W \geq Up + \left(n + 4 + t_s\right)\lambda \tag{6.21}$$

$$L = 2d_{pp} + (2k + n + 4 + 2t)\lambda \tag{6.22}$$

From these minimum dimensions we can analyze the range and the potential reflections and estimate the absorber requirements for these types of range geometries.

6.3.6 Absorber Layout

Ideally, as the probe is positioned the reflected energy entering the QZ remains constant or as close to constant as possible. Let us consider the two surfaces that are parallel to the plane that contains the movement of the probe. Figure 6.21 shows the ray tracing that is done to estimate the absorber thickness on the side walls. If the left side of Figure 6.21 is compared to Figure 6.14, it can be seen that the angles of incidence are the same; hence, the worst angle β is given by (6.17), and t is roughly 3 or 2.5, yielding absorber treatments that are 3λ to a minimum of 2.5λ.

For the gantry case, as shown on the right side of Figure 6.20, it will appear that the gantry arm itself should be treated with 3- or 2.5-λ absorber, but this will depend on the dimensions of the gantry. If the gantry arm is wider than λ, typically we want to treat it with thick absorber as is the case in Figure 6.17. For smaller widths (under $\lambda/4$), the arm can be treated with a shaped piece of lossy foam block, which is enough to reduce the effects of the arm.

The absorber on the wall, behind the gantry could be shorter since it is usually further away than the opposite wall so sometimes 2λ is sufficient to provide a −40-dB reflectivity.

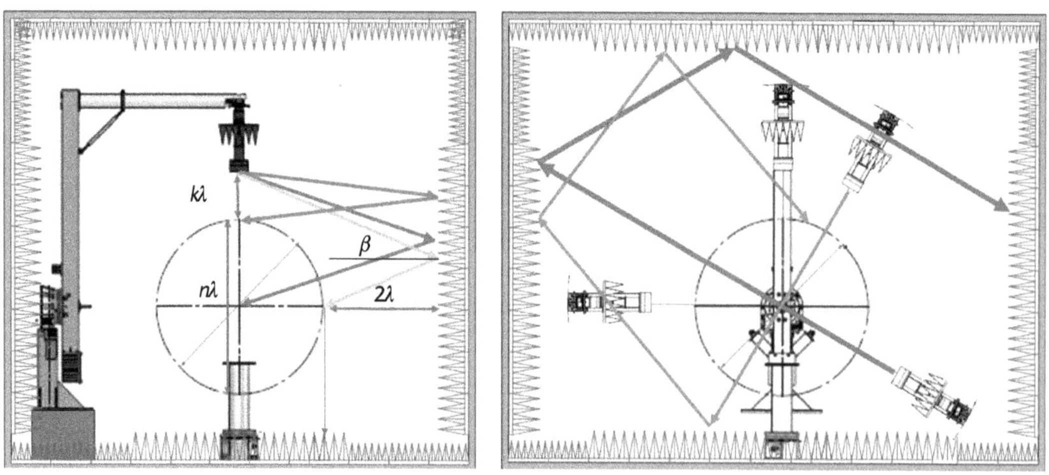

Figure 6.21 Ray tracing used to estimate angles of incidence and critical areas.

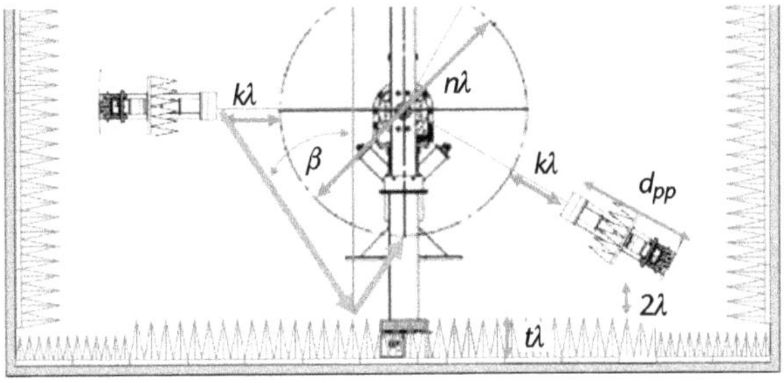

Figure 6.22 Estimating the worst-case angle of incidence onto the floor.

The right side of Figure 6.21 shows some rays for a couple of orientations. For those two orientations, the ray will undergo several reflections before entering the QZ. Additionally, it can be see that for the $\theta = 0°$ and the $\theta = -90°$, the range looks like the range geometry in Figure 6.11, hence the floor, and the two other walls that are perpendicular to the plane on which the probe moves can be treated with at least 2-λ absorber. This will provide −40 dB at normal incidence. However, for the $\theta = -90°$, the floor absorber is also a lateral surface absorber where the worst angle of incidence may vary depending on the height over the ground of the range.

If the height is based on (6.20), then the worst angle of incidence is given by (6.17). However, if the height is given by (6.19), the equation for the worst angle is a bit more complicated. Looking at Figure 6.22, the distance from the center of the QZ to the absorber tips is given by

$$D = 2\lambda + \left[\left(\frac{n}{2} + k\right)\lambda + d_{pp}\right]\sin\varphi \tag{6.23}$$

Then the angle of incidence is given by

$$\beta = \tan^{-1}\left(\frac{\left(\frac{n}{2} + k\right)\lambda}{2D - \frac{n\lambda}{2}}\right) \tag{6.24}$$

As with all the analysis done in this chapter, the wavelength is assumed to be at the lowest frequency of use for the range. In general the angle is going to be more acute for (6.24) than for (6.17) so choosing an absorber that provides the required reflectivity using (6.17) will meet the levels for (6.24) provided that φ is large enough and that k and n are the same in both cases.

As in other cases, one of the purposes of going through these sets of dimensional analysis of the range is to allow the user of these ranges to evaluate them and estimate the reflectivity of the range. In the IEEE Standard 1720 [4], an 18-term error analysis is suggested to evaluate the uncertainty. One of the terms is room scattering. The approach suggested by the standard is to reposition the probe and AUT within the room and check the difference. Since this causes the angles of incidence to change, it can provide a guess of the effects from the room. This may be easy to do on a PNF, and a procedure is given in [4], but it is more challenging on a SNF system like a gantry system that is usually fixed within the room. Knowing the reflectivity expected in the QZ it is possible to estimate the error in the measurement as shown for the far-field range in Chapter 5. That error can be treated as the potential uncertainty from the room scattering and used in the 18-term error assessment.

6.4 The CNF Range

The CNF range is the least common of the near-field ranges. It is usable for a particular set of antennas with omnidirectional (or very broad beam) patterns on one principal plane while having highly directive patterns in the orthogonal plane.

The typical CNF range uses a vertical scanner, where the AUT is rotated on an azimuth positioner. Figure 6.23 shows the typical cylindrical near-field scan. As it can be seen, the end caps of the cylindrical surface are not measured. The assumption is that almost no energy radiates in that direction.

The length of the probe scan is estimated in the same way that the scan size is estimated in a PNF system; hence, (6.1) and (6.2) are used to yield the scan length and the angle of interest. This means that the range is partially sized using the PNF range approach and partially using the SNF approach.

6.4.1 Sizing the Chamber

The height of the range is calculated using (6.6) since the height follows the geometry shown in Figure 6.3. The width and length of the range can be estimated using (6.13) and (6.14) given that the length and width follow the same geometry as shown in Figures 6.12 and 6.13. There are a couple of differences. The antennas on a CNF do not need to be rolled; hence, typically because only an azimuth positioner is required, the value for $d_{\mathrm{AUT}p} = 0$. On the other side, the size of the probe positioner $d_{\mathrm{pp}} = S_{\mathrm{clr}}$ since the probe positioner is typically the vertical stage of a PNF scanner.

6.4.2 Absorber Layout

The floor and ceiling absorbers are typically short. In general, there is very little radiation toward the ceiling and the floor (as is the case in the PNF range). Thus,

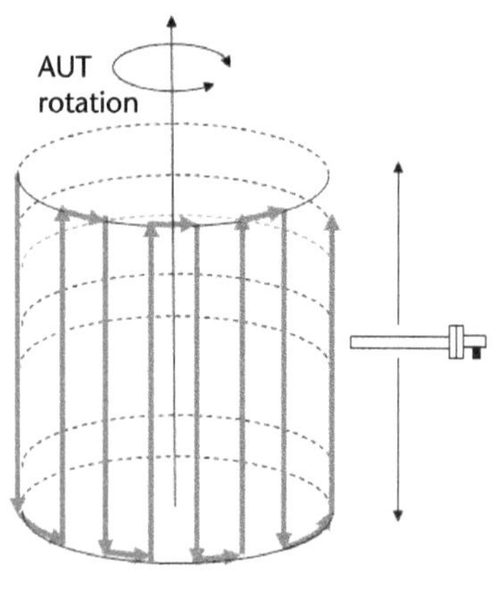

Figure 6.23 The CNF scan.

an absorber that is one wavelength at the lowest frequency of operation is sufficient for the floor and the ceiling.

For the end walls behind the vertical scanner and across from it, the absorber is the same as in a fixed-probe SNF range; hence behind the scanner one wavelength is sufficient, while on the opposite wall, two-wavelength absorber is required if we desire that –40-dB level. Accordingly, for the lateral walls, the 3- to 2.5-λ absorber, like those in the fixed probe SNF range, is recommended.

6.5 Error and Uncertainty in a Near-Field Range

Recall the 18-term error approach to uncertainty analysis for near-field systems. The sources of uncertainty for near-field systems are listed as follows [2]:

1. Probe relative pattern;
2. Probe polarization ratio;
3. Gain standard;
4. Probe alignment;
5. Normalization constant or gain standard far-field peak;
6. Impedance mismatch factor;
7. AUT alignment;
8. Data point spacing;
9. Measurement truncation;
10. Probe transverse position errors;
11. Probe orthogonal position errors;
12. Multiple reflections between probe and the AUT;
13. Receiver amplitude nonlinearity;
14. Phase errors related to flexing cables and rotary joints, temperature effects, and receiver errors;
15. Receiver dynamic range;
16. Room scattering;
17. Leakage and cross-talk;
18. Random errors.

Not all of these errors carry the same weight. Flexing of cables is more critical on planar scanners than on fixed-probe SNFs, where the rotary joints may be a more important source of uncertainty. The standard IEEE 1720 provides some guidance on how to assess some of these terms [2]. The effects of the multiple reflections, for example, can be determined by acquiring the data at different k values. Leakage levels can be measured by terminating either the probe or the AUT with a 50-Ω load. For room scattering, the document [2] provides a methodology to follow.

For PNF systems, the approach is to measure the same AUT with a given scan size and reposition the AUT and probe scan by a given distance (on the order of a number of wavelengths) in x, y, and z. Then after repositioning, the AUT is measured again. The difference in the far-field patterns provides an estimate of the room-scattering effects.

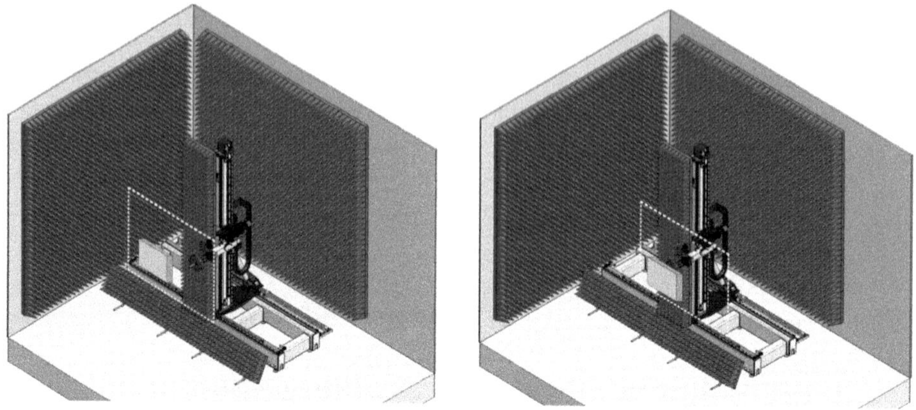

Figure 6.24 An AUT can be placed at different locations within the main scan area and a smaller sized scan performed. The probe may also be shifted with the AUT to maintain the $k\lambda$ distance.

This approach is relatively easy to do on a PNF. As Figure 6.24 shows, an AUT can be placed at different heights over the floor and at different distances with respect to the lateral walls. Some planar scanners also provide a linear motion of the probe in the direction normal to the scan plane. Thus, the probe can also be repositioned and the antenna moved away from the end wall so that the test distance $k\lambda$ is kept constant.

However, translation of the probe and the AUT within a room for spherical and cylindrical is very difficult, if not impossible, for typical SNF range configurations.

6.5.1 Case Study: A PNF Range

Using the range analyses presented in this chapter, it is possible for a range user to estimate the reflectivity from different surfaces using the Matlab routine provided in Chapter 3 to obtain the potential room scattering in order to estimate the potential effect of the absorber. To demonstrate this, let us follow a simple example for a PNF. Consider the PNF range shown in Figure 6.25. The scanner can move the probe on a 2.49m-by-2.47m plane. The absorber in all the surfaces is 12-inch pyramidal. The room housing the scanner is 4.27-m-tall by 4.27-m-long by 5.5-m-wide. Let us assume an antenna that operates at 2.4 GHz. The antenna, placed at 3λ from the probe ($k = 3$) or 37.5 cm, is a moderate-gain array with an aperture that is 32 cm by 24 cm. The entire scan plane is used to facilitate obtaining valid data for the antenna out to 70°. Using (6.1), it can be seen that this goal is achievable.

Notice in Figure 6.25 that although the scanner frame is off-center on the room, the scan plane is aligned with the centerline of the room's width. The distance from the edge of the scan plane to the floor is 79 cm; thus it is about 48 cm to the tips of the 12-inch absorber. This places the scan plane at 1.01m from the ceiling or about 70 cm from the ceiling absorber tips. For the width, the edge of the scan plane is about 1.5m from each of the side walls. The tips of the absorber are 1.18m away.

Using those dimensions, we can now estimate the angle of incidence onto the lateral surfaces of the range. Looking at Figure 6.7 we have the $k\lambda$ being 37.5 cm.

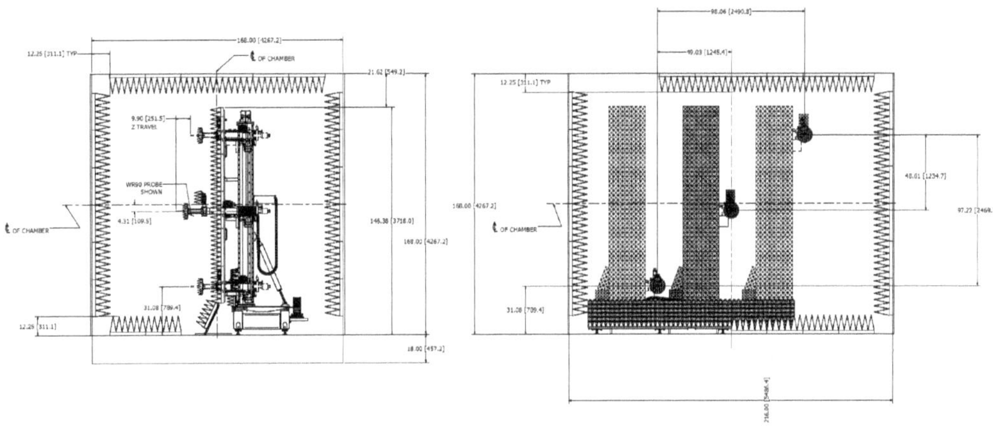

Figure 6.25 A PNF range; different locations of the probe are shown to demonstrate the range of motion of the probe on the scanner.

If the antenna under test is mounted so that the wider dimension (i.e., 32 cm) is on the horizontal plane, then the distance from the edge of the AUT to the edge of the scan plane is 1.245m – 0.16m or 1.085m. Recall that from the edge of the scan area to the tips of the absorber, it is roughly 1.18m. Taking the radiation from the edge of the antenna toward the specular point, the angle of incidence is given by

$$\beta = \tan^{-1}\left(\frac{k\lambda}{\frac{L_x - \mathrm{AUT}_x}{2} + 2\Delta_{\mathrm{Abs}x}}\right) \tag{6.25}$$

where $k\lambda$ is the test distance at the frequency of interest (37.5 cm), L_x is the scan plane on the horizontal ($L_x = 2.49$m), AUT_x is the AUT size on the horizontal (AUT_x = 32 cm), and $\Delta_{\mathrm{Abs}x}$ is the distance from the edge of the scan plane to the absorber (which should be, and is, larger than 2λ): $\Delta_{\mathrm{Abs}x} = 1.18$m.

Hence, the worst-case angle of incidence is given by $\tan^{-1}$ (37.5 cm/344 cm), which equals 6.22°. The absorber treatment is 2.4λ in thickness so for the angle of incidence (6.22°), the reflectivity at is better than –43 dB. Since the scan plane is centered, this is the same for the opposite wall.

In addition to that reflectivity, since an OEWG is being used, this means that the probe is receiving the reflected energy from around 84°, which, from the typical probe pattern (as those superimposed on Figure 6.13), means an additional –4 dB. This brings the reflectivity level down to better than –47 dB.

The AUT vertical dimension is $\mathrm{AUT}_y = 24$ cm, and the scan plane is $L_y = 2.47$m. The distance from the edges of the scan plane to the floor are $\Delta_{\mathrm{Abs}y} = 48$ cm, while to the ceiling it is 70 cm; hence, using a similar equation to (6.25), the angle to the ceiling absorber is $\tan^{-1}$ (37.5 cm/251 cm) = 8.5° while the angle to the floor is $\tan^{-1}$ (37.5 cm/207 cm) = 10.2°. Either of those angles provides a reflectivity of about –43 dB, and with the probe pattern we are in general at –47 dB. This is equivalent to a potential error from reflected energy of ±.04 dB.

This is a worst-case estimate for the potential error from the range for this antenna at this frequency. We did not consider the absorber on the wall behind the scanner; that wall will be at normal incidence, and the levels will be at about −43 dB, but with the probe front-to-back ratio, its effect will be very low compared to the lateral surface absorber. Similarly, this range has absorber on the wall opposite the scanner. That absorber will absorb the back lobe from the AUT, which will be easily 10 to 15 dB below the main beam, making the reflected levels in the −53-dB range or higher.

References

[1] Hess, D., "Near-Field Measurement Experience at Scientific Atlanta," *Proceedings of the 1991 Symposium of the Antenna Measurement Techniques Association*, Boulder, CO, October 7–11, 1991, pp. 5-3–5-8.

[2] "IEEE Recommended Practice for Near Field Antenna Measurements" IEEE Std 1720-2012, December 5, 2012.

[3] Parini, C., et al., *Theory and Practice of Modern Antenna Range Measurement*, London, UK: IET, 2015.

[4] "IEEE Standard Test Procedures for Antennas," IEEE Std 149-1979, August 8, 1980.

CHAPTER 7

Anechoic Ranges for Compact Range Measurements

7.1 Plane Wave Generators

The goal of an antenna measurement is to obtain the far-field radiation pattern. There are various ways to accomplish this in indoor ranges. One way is to be far enough so that the phase deviation across the AUT aperture is less than 22.5°. This deviation yields the $2D^2/\lambda$ used in the indoor far-field ranges. The approach used in Chapter 6 is a mathematical technique by which the true far field is obtained after measuring the near field of the antenna. This chapter presents an additional approach to measuring the far-field pattern of antennas.

This technique consists of generating a plane wave at the aperture of the antenna. Thus, the AUT will appear to be in the far-field environment. Generating this far field is done so that the distance at which the plane wave is generated is shorter than the distance required to generate that plane wave behavior in free space. That is, we do not increase the distance between the range antenna and the AUT, but we use a structure or device that will create or generate that plane wave in the proximity of the range antenna.

There are three main approaches used to accomplish this: lenses, both dielectric and metallic; parabolic reflectors; and arrays of sources. The compact range reflector is the most common of these approaches.

Dielectric lenses can be used to create the plane wave behavior in front of the lens, but they can be heavy as they need to be thicker if we want the range antenna (referred to as "the feed" in this chapter) close to the lens (thus reducing the overall antenna range length). Figure 7.1 illustrates the use of a lens to generate a plane wave. A simulation was performed using a commercial computational electromagnetics software package that used a FDTD-type technique (Simulia's CST Suite™). Figure 7.1 shows the amplitude and the phase at 3 GHz; the plane wave behavior of the fields close to the lens is clearly seen. Recently artificial materials (metamaterials) have been used to create light materials with a dielectric constant that is uniform across some frequency ranges. These artificial material lenses created a short renaissance for the idea of using a dielectric lens for antenna measurements [1]. These dielectric lenses have been successfully used in taper ranges to increase the usable QZ as the frequency increases. In these configurations the focal length of the lens is long enough (given by the taper length of the range) for a thinner lens to be used. The thinner lens translates in a lower weight and less material [2].

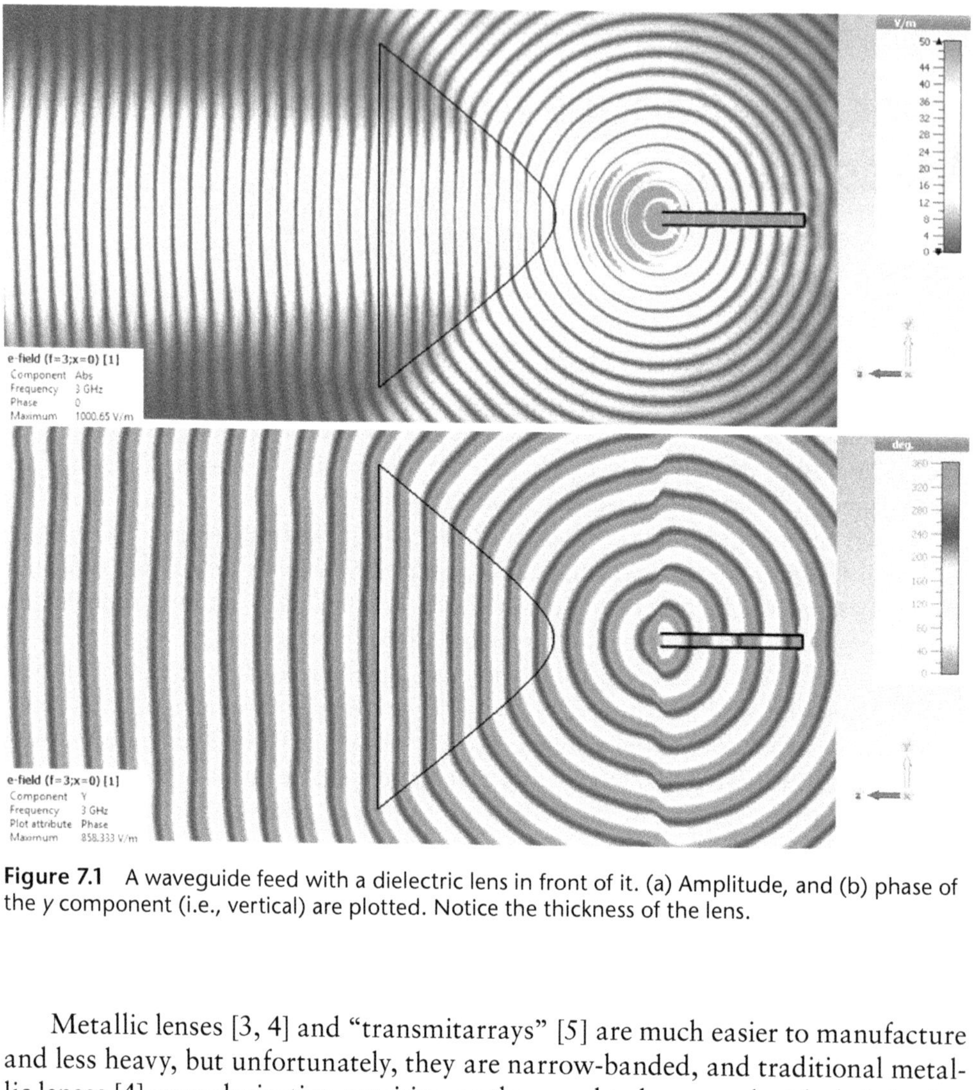

Figure 7.1 A waveguide feed with a dielectric lens in front of it. (a) Amplitude, and (b) phase of the *y* component (i.e., vertical) are plotted. Notice the thickness of the lens.

Metallic lenses [3, 4] and "transmitarrays" [5] are much easier to manufacture and less heavy, but unfortunately, they are narrow-banded, and traditional metallic lenses [4] are polarization-sensitive, so they need to be rotated with their feed to look at the orthogonal polarization. Figure 7.2 shows the numerical model results for a transmitarray in the presence of a point source; the results are obtained using the same numerical package.

The use of arrays to generate the plane wave has been demonstrated [6], but these can be complex and may require the entire array to be rotated to measure the orthogonal polarization [7]. Additionally, they are also frequency band–limited. Lastly, depending on the type of feed network, they may not be reciprocal as the other methodologies are.

Because of limitations in the use of other approaches, the plane wave generator using a parabolic mirror is the most popular approach. There are many incarnations of this technology, some of which use multiple mirrors to achieve a phase variation that is within a couple of degrees of the true plane wave. This chapter concentrates on ranges using a single mirror or reflector, which is the original geometry introduced by Johnson when he filed his patent in 1964 for an "antenna range providing

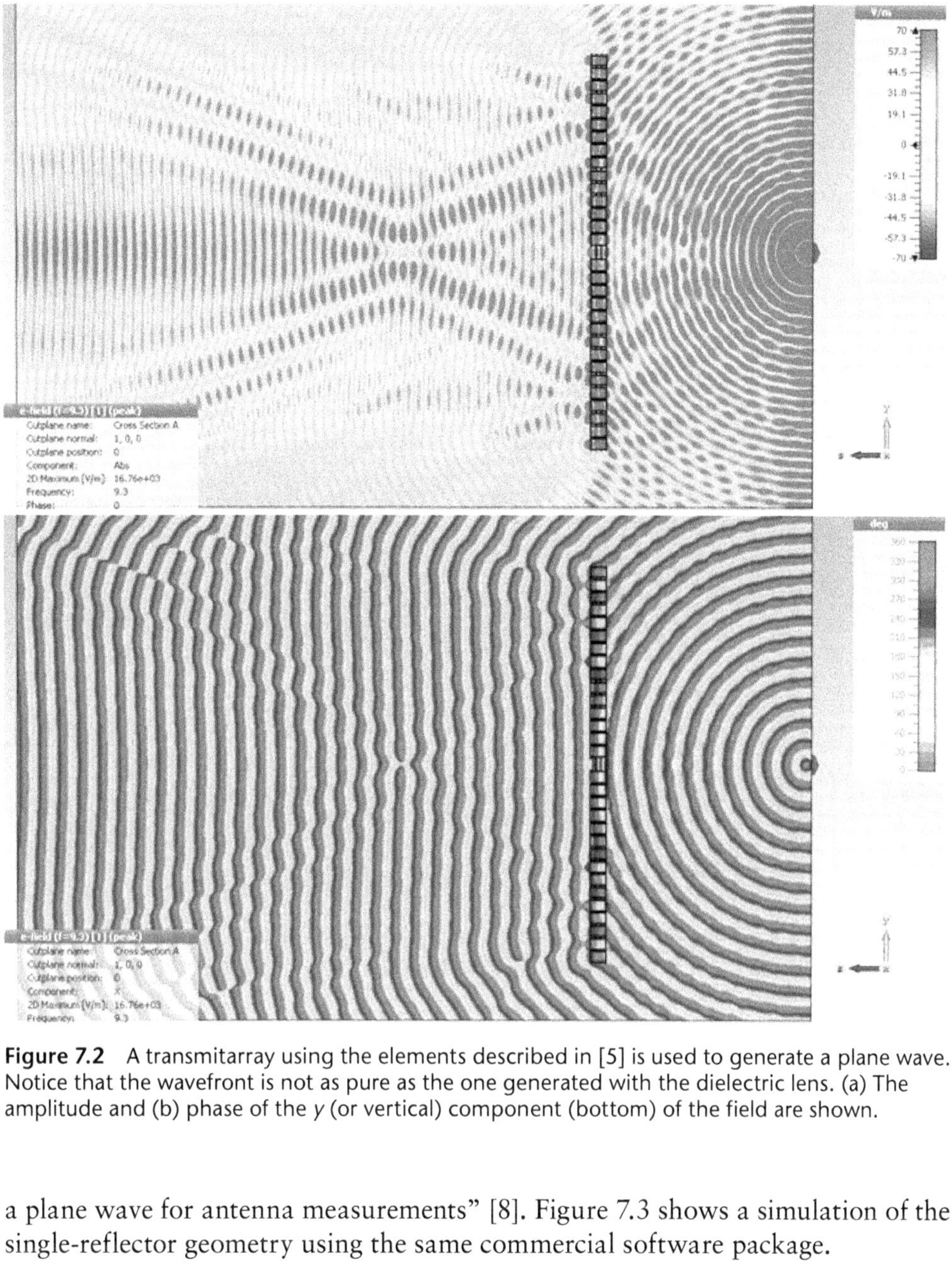

Figure 7.2 A transmitarray using the elements described in [5] is used to generate a plane wave. Notice that the wavefront is not as pure as the one generated with the dielectric lens. (a) The amplitude and (b) phase of the *y* (or vertical) component (bottom) of the field are shown.

a plane wave for antenna measurements" [8]. Figure 7.3 shows a simulation of the single-reflector geometry using the same commercial software package.

7.2 The Compact Range

The compact range is one of the most popular techniques to measure electrically large antennas indoors. Unlike mathematical transforms used in near-field to far-field measurements, it does not require a full data acquisition to obtain the far field. It does provide real-time measurements of the antenna so principal plane cuts can be obtained relatively fast. However, typically, for a reflector to be efficient, it

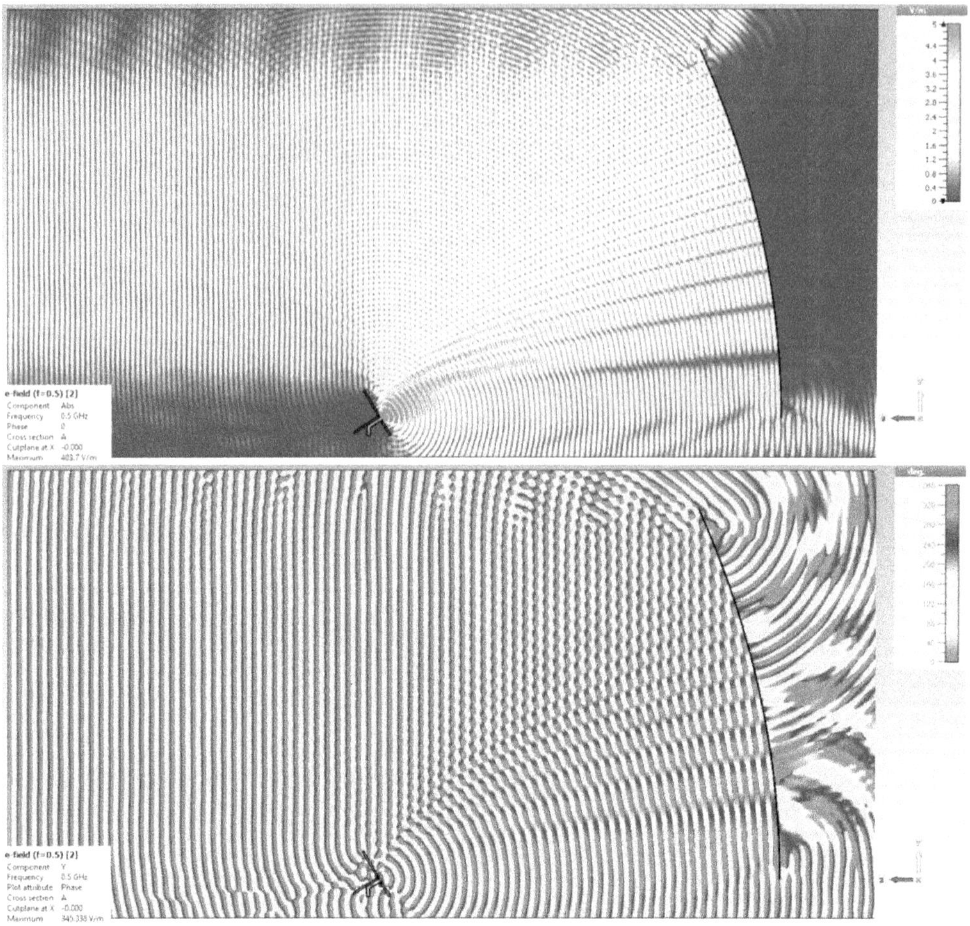

Figure 7.3 A single-reflector compact range. The feed illuminates a parabolic reflector from the focal point of the paraboloid. (a) The amplitude of the E-field and (b) the phase of the y component are shown.

should be at least 10λ in size, so operating at low frequencies requires physically large reflectors and ranges to house them.

7.2.1 History

There are several different configurations of compact ranges. These can be single- or dual-reflector systems. The most common geometry is the single-reflector configuration. Indeed, that single-configuration system is the one introduced by Johnson in his patent document [8] and the one that he describes in his original paper [9]. While Johnson's paper was published in 1969, the work on the compact range idea goes back to the early 1960s. The patent was filed in 1964, and as early as 1966, Johnson and his colleagues were publishing reports [10] about the reflector technique. Reading the report, Johnson acknowledges that the idea of doing indoor antenna measurements by generating plane waves is not new. In his report he refers to the work performed at McGill University in the 1950s that, according to Johnson's

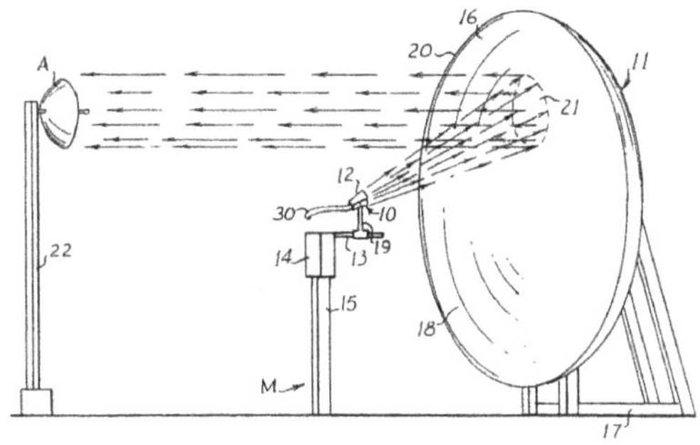

Fig. 1

Figure 7.4 Figure 1 from Johnson's patent [8] showing the feed and a full parabolic reflector.

account and based on private communications with McGill University personnel, did not had satisfactory results. A master's thesis by J. H. Crysdale presented at McGill University deals with a parabolic reflector [11].

All these early exercises deal with a single reflector and a feed located at the focal point of the parabola. As Figure 7.3 illustrates, spherical waves emanate from the range antenna, the feed, as we will refer to it, and are reflected by the parabolic surface generating a plane wave in the near field of the reflector. Figure 7.3 shows the typical configuration where a portion of the parabolic surface is used to reflect the wave. This is different than the original approach by Johnson, as shown in Figure 7.4. However, Johnson had the idea of a section of the parabolic surface as his figures for a two-dimensional approach show. Those figures from the patent document are shown in Figure 7.5

Interestingly, to create a "line-source" for the 2D approach, Johnson uses a second parabolic reflector, making his approach a dual-reflector system. Figure 4

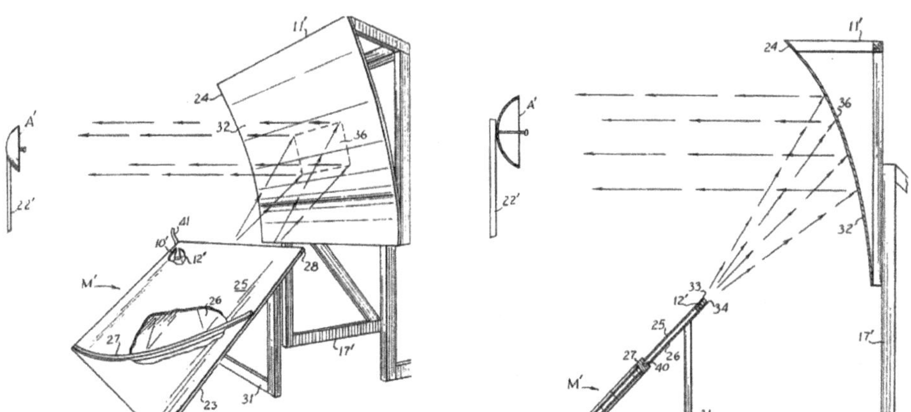

Fig. 3 Fig. 4

Figure 7.5 Johnson's figures for a "line source" and a 2D partial parabolic reflector [8]. Figure 4 on the left shows a configuration that is similar to what was modeled in Figure 7.3.

of the patent documents, shown in Figure 7.5, shows a configuration very similar to the one that was simulated and presented in Figure 7.3.

One thing that is apparent in the figures from [8, 9] is that there is no edge treatment. Johnson uses a very large reflector compared to the size of the AUT. This is not very efficient, and even with a large reflector the edges will give rise to diffraction, and the diffracted fields will propagate into the QZ, interfering with the plane wave and creating a ripple onto the wavefront.

7.2.2 Edge Treatments

There are different approaches to mitigate the edge diffraction. Of the most commonly used, one consists of serrating the edges of the reflector so that the diffraction is directed away from the QZ and in different directions. The other commonly used approach is to roll the edges of the reflector to have the diffraction on the back of the main reflecting surface (illustrated in Figure 7.6). In addition to placing the diffraction edge on the back, the smooth roll slowly radiates the currents traveling on the surface, reducing their magnitude as they approach the edge.

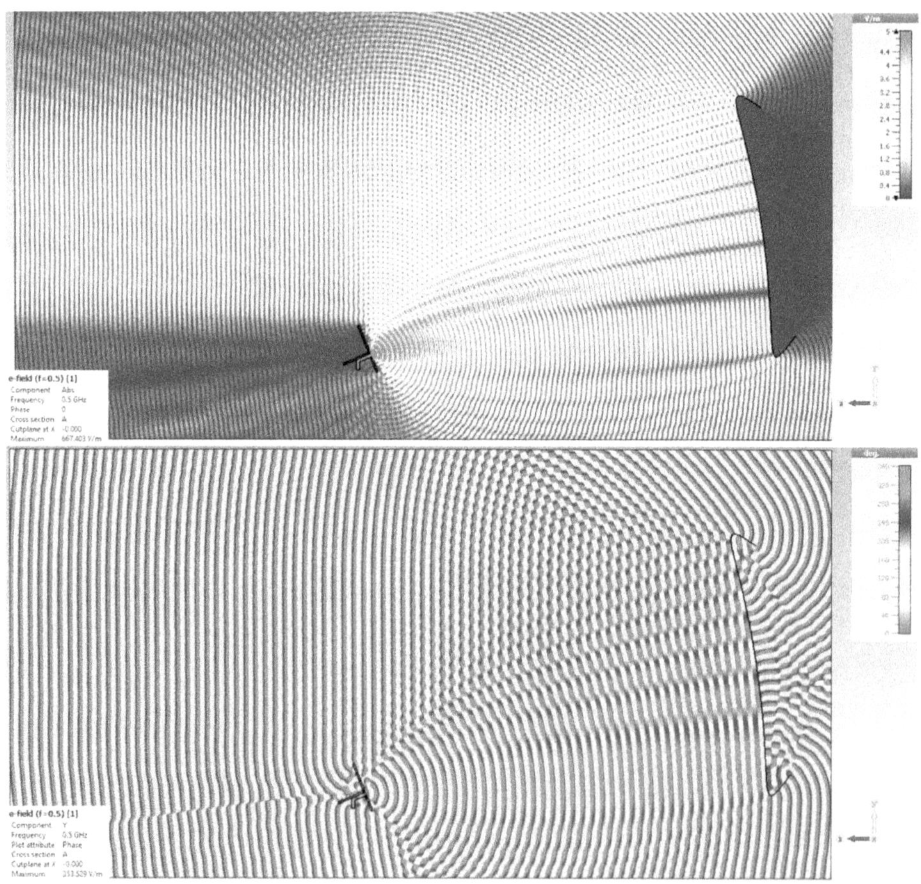

Figure 7.6 A rolled-edge reflector geometry simulation: the field amplitude (top) and the phase of the vertical component (bottom).

There is a controversy over which edge treatment is better. The first compact range system that Johnson used was a standard 3-m dish antenna with a circular rolled edge. In his 1973 invited contribution to the proceedings of the IEEE [12], Johnson describes a rolled-edge solution to attenuate the diffraction from the edge of the reflector. Johnson suggests a radius of a wavelength at the lowest frequency and an arc length of at least 180°. This cylindrical roll should not be confused with the cosine blended-edge rolls that are common today and that were introduced by Burnside et al. [13]. The rolled edge shown in Figure 7.6 is a blended rolled-edge type. By 1974, Scientific-Atlanta had introduced the model 5751 as reported by Pape in 1980 [14]. The 5751 was the first commercial compact range reflector. That reflector showed three different types of edge treatment [15]. The reflector illustrated in [15] showed a cylindrical roll on the top and serrated edges on the two sides, and the bottom was treated with RF absorber. Incidentally, in Johnson's 1969 paper [9] RF absorber is placed on the reflector edge to cut the edge effects. Placing absorber onto the reflector to improve performance is a technique that has been used several times. Baggett described cutting flat absorber to recreate serrations on the bottom of a reflector [16]. An absorber treatment to treat the bottom of the reflector that extended onto the floor was done in New Mexico in 1986 (see Figure 7.7). Baggett also describes extending reflectors to the walls and having holes cut into the wall, to allow serrations poke through, at an installation in Missouri around 1990. Both cases described by Baggett performed within specifications and were done 22–24 years prior to the patent awarded to Liu, which presents a reflector extending into the absorber treated wall as a technique to reduce the edge diffraction [17].

It is not the intent of this book to determine the best or worst approach to treat the edge. Serrations are easy to manufacture and lower cost than a blended edge; even Burnside et al. [13] acknowledge this. However, typical serrations are about

Figure 7.7 An early blended rolled-edge reflector where the bottom runs into the floor and into the absorber floor treatment as suggested in [17]. The reflector did not meet RCS requirements and was modified with the LFB serrations shown [16]. (*Source:* Marion Baggett. Reprinted with permission.)

four to five wavelengths at the lowest frequencies of operation, typically limiting the lower frequency of use of serrated edges to 500 MHz. Blended rolled-edge reflectors tend to be smaller than serrated-edge ones, especially at those low frequencies; however, as Lee and Burnside state [13], the rolled edge will "illuminate the chamber walls more significantly. Fortunately, this is not a problem, provided that the appropriate absorber materials are used to line the chamber." There lies the importance of the edge treatment. The chosen edge treatment for the compact range reflector will have implications for the construction of the indoor range and the absorber that will be used to treat the surfaces of the range.

7.3 Sizing the Chamber

The size of the indoor anechoic range for a compact range is determined mainly by the size of the reflector. While we are not focusing on other plane wave generators, the size of the array will also be the determining factor for lenses and arrays. For lenses and reflectors, the other factor is the focal length of the optical apparatus (i.e., lens or mirror).

Of the various configurations of compact ranges, we focus here on the simple geometry based on Johnson's original work, the single-reflector centerline fed approach, which is the most common implementation of the compact range reflector.

7.3.1 The QZ

The QZ of compact ranges is usually defined in the shape of a horizontal axis cylinder. The axis is aligned with the range length. Additionally, the cylinder can be circular or elliptical in cross-section. Thus, we will define the QZ as a cylinder with dimensions QZl, QZw, and QZh, where QZl is the length of the cylindrical QZ, QZw is the width, and QZh is the height. Figure 7.8 illustrates the geometry of the typical compact range QZ.

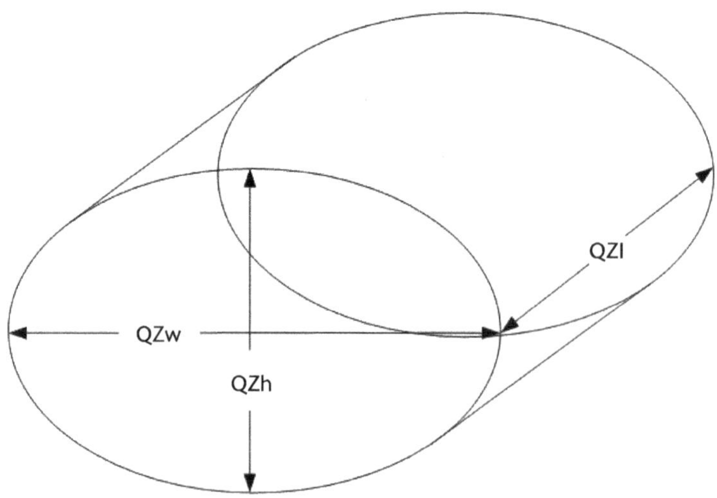

Figure 7.8 Typical compact range QZ.

For plane wave–generating systems the range will be evaluated based on how well the field on the quiet zone approaches a plane wave. Thus, during range evaluation the field is measured across the QZ, and some parameters are extracted to provide a measurement of how close the illumination is to a plane wave.

The fields are measured in both amplitude and phase, and parameters that define the quality of the plane wave are extracted from both the amplitude and the phase distribution. Figure 7.9 shows the typical parameters measured on a compact range QZ.

The most common parameter is the amplitude taper (Figure 7.9(a)). The amplitude taper is obtained from the acquired amplitude of the field across the QZ at a given frequency along a given line (perpendicular to the range length). After the data is acquired it is curve-fitted to a polynomial (typically a second-order polynomial); the amplitude taper is the variation in decibels of that polynomial approximation across the QZ. The amplitude taper is mainly related to the pattern of the

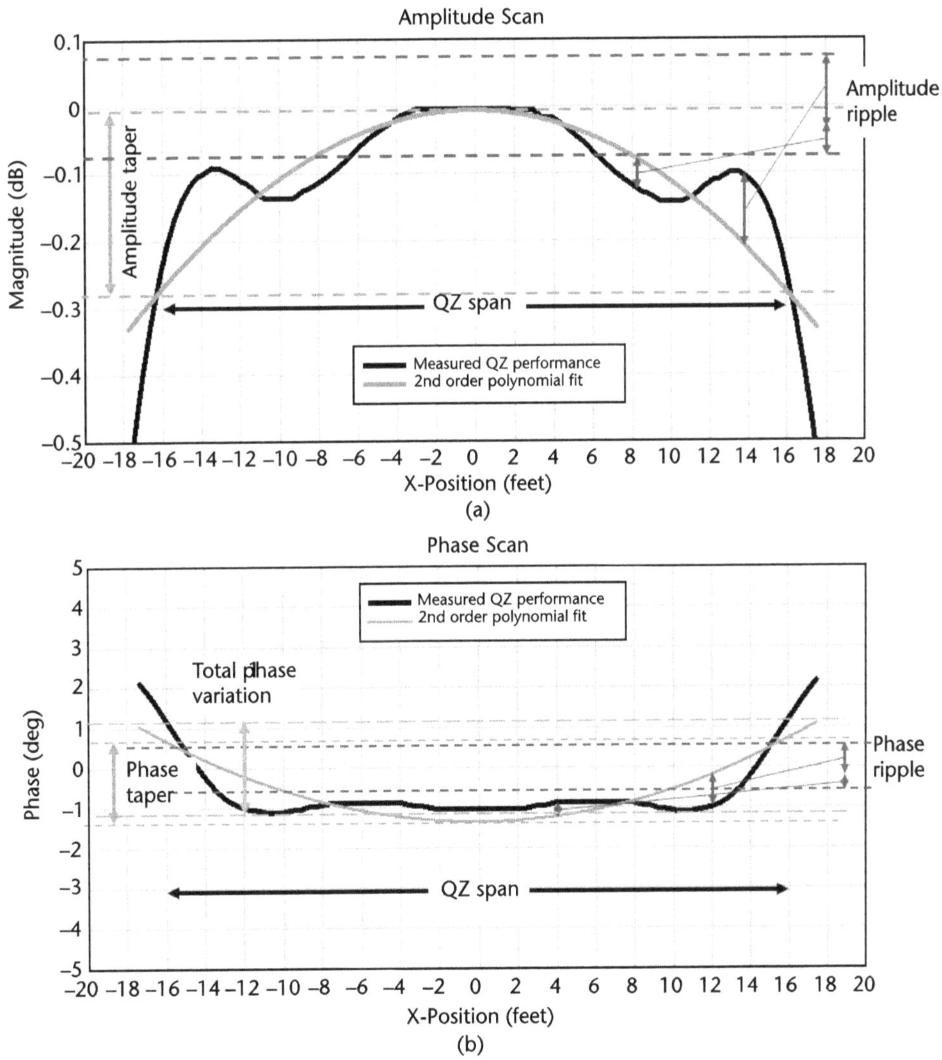

Figure 7.9 QZ parameters for a plane wave–generating antenna measurement system.

illuminating feed. We want a pattern that is wide enough to illuminate the body of the reflector with a very small amplitude taper. Physically this causes the edge treatment of the reflector to get illuminated and increase the diffraction, which shows on the QZ as a ripple riding on the taper. That amplitude ripple is another of the QZ parameters that define the purity of the plane wave generated on the QZ. The phase is important. If everything was perfect, the phase would be perfectly flat across the QZ, giving us a plane wave. Imperfections and diffractions from the edges cause a variation on the phase. Typically, the phase across the QZ can be postprocessed as the amplitude, and a phase taper and phase ripple can be extracted; however, to simplify the analysis the total phase variation is given (see Figure 7.9(b)).

7.3.2 Room Size

Let us start with the length of the range. The length is still mainly related to the reflector use. The reflector will have a given parabolic curvature. The parabola will have a focal point, and at that focal point the range antenna, the feed, will be located. The quality of the plane wave generated by the reflector is closely related to the feed. To achieve an adequate phase distribution across the QZ, the feed must be at the focal point. To be more specific, the phase center of the antenna should be at the focal point. Thus, antennas where the phase center shifts with frequency are not ideal for illuminating a compact range. For example, LPDAs, Vivaldi-printed antennas, and ridged horns are not ideal. This does not mean these antennas cannot be used, only that there will be a deviation on the phase from the ideal plane wave. This deviation may be acceptable given the advantages of using a broadband antenna. The decision has to be made based on what uncertainty level is acceptable for the measurements.

Let us consider the geometry shown in Figure 7.10, which illustrates a blended rolled-edge reflector, where the edge treatment is irrelevant when considering the

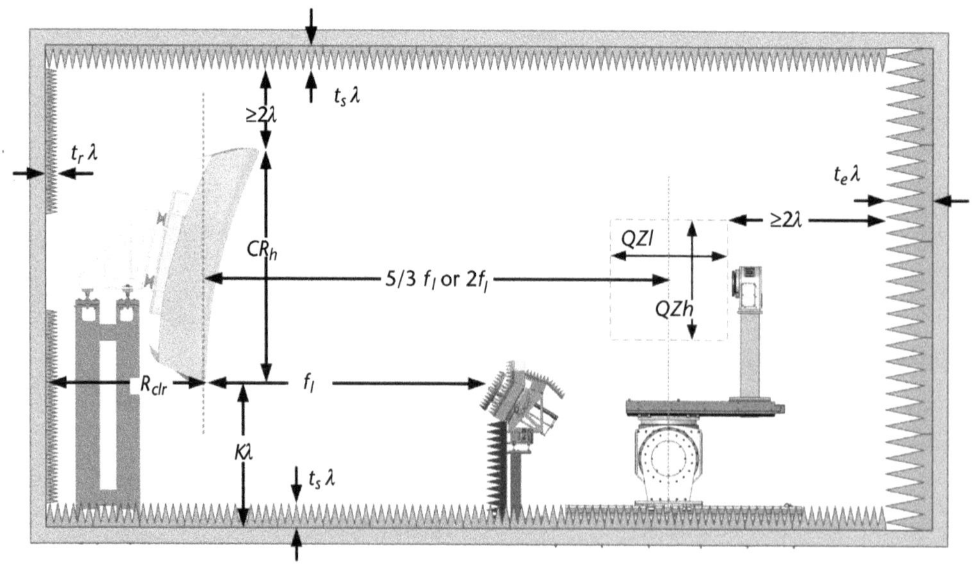

Figure 7.10 A typical range geometry showing the elevation for a compact range reflector.

rough size of the range. The edge treatment will affect the lateral surface absorber as shown later in Sections 7.4.3 and 7.4.4.

Regardless of the treatment, the reflector will have a given focal length. The reflector is potentially a very heavy structure that needs to be accurately positioned. Additionally, it is crucial that the reflector does not deform. Deformation may occur due to temperature variations in the indoor range, so it is important to maintain a constant temperature. Some reflectors are physically very large, providing a QZ that can be 5m (16.5 ft.) by 5m by 5m in size. Such a reflector cannot be machined out of a single piece so multiple panels are machined. These panels are mounted to a structure that allows the panels to be adjusted for perfect alignment during installation. The point that we are trying to get across is that there can be a significant structure behind the reflector to help support it, align it, and keep it stable.

Figure 7.11 shows a picture of a factory preliminary assembly of a large reflector (5m by 5m by 5m QZ). Only the center section of the reflector is being assembled. In Figure 7.11, each section of the heavy structure that supports the reflector has a mechanism that allows for alignment of each panel. Each panel (roughly 2m by 2m in size) can be moved linearly along three orthogonal axes as well as rotate around each of these orthogonal axes. The structure requires space, and that is given by the parameter R_{clr} shown in Figure 7.10. Typically, R_{clr} varies from 90 cm for a small 50-cm QZ reflector to up to 2m for large multipanel installations.

Let us go back to the focal length. As noted, one parameter that defines the purity of the QZ is the amplitude; it is desirable to have a very small variation of the illuminating pattern across the reflector. This means that, typically, the focal length is selected such that the illuminating feed has a very small taper across the QZ. The most common feed is the circular aperture with corrugations (see Figure 7.12). These feeds typically have a 1-dB beamwidth of about 30 degrees (see Figure 7.13).

Simple trigonometry can be used to estimate the required focal length for a

Figure 7.11 A multipanel compact range reflector (5-m QZ) during preliminary factory assembly. Notice the large structure supporting the reflector. The blended rolled-edge treatment is not attached except for a small portion on the lower right side. (Photo courtesy of NSI-MI Technologies.)

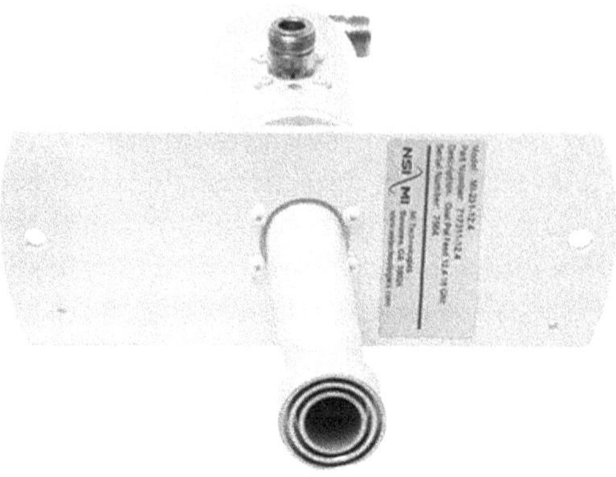

Figure 7.12 A circular aperture with corrugation typically used as a compact range feed—a dual-polarized (two orthogonal inputs) version. (Photo courtesy of NSI-MI Technologies.)

given QZ size. If a 1-dB amplitude taper is required over a 1.82-m (6-ft) QZ, then the focal length should be at least 3.4m (11.15 ft.). Basically, the equation is

$$f_l = \frac{QZ_{max}}{\tan BW_{1dB}} \tag{7.1}$$

where QZ_{max} is the largest of the width and height of the QZ and BW_{1dB} is the 1-dB beamwidth. A long focal length is important because the geometry will reduce the cross-polarization of the reflector illumination. Stutzman and Thiele provide a good, simple discussion of the offset reflector geometry cross-polarization in [18]. It is important to understand the limitations of cross-polarization in offset center-fed reflectors. If cross-polarization is a driver or a critical parameter in the antenna being measured, near-field systems, such as those discussed in Chapter 6, are a better approach.

The QZ for a compact range is located such that its center is about 2/3 away from the focal point or a full focal length away from the focal point. Hence, from

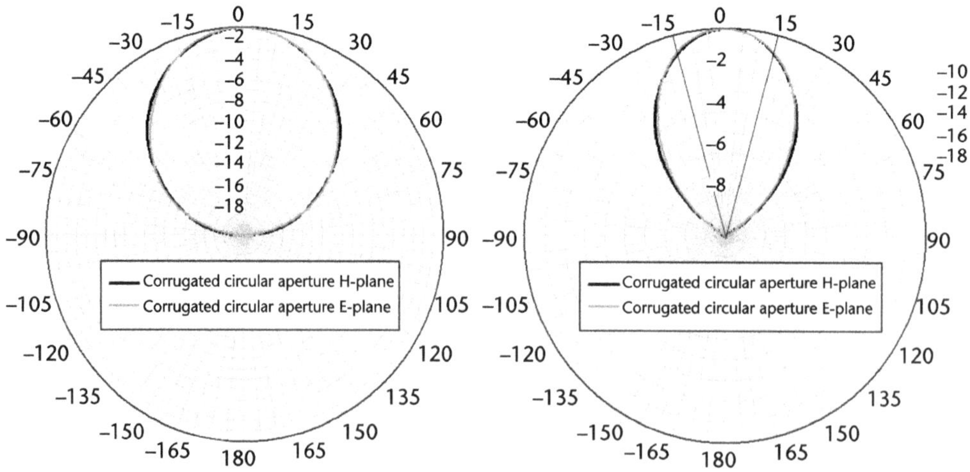

Figure 7.13 Radiation pattern for a typical corrugated feed like the one shown in Figure 7.12. On the right is the 1-dB beamwidth.

the reflector to the center of the QZ, there is either $5/3f_l$ or $2\ f_l$. The end wall absorber (as is the case with far-field ranges) is typically located at least 2λ from the QZ, where λ is the wavelength at the lowest frequency. In general, the length of a compact range is given by

$$L \geq R_{clr} + \frac{5}{3}f_l + \frac{QZl}{2} + \left(2 + t_e\right)\lambda \tag{7.2}$$

where t_e is the electrical size of the end wall absorber at the lowest frequency of operation. Section 7.4.1 discusses the electrical size recommended for that end wall absorber. Equation (7.2) provides a minimum size. As we will see, positioning equipment may have an effect on the length because the positioning system may be used to help the loading of the AUT.

The height and width of the range are mainly driven by the reflector size. The reflector is typically placed a certain distance from the absorber, and while there are known cases where the reflector has been placed all the way into the absorber treatment, it is desirable for installation purposes to allow some clearance to help with the installation. Figure 7.14 shows the on-site installation of the reflector shown in Figure 7.11. Space above the reflector is needed for the cranes to be able to lift the panels to their required positions.

The height is typically given by

$$H \geq CR_h + \left(2 + k + t_s\right)\lambda \tag{7.3}$$

where CR_h is the overall vertical size of the reflector and t_s is the electrical size of the ceiling absorber. As is typical, the absorber is kept 2λ away from the reflector

Figure 7.14 Installation of a large reflector in an indoor range. Notice the space above the reflector; that space is required for installation of the different panels. (Photo courtesy of NSI-MI Technologies.)

edges (see Figure 7.10). The other parameter in (7.3) is k. The parameter k is such that $k \geq (2 + t_s)$. If k is equal to $2 + t_s$, the range is symmetrical in the elevation cut. This is typically the case for large compact ranges operating down to frequencies on the order of 500 MHz, where $2\lambda = 1.2$m, thus providing enough space for the feed positioner. However, for ranges with a higher lower-end frequency of operation typically the factor $k > (2 + t_s)$.

In some cases, the floor absorber may be thicker than the ceiling absorber in order to reduce the floor reflection from the feed. Because of its location close to the floor and its wide beamwidth, the feed may illuminate the floor with high levels, and the reflection from the floor may be reflected to the QZ by the reflector. This will be discussed in more detail below.

Like the height, the width of the range depends mainly on the size of the reflector. Figure 7.15 shows a typical geometry for an indoor compact antenna test range, where the width of the range is easily obtained using

$$W \geq CR_w + \left(4 + 2t_s\right)\lambda \tag{7.4}$$

where t_s is the electrical size of the lateral wall absorber at the lowest frequency of operation. Equations (7.2) to (7.4) provide a good rough order of magnitude to estimate the overall required area to implement a compact range.

Typically, compact ranges are for electrically large antennas where the far-field condition makes it not feasible to measure the antenna in a far-field range. Generally, this means that the antennas need to be physically large; this means that the size of the range may need to be increased to allow for handling the AUT.

7.3.3 Positioners and Size

It is common to have an elevation stage in the AUT positioning stack-up to allow for the antenna to be mounted to the positioner. When tilting the positioner back, there

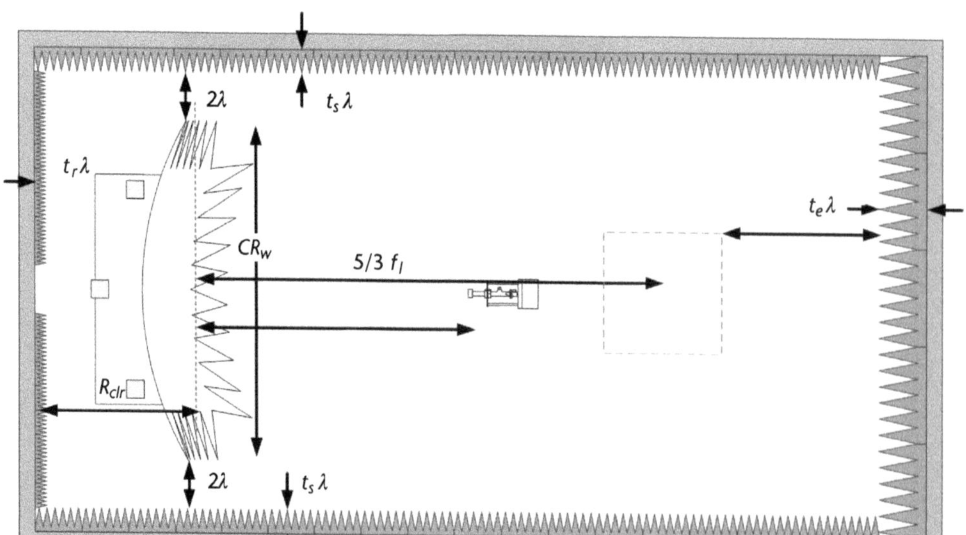

Figure 7.15 Typical floor plan of an indoor antenna compact range.

is a need for additional space in the range length. Figure 7.16 shows the geometry of a range with the positioner locating the antenna in the QZ and with the positioner tilted backward to load the antenna. To allow for the loading position the length of the range is given by

$$L \geq R_{clr} + \frac{5}{3}f_l + S_{uld} + t_e\lambda \tag{7.5}$$

where the parameter S_{uld} provides the space for the positioner to tilt back an antenna that fits in the QZ. We can approximate S_{uld} using

$$S_{uld} = \frac{QZh}{2} + \frac{CR_h}{2} + (k+2)\lambda \tag{7.6}$$

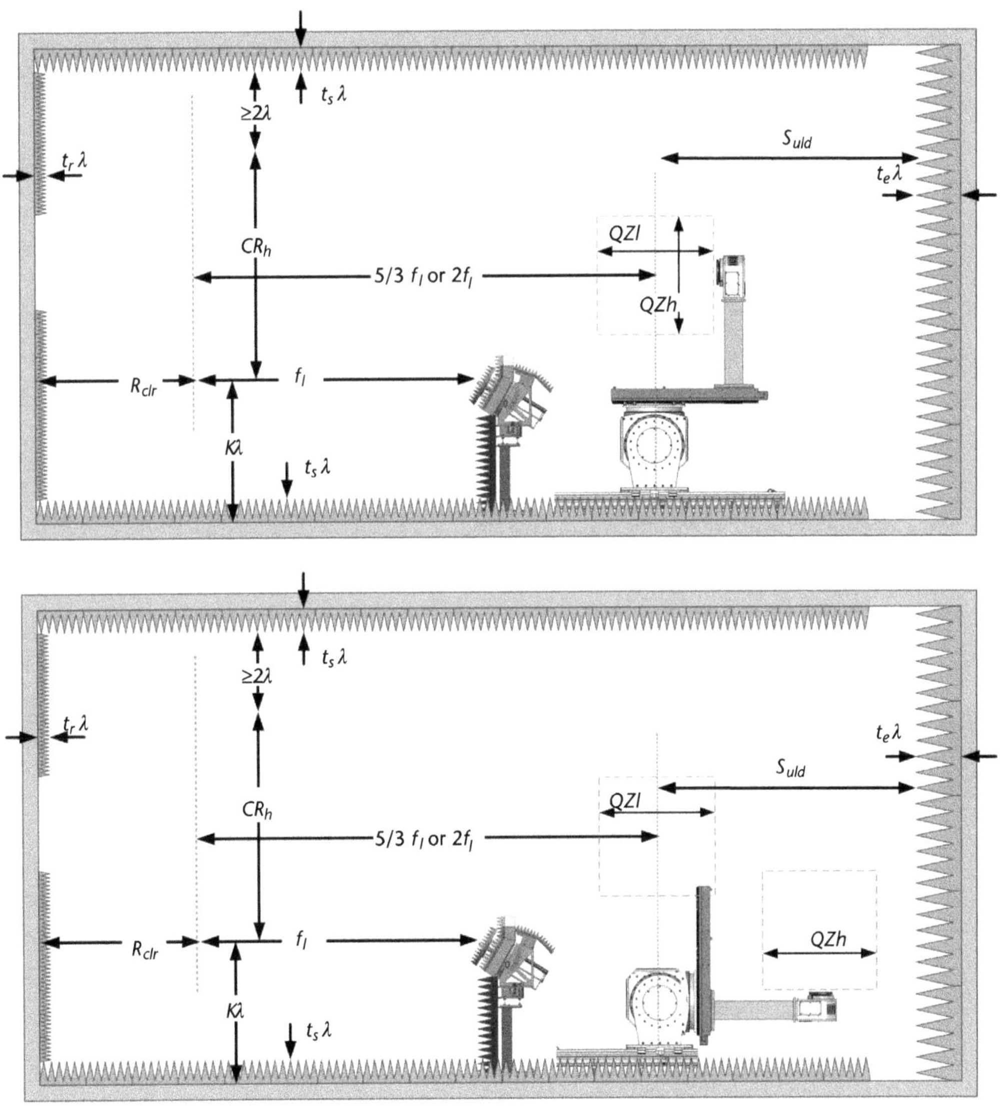

Figure 7.16 Compact range length accounting for loading and unloading of the AUT.

which turns (7.5) into

$$L \geq R_{clr} + \frac{5}{3}f_l + \frac{QZh + CR_h}{2} + \left(k + 2 + t_e\right)\lambda \tag{7.7}$$

Equation (7.7) provides enough internal room in the range for the AUT positioner to tilt back for loading the antenna. There are, however, several parameters in these equations for the range size that depend on the required absorber. Section 7.3.4 examines the absorber size requirements to obtain a given reflectivity from the different surfaces in the range.

7.3.4 Rules of Thumb for Reflector Sizing

One question that needs to be answered is the relative size of the reflector width CR_w and height CR_h. Examining a series of commercially available reflectors, some rules of thumb can be extracted. These reflectors are designed to meet certain levels of amplitude taper, ripple, and phase taper. Let us start by defining λ_{low} as the wavelengths at the lowest frequency of interest.

Per (7.1) it is expected that $f_l \geq 1.732\text{QZ}$. Looking at the focal length of available reflectors, the focal length ranges between $21.34\lambda_{\text{low}}$ and $24.4\lambda_{\text{low}}$ for QZ sizes between $8\lambda_{\text{low}}$ and $12.2\lambda_{\text{low}}$. The reflector body width for a serrated-edge reflector follows $CR_w = F \cdot QZw + 10\lambda_{\text{low}}$, where $1.83 \leq F \leq 2.3$; and the $10\lambda_{\text{low}}$ accounts for the length of the serrations.

The height of the reflector follows $CR_h = M \cdot QZh + 10\lambda_{\text{low}}$, where $1.4 \leq M \leq 1.83$. These are approximate rules. Serrations can vary in length depending of the shape of the "teeth" of the serration. In addition, the focal length may vary depending on the acceptable or desired levels of cross-polarization.

Rolled-edge reflectors tend to be smaller, because the factors M and F are equal to 1; that is, the body of the reflector is equal to the size of the QZ. The rolls are approximately between $5\lambda_{\text{low}}$ and $6\lambda_{\text{low}}$ in projected length [13]. As a rule of thumb, this makes the blended rolled-edge reflector $CR_h = QZh + 10\lambda_{\text{low}}$ by $CR_w = QZw + 10\lambda_{\text{low}}$ at a minimum.

The sizes recommended in this section are minimum sizes. The larger the indoor range, the better the potential performance. There is, however, a balance between the economical cost of the range and the performance.

7.4 Absorber Layout

Typically, compact ranges are set in rectangular chambers. However, unlike the rectangular chambers discussed in Chapter 5, in the compact range chamber, the critical wall is the end wall opposite the reflector. This section discusses the different internal surfaces of the range and estimates the best absorber size for each of them.

7.4.1 The End Wall

The critical wall is the range end wall opposite the reflector. As the simulations show in Figure 7.3 and Figure 7.6, the bulk of the reflected energy travels from the reflector to the end wall as a plane wave. It is important to understand that it travels as a plane wave; this means that it travels without attenuating. There is no path loss as is the case with a spherical wavefront that spreads over a larger surface as it propagates, reducing the power density as it travels away from the source. This unattenuated energy will hit the end wall and reflect toward the QZ. To avoid a significant standing wave, it is desirable to have pyramidal absorber on the end wall that is at least 2λ (where λ is the wavelength at the lowest frequency) on that wall. Thus, it follows that $t_e > 2$. Ideally, if possible, $t_e \geq 3$.

7.4.2 The End Wall Behind the Reflector

The opposite wall to the range end wall discussed in Section 7.4.1 is the wall behind the reflector. Figures 7.10, 7.15, and 7.16 show that wall having only absorber in the periphery of the wall. In the size equations that were presented above, the parameter t_r is not mentioned. This parameter defines the electrical size of the reflector side end-wall absorber at the lowest frequency. The reflector support clearance parameter, R_{clr}, is shown to include the reflector side end-wall absorber.

7.4.2.1 Serrated-Edge Treatments

In Figure 7.3, it appears that there is very little radiation going toward the reflector side end wall, but this is a bit misleading. The cut is done at the middle of a serration, and the simulation shown in Figure 7.3 is a quasi-two-dimensional simulation. The reflector simulated is one full serration in width, and using a perfect magnetic conductor boundary on the sides of the simulation makes it appear as an infinitely wide serrated reflector. It is an approximation, but it provides a good visualization of the problem. If the field is obtained at a valley of a serration for the same vertically polarized feed, we see that there is not much difference, as shown in Figure 7.17, which illustrates the field at planes located at the valley and the peak of a serration.

This validates what Joy and Wilson [19] called a field transition point-of-view for the serrations instead of the quasi-GTD approach. That is, the serration area provides an area where the reflectivity changes from that of PEC (or metal) at the center of the reflector to that of air at the edge. The serrations provide a smooth transition for the reflected energy from the main body to the edge. Notice that the fields behind the reflector are in the worst case, a factor of 0.1, of the fields from the feed illumination. That is about −20 dB lower. Hence, the absorber on that wall can have a reflectivity that is 20 dB higher than that of the opposite end wall. Typically, the reflector end-wall absorber can be of a size such that $t_r \geq 0.7$. This means that the absorber on the reflector end wall can be roughly one-third the length of the opposite end-wall absorber.

7.4.2.2 The Base of the Reflector

Figure 7.17 shows that the field under the reflector is higher in magnitude than on the rest of the wall. However, the heavy support structure is typically treated with an absorber that usually has the same thickness as the lateral walls, ceiling, and floor. Figure 7.18 shows a serrated-edge reflector with the base of the structure clearly masked by an absorber.

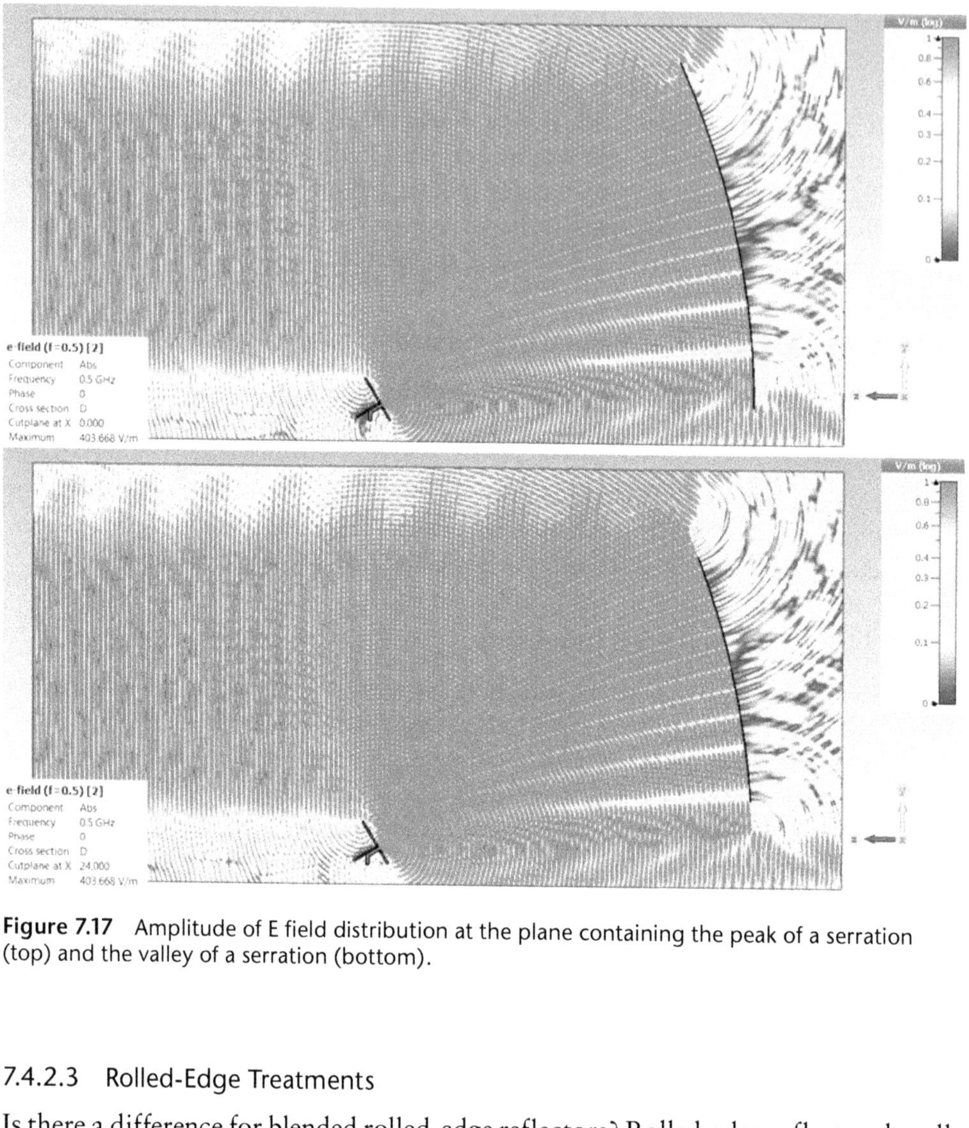

Figure 7.17 Amplitude of E field distribution at the plane containing the peak of a serration (top) and the valley of a serration (bottom).

7.4.2.3 Rolled-Edge Treatments

Is there a difference for blended rolled-edge reflectors? Rolled-edge reflectors handle the edge diffraction in a different way. Figure 7.19 shows levels that are in some cases a factor of 0.3 the field from the source. That means that we are looking at about −13 dB lower. Hence for the rolled-edge case the absorber on the end wall behind the reflector should be about a wavelength, $t_r \geq 1$, or at least about half the thickness of the opposite end wall treatment.

Figure 7.19 also shows that the field around the reflector presents a very coherent phase; this was also apparent when looking at the phase plots in Figures 7.3 and 7.6. The amplitude plots shown in Figures 7.17 and 7.19 show the chaotic fields on the serrated edge versus a very coherent wave on the rolled-edge case.

7.4.3 The Lateral Surfaces of the Range

We are left with the treatment of the lateral surfaces of the range. Despite the concept of the reflector collimating the energy, we have to treat the lateral surfaces of

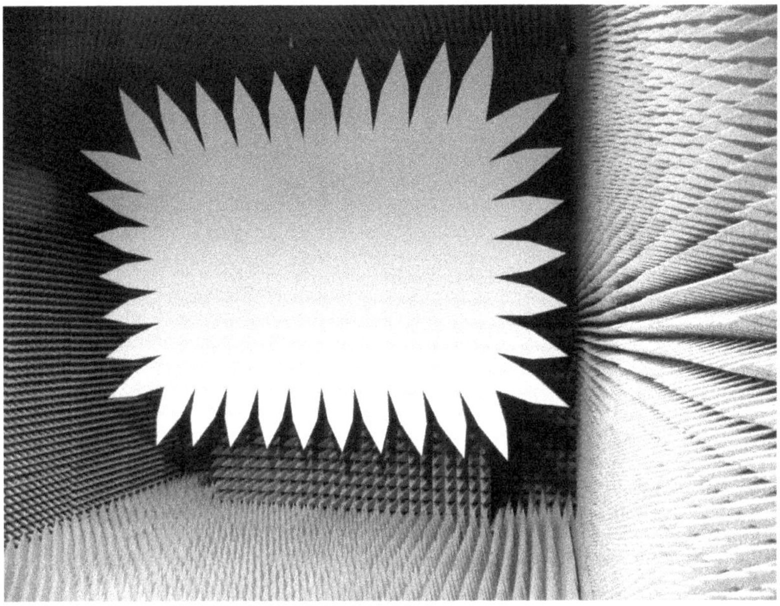

Figure 7.18 A serrated-edge reflector showing the base of the structure covered with an absorber to hide it from the feed direct illumination. (Photo courtesy of NSI-MI Technologies.)

the range. Even Johnson in his classical paper [9] shows a figure where the lateral surfaces of the range have been covered with an absorber.

7.4.3.1 The Feed Antenna

One of the reasons for covering the lateral sides is the feed. Figures 7.17 and 7.19 show the feed illuminating the ceiling and the floor. (Similarly, although not shown, the feed illuminates the lateral side-walls.) The ray-tracing diagram in Figure 7.20 shows that the main concern that we have is the potential for energy from the floor to be reflected off the reflector and interact with the main collimated wave to create amplitude ripple across the QZ.

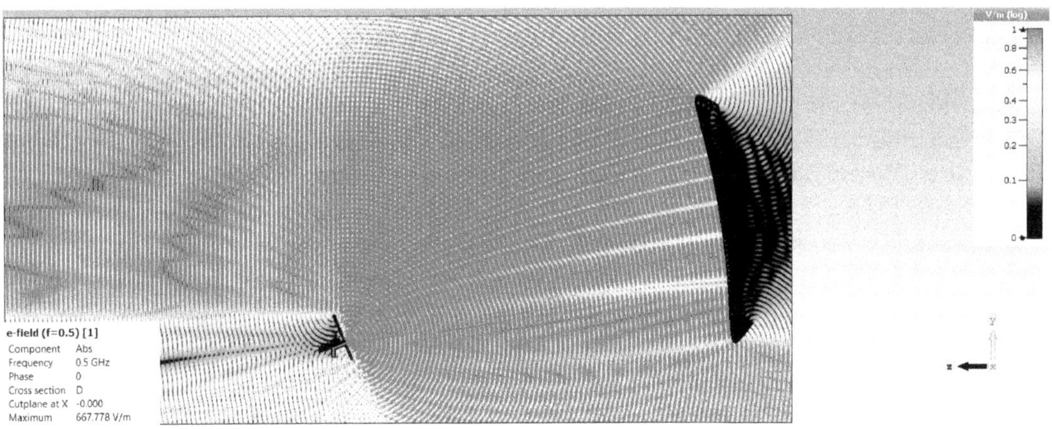

Figure 7.19 Fields behind the reflector body for a rolled-edge geometry.

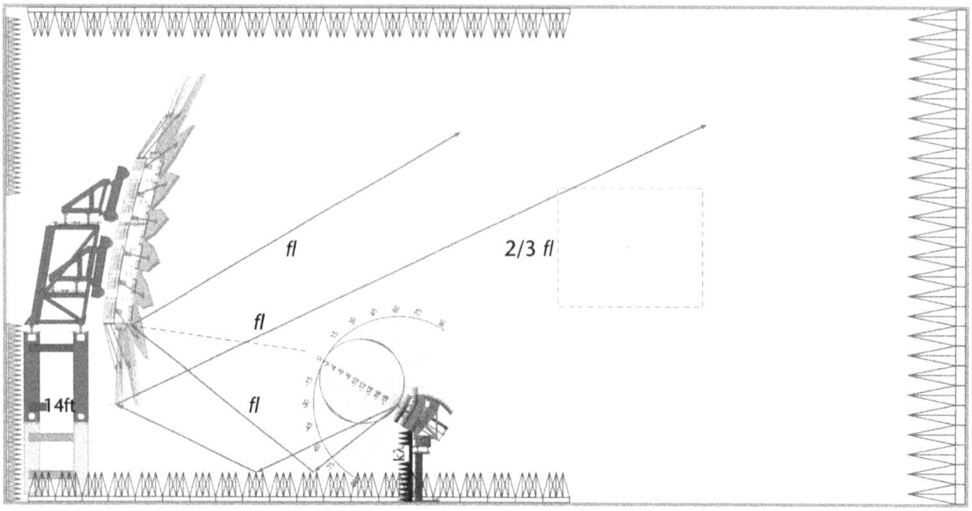

Figure 7.20 Simple ray tracing showing a potential for floor reflection to be reflected into the QZ. This is more likely in a large reflector with long focal lengths.

When looking at setting up a compact range it is important to look at the geometry to ensure that the floor between the reflector and the feed is properly treated. In general, the levels of radiation toward the specular areas of the floor are in the −10-dB level because of the tilting of the feed. (This will, of course, depend on the feed being used.) Usually, an absorber that is one wavelength is sufficient usually. However, in some cases for very large reflectors with QZ exceeding 4m, a longer absorber may be required as a longer focal length will cause a larger angle of incidence onto the floor absorber.

The field simulations shown in Figure 7.17 and 7.19 show a very high level field on the ceiling between the reflector and the QZ. Looking at another ray-tracing exercise illustrated in Figure 7.21 we can see that the location of the feed causes a large amount of energy that may interfere with the direct illumination from the reflector. The most critical will be the direct path between the feed and the QZ, which is noted as triangular area with vertical lines across it in Figure 7.21. The feed fence (discussed in Section 7.4.6 usually takes care of that direct coupling between the feed and the QZ. All other paths involved reflected signals that have been already attenuated by the lateral surface absorber treatment.

Studying the propagations of a modulated pulse in a compact range may provide some insight. Figure 7.22 shows a total of 10 time frames taken every 25 ns, starting at 25 ns and continuing to 250 ns. The compact range reflector shown is a 75-ft (22.86-m) focal length reflector intended to provide a cylindrical QZ with QZh = 24 ft. (7.31m) down to 500 MHz. The center of the QZ in this case is at 5/3 *fl* so basically it is 125 ft. (38.10m) from the vertex of the reflector or 50 ft. (15.25m) from the focal point. The reflector itself has a 14.6-m (47.8-ft.) body (and thus a factor M of almost 2), and 3-m (10-ft.) serrations (i.e., $5\lambda_{low}$). Thus the CR_h = 20m (68 ft.).The plots for the 25-ns, 50-ns, 75-ns, and 100-ns time frames illustrate the field from the source propagating. Like Figure 7.21, the ceiling area above the feed is a critical one; reflection from there may reflect to the reflector or to the QZ itself. The simulation in Figure 7.22 does not show the reflections since

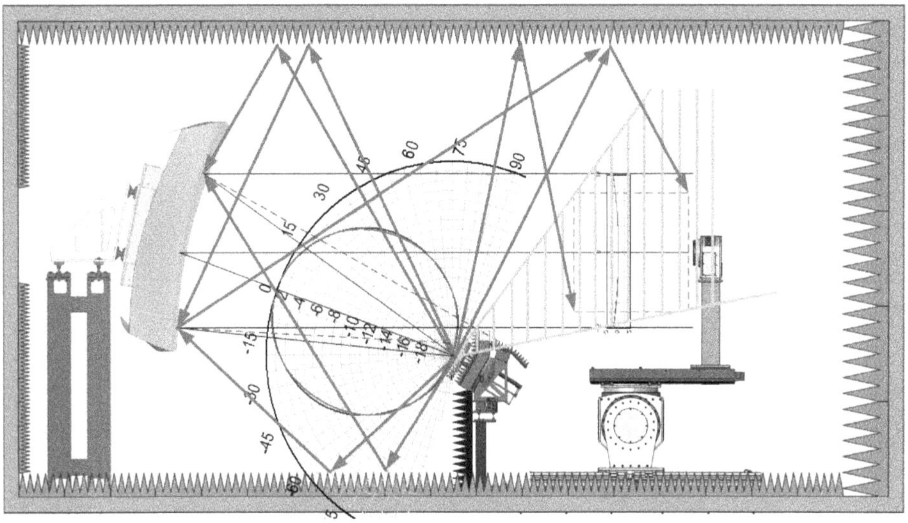

Figure 7.21 Ray tracing showing potential paths from the feed that could negatively interact with the QZ. The feed fence should block the direct paths to the QZ indicated by the shaded area.

the computational domain is terminated by a numerically perfect absorber. Looking at the feed pattern superimposed on the sketch in Figure 7.21, the amplitude of the fields propagating to the ceiling compared to the ones illuminating the reflector are 7 dB lower at the worst case. The amplitudes that can be reflected into the QZ are more than 20 dB lower. The plot in Figure 7.22 at 100 ns shows the levels at the ceiling to be in a similar range. Based on those levels and given that we are close to normal incidence, the value for the electrical size of the absorber is $t_s \geq 1.5$. This value will take care of feed radiation illuminating the lateral surfaces. This value should be sufficient for the floor between the reflector and the feed, given that the feed is tilted toward the center of the reflector, and the levels radiated toward the floor will be lower than those radiating toward the ceiling. However, it is important to check that there is no potential issue from floor reflections due to angles of incidence as pointed out above.

7.4.3.2 Estimating the Illumination of the Lateral Surfaces by the Reflector

While the reflector collimates the energy radiated from the feed, the edge treatment and surfaces tolerances (as well as reflection from the feed) cause some illumination of the lateral surfaces. Chapter 5 discusses an approach in which the far-field pattern of the range antenna is used to estimate the magnitude of the rays incident onto the specular areas of the lateral surfaces of the ranges. I suggest a similar approach for compact ranges in [20]. This approach is a rough approximation, as the reflector is being used in the extreme near field where the plane wave is being generated. Studying the propagation of a modulated pulse in a compact range may provide some insight into the accuracy of the far-field approximation.

7.4.3.2.1 Time Domain Analysis of the Reflector Near Fields

Recall that Figure 7.22 shows a total of 10 time frames. These time frames are taken every 25 ns starting at 25ns and continue to 250 ns.

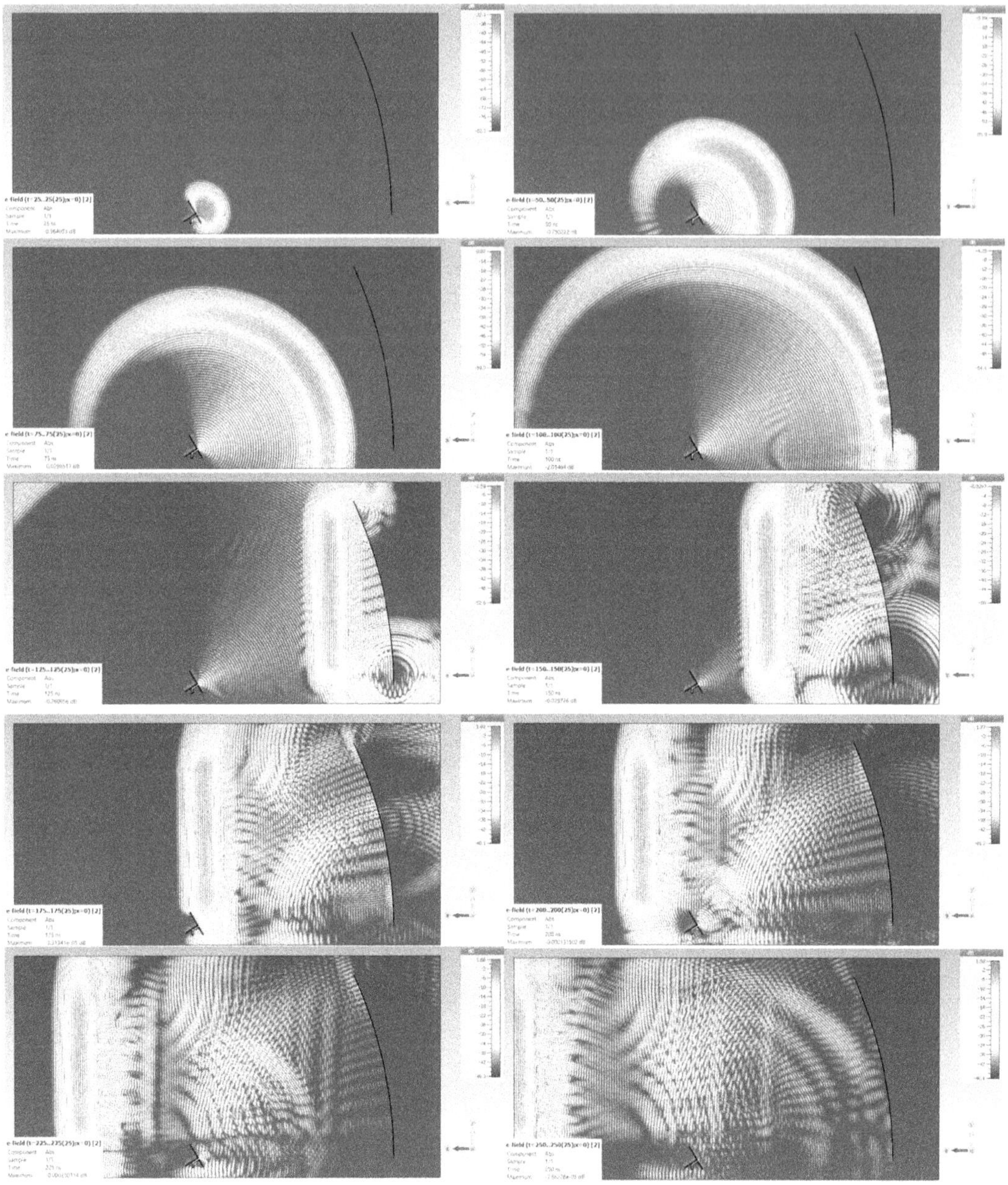

Figure 7.22 Time domain slides of a compact range. Frames show every 25 ns from 25 ns to 250 ns. The focal length is 75 ft. (22.86m).

The distance from the reflector vertex to the center of the QZ is roughly 38.10m (125 ft.) and so the specular point is about 19.05m (62.5 ft.) from the reflector vertex. The slides at 150 and 175 ns are the ones that may provide some insight into the lateral surface illumination. The range shown in Figure 7.22 is 82-ft. (25m) high. The center of the QZ is located centerline of the range, so at 41 ft. (12.5m) over the floor. Using trigonometry we can estimate the angle of incidence for a ray departing the centerline of the reflector and reflecting on the ceiling to the center of the QZ. The angle is given by $\tan^{-1}$ (38.1/25), which is 56.73°.

Figure 7.23 details the 150-ns and 175-ns plots presented in Figure 7.22. Figure 7.23 shows the location of the QZ, the focal point, the feed location, and the reflector vertex. A ray emanating from the center of the reflector and propagating to the ceiling where it is reflected into the QZ is also shown superimposed on the plots on Figure 7.23. In [20], the argument is that the far-field pattern provides an indication of the radiation levels that propagate from the near field, and thus, the far-field pattern provides an estimate of the incident ray magnitude. Looking at the spherical wavefront on the edge of the reflected pulse, it appears that the amplitude incident onto the ceiling is around 13 dB below the magnitude of the plane wave traveling to the QZ. Examining the pulse at 175 ns, it shows that the amplitude of the spherical wavefront on the edge has decreased as it traveled away from the origin, but there is no relation between the far-field pattern and the near-field behavior as assumed in [20].

7.4.3.2.2 Frequency Domain Analysis of the Reflector Illumination

If we look at the field distribution for the reflector and feed in the frequency domain, at 500 MHz to be more specific, and we compute the field distribution for the feed antenna without the reflector at the same frequency, it is possible to subtract the fields from the feed source from the total fields of source and reflector. Doing this eliminates the source illumination of the ceiling, and we can look at the amplitude of the field from the paraboloid reflector incident onto the ceiling. This process is illustrated in Figure 7.24. If we superimpose the far-field pattern of this reflector at the frequency of interest it is clear that using the far-field pattern is overly optimistic. The far-field pattern estimates a −35-dB field versus the peak, versus levels

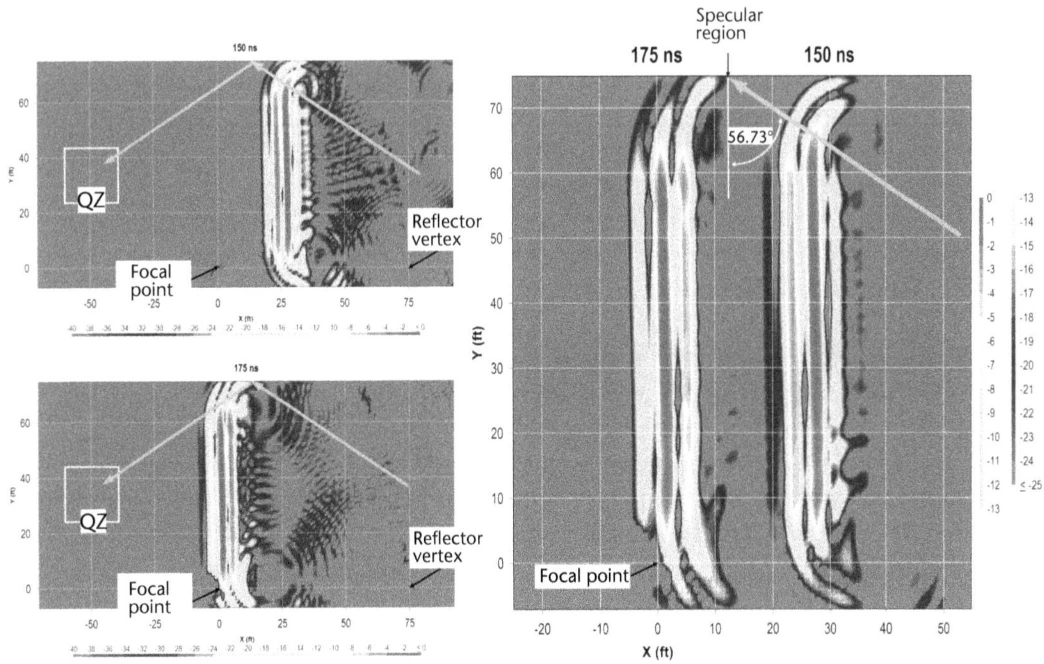

Figure 7.23 Detail of the propagation of a cosine modulated Gaussian at 150 ns and at 175 ns, showing the features of the range superimposed on the plots (left). On the right, the pulse at 175 and at 150 ns is superimposed on the same plot, showing the angle of incidence of a specular ray.

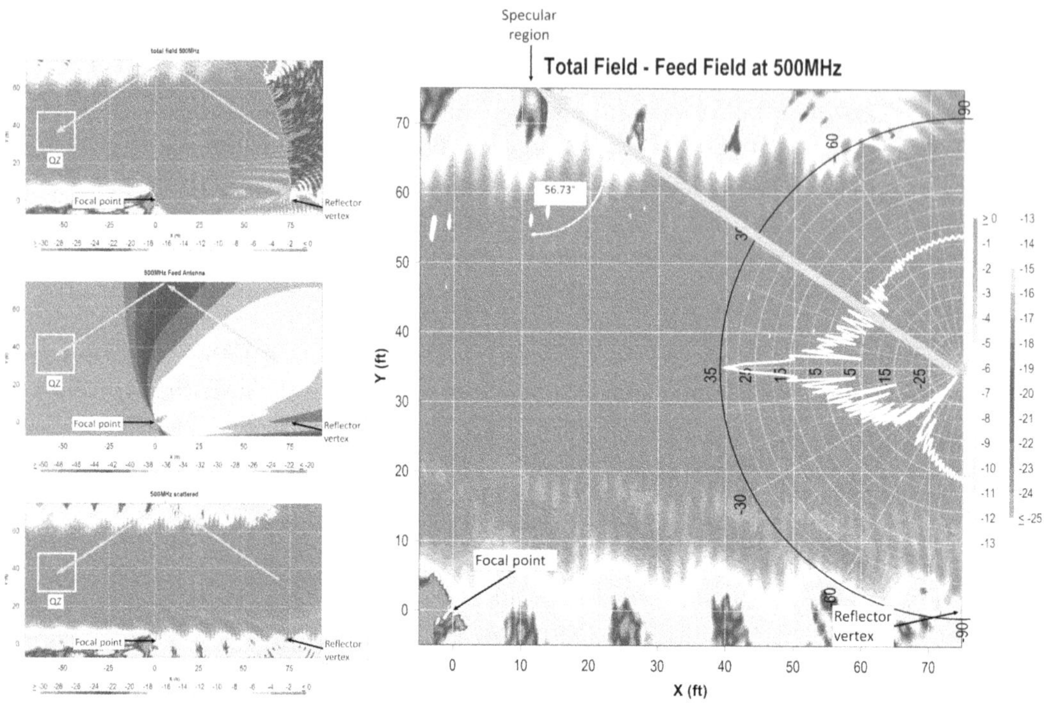

Figure 7.24 Field distribution without the feed fields. The fields from the feed are subtracted from the total field distribution of feed and reflector to obtain the "scattered" fields from the reflector (left). The estimated reflector pattern is superimposed on the "scattered" fields from the reflector (right).

in the −12 to −13 dB range. These are similar to the levels that are seen in the time domain plotted in Figure 7.23.

7.4.3.2.3 Power Flow Analysis of the Reflector Illumination

Let us look at the power flow, or Poynting vector. In determining the Poynting vector, we obtain the direction in which the power travels in the compact range. We are interested in the power traveling in the direction of the specular ray, which is at 146.73° from the x-axis (see Figure 7.24). The unit vector that describes the direction of the ray incident onto the specular area of the ceiling is $\hat{a}_{spec} = -0.8361\hat{a}_x + 0.548\hat{a}_y$; thus, if we perform a dot product of the Poynting vector with $\hat{a}_{spec}$ the result is the amplitude of the power density traveling in the direction of $\hat{a}_{spec}$. Let us consider the Poynting vector on the compact range described above. Figure 7.25 illustrates with vectors the power flow in the compact range. We see a large magnitude emanating from the focal point where the feed is located. We can perform the dot product of the pointing vector on the top lateral surface (i.e., the ceiling), with $\hat{a}_{spec}$, and that will give us the power density traveling in the specular ray direction $\vec{P}_{spec} = (\vec{P} \cdot \hat{a}_{spec})\hat{a}_{spec}$. The next step is to integrate the Poynting vector at the $x = -70$ ft. surface, $\vec{P}_{x=-70\text{ ft}}$ and the $\vec{P}_{spec}$ throught the top surface $y = 75$ ft. Thus

$$P_{180} = \oint \vec{P} \cdot -\hat{a}_x \, ds_{\text{end wall}} \tag{7.8}$$

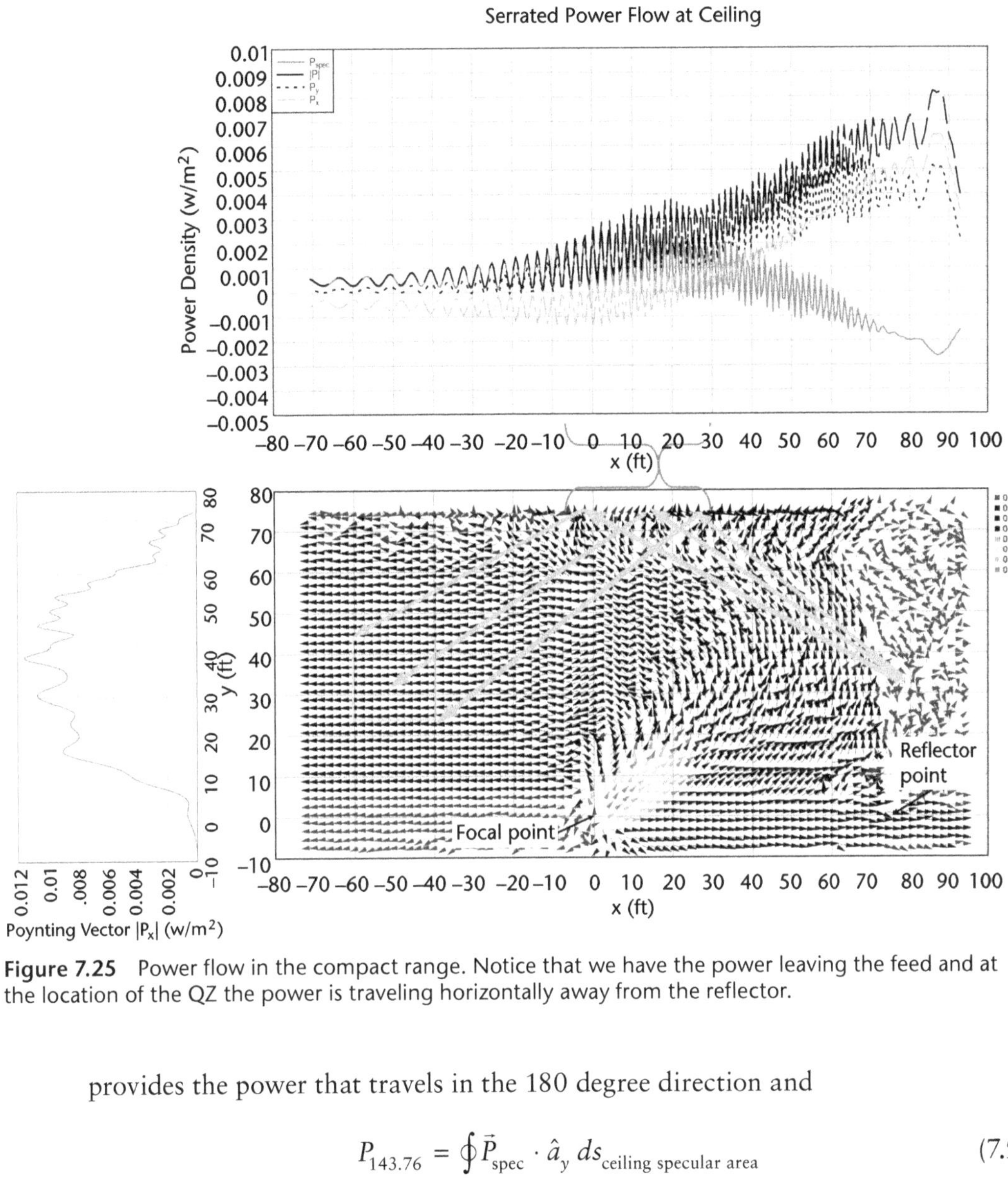

Figure 7.25 Power flow in the compact range. Notice that we have the power leaving the feed and at the location of the QZ the power is traveling horizontally away from the reflector.

provides the power that travels in the 180 degree direction and

$$P_{143.76} = \oint \vec{P}_{\text{spec}} \cdot \hat{a}_y \, ds_{\text{ceiling specular area}} \tag{7.9}$$

Performing these integrals yields power values that we can compare to one other. However, for the ceiling let us only consider the specular region (roughly from $x = -10$–30 ft. as shown in Figure 7.25), where reflected rays enter the QZ. This exercise is an approximation, but the Poynting vector provides the direction of the power flow so we can get an estimate of the power incident onto the lateral surface traveling in the direction of the specular ray.

Figure 7.25 shows $|\vec{P}_{\text{spec}}|$ on the top surface; in this quasi-2D case being simulated the top surface reduces to a line. $|\vec{P}_x|$ at the end wall is also shown. Evaluating those integrals in (7.8) and (7.9), we get a difference of −15 dB between the power going through the end wall and the power that exits the specular region. This gives a value of the expected fields into the specular region. Unlike the analyses done using the E-fields in the time or frequency domain, the Poynting vector analysis provides information on the magnitudes traveling in the direction of interest.

7.4.3.2.4 Other Methods for Estimating the Lateral Surface Illumination

Physical optics approaches or similar quasi-optic methodologies to solve for the near-field distribution of the reflector field can be used to determine the field levels at the edges of the lateral surfaces. These can provide an estimate of the incident field levels at the walls. GRASP™ and similar commercial packages can be used to obtain the reflector solution. Such analyses are performed on the reflector in Chapter 8 to estimate the incident energy onto the lateral walls.

7.4.3.2.5 Rule of Thumb Estimation: The Ceiling and the Floor

We have considered how complex it can be to predict the illumination of the lateral walls on a compact antenna test range. We can, however, assume a level based on experience.

Over the past 40 years, similar analyses and actual measurements, like those reported by Brumley [21], have been performed. Empirically it has been shown that the lateral wall illumination is approximately between −15 and −20 dB below the levels at the center of the range.

Using the rule of thumb for the reflector size, a range of angles of incidence onto the lateral surfaces can be estimated. With that angle and the estimate for the illumination of the lateral surfaces, a recommendation for the minimum lateral wall absorber can be made.

Let us assume $10\lambda_{low}$ for the height of the QZ. Choosing a $24\lambda_{low}$ focal length and a $CR_h = M \cdot QZh + 10\lambda_{low}$ where $M = 1.4$ we have $CR_h = 24\lambda_{low}$. Per (7.3) and following Figure 7.10 we have that the angle of incidence to the rear top edge of the QZ, which would be the worst-case angle (i.e., the broader angle), is given by

$$\theta_{ceiling} \leq \tan^{-1}\frac{45\lambda_{low}}{25\lambda_{low}} = 61° \tag{7.10}$$

Hence, the worst angle in a case with the minimum $2\lambda_{low}$ separation between the reflector and the absorber and with the minimum $M = 1.4$ and a $10\lambda_{low}$ QZ, the worst angle on the ceiling will be 61°. Figure 7.26 illustrates how the angle is determined. The ceiling absorber treatment with $t_s = 1.5$ will yield about −36 dB by combining the absorber reflectivity (see Figure 3.9) and the lateral wall illumination of −15dB, to which we have to add the path loss. (The energy going to the lateral surfaces does attenuate as it travels unlike the plane wave traveling to the QZ.). Note that this is the worst case; the angles vary from 46.7° for the best case to 61° for the worst case. Analyzing possible combinations of reflector height size and focal length and QZ size gives us an estimate of the worst possible angles of incidence. The results of this analysis are shown in Figure 7.27.

The value of $t_s = 1.5$ for the ceiling can also be applied to the floor; in general the factor k for the position of the reflector in relation to the floor is larger than 2, that is $k > 2$. Thus, the angles of incidence into the floor will be better than those in the ceiling. Additionally, recall that $t_s = 1.5$ is the recommended size for the floor treatment between the feed and the reflector.

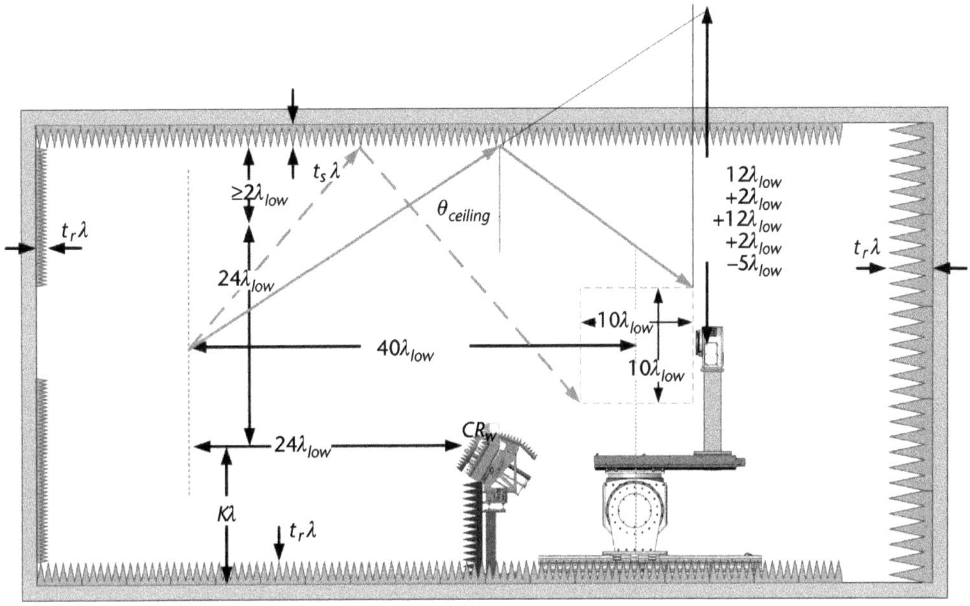

Figure 7.26 Worst-case angle of incidence onto the ceiling.

7.4.3.2.6 Rule of Thumb Estimation: The Side Walls

A similar approach can be followed for the lateral surfaces. Recall that the reflector width is given by $CR_w = F \cdot QZw + 10\lambda_{\text{low}}$, where $1.83 \leq F \leq 2.3$. It should be obvious that since the minimum factor F is the same as the maximum factor M, serrated reflectors tend to be wider rather than taller, so the width of the room is

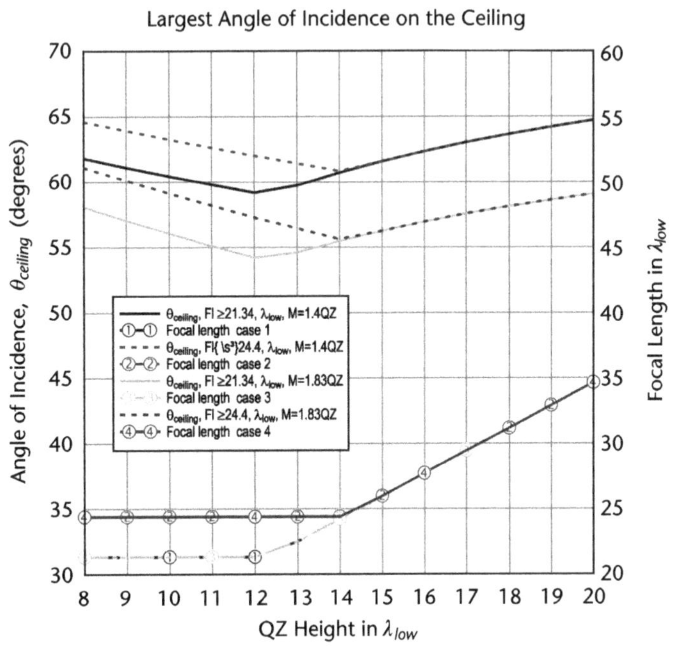

Figure 7.27 Analysis of some combinations of reflector height, QZ size at the lowest frequency, and focal length to show the worst-case angle of incidence.

Table 7.1 Absorber Treatment Recommendations for a Serrated-Edge Compact Range

Surface	*Absorber Size in* λ_{low}
Wall opposite reflector	$t_e \geq 2.5$
Wall behind reflector	$t_r \geq 0.7$
Floor and Ceiling	$t_s \geq 1.5$
Side walls	$t_s \geq 1.5$

generally wider. Thus, the curves shown on Figure 7.27 for $M = 1.83$ will be the same as the ones for $F = 1.83$. As F increases to $F = 2.3$, the angles of incidence will decrease. Thus, the lateral wall absorber can also be of a size such that $t_s = 1.5$. Hence, we arrive at the following recommendations, listed in Table 7.1, for the absorber treatment in a compact range with a serrated-edge reflector. These are recommendations, and the reader may want to follow a similar approach to define the absorber in a given range or to estimate the potential reflected signal entering the QZ given the range geometry and reflector performance.

This is a recommendation for serrated-edge reflectors. Remember that rolled-edge reflectors require a different absorber size on the wall behind the reflector.

7.4.4 Rolled-Edge Lateral Surface Absorber

Rolled-edge reflectors deal with the edge diffraction in a different way. Thus, the radiation from the edge of the reflector propagates differently from the serrated-edge treatments. Figure 7.28 shows the plot of the Poynting vector for a rolled-edge compact range with geometry similar to the serrated edge plotted in Figure 7.25. For this case the focal length is 24.4m (80 ft.). Thus the projected distance from the reflector vertex to the center of the QZ is 40.65m (133.33 ft.). The reflector body is the same size as QZh. Thus, the body is 7.3m (24 ft.). The rolls of the edge treatment are 3.96m (13 ft.) in the projected area, which is about $6\lambda_{\text{low}}$. Thus $CR_h = 15.22$m. Thus the rolled-edge reflector is 4.78m smaller than the serrated one. The computational area is the same for both reflectors, so the height of the computational domain is 25m (82 ft.). However, let us sample the pointing vector on the ceiling as if the range height has been reduced by 4.78m to 20.22m overall height. The range will remain at midheight so the center of the QZ is at 10.11m over the ground. The specular point for a ray traveling to the center of the QZ is at $x = 4$m (13.33 ft.), or 4m in front of the focal point. The angle of incidence to the specular area is $\theta = 63.55°$. The plot in Figure 7.28 illustrates the difference between the two reflectors' styles when compared to the results illustrated in Figure 7.25.

The Poynting vector behind the reflector is more organized on the rolled edge (Figure 7.28) than it is on the serrated edge (Figure 7.25).

If we perform the integrals on (7.8) and (7.9), we get that the difference (in this case) is less than 1 dB; however, what changes is that the angle of incidence at which the reflected energy will have power traveling in the direction of the QZ. For $t_s = 1.5$ the absorber degraded from the 56° for the serrated to the 64° by about 5 dB. It may be necessary to get an absorber that is $t_s = 2.5$ to achieve those extra 5

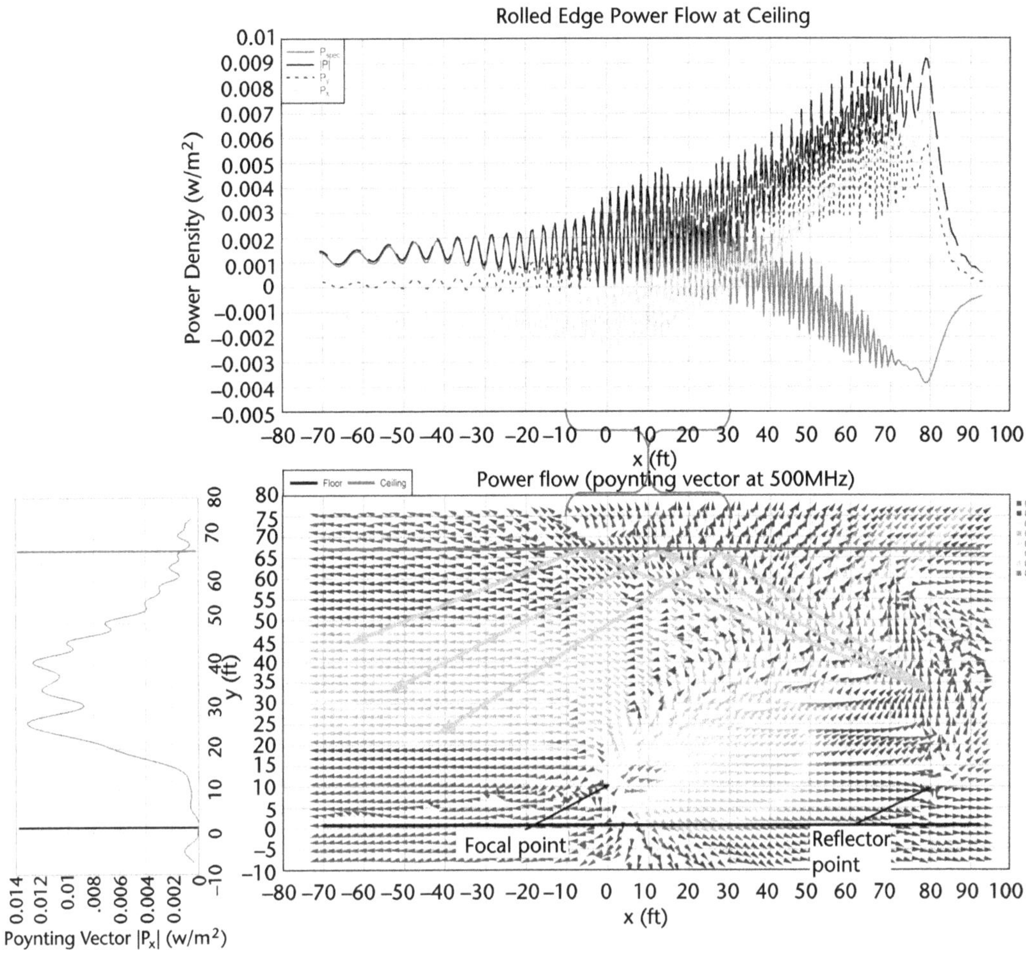

Figure 7.28 Power flow in rolled-edge compact range. Notice that the power flow to the wall behind the reflector is more organized.

dB of reflectivity. It may be a good idea to increase the height to improve the angle of incidence. In general, as mentioned by Lee and Burnside [13], it may be a good idea to increase the lateral surface absorber or to increase the range.

Notice that when we compare Figures 7.25 and 7.28 the power flow into the lateral surfaces at the location of the QZ is much higher for the rolled edge than it is for the serrated-edge. While those are noncritical areas for antenna measurements, they *are* critical in RCS measurements (as covered in Chapter 8), making serrated-edge reflectors a better choice for RCS chambers. Figure 7.29 illustrates the difference in illumination of the ceiling. Being higher, the rolled-edge feed has a smaller angle of tilt versus the horizontal; hence there is less illumination by the source compared to the serrated. The time domain allows us to see the difference between the feed illumination and the reflector fields. Notice that the rolled-edge reflector illuminates the ceiling specular point with amplitudes that are twice those of the serrated. This clarifies further the statements by Lee and Burnside about the rolled edge illuminating more of the lateral walls. Illuminating the lateral walls is a key issue in RCS chambers as discussed in Chapter 8.

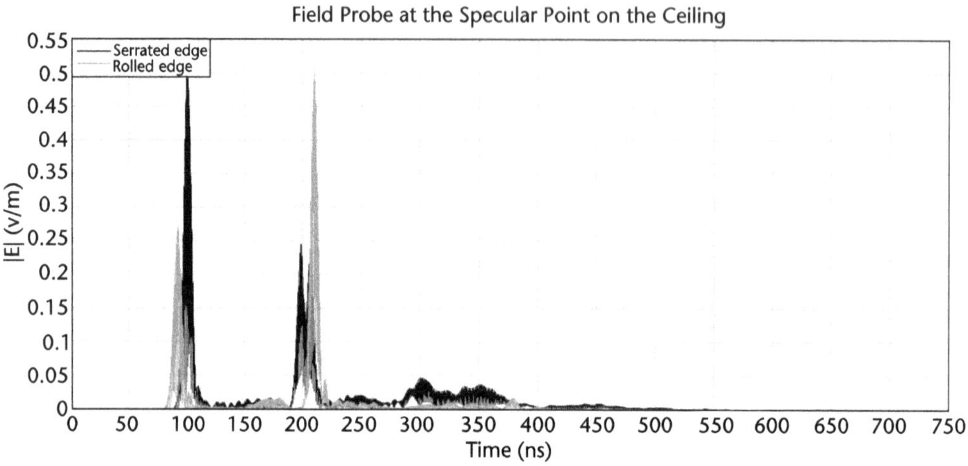

Figure 7.29 Time domain of the total electric field at the specular point.

7.4.4.1 Recommendations for Rolled-Edge Compact Ranges

The recommended absorber layout for rolled-edge reflectors is given in Table 7.2.

These are only recommendations, and it is important to understand how the reflector radiates to properly treat all surfaces. Making the room taller and wider for a rolled edge may reduce the lateral surface absorber. Also recall that since the illumination by the serrated edge is much smaller behind the QZ, it may be possible to reduce the absorber coverage in that area. Such reduction may not be possible in a rolled edge since the illumination of those areas is higher in magnitude.

It should be noted that in the two samples we have considered the absorber thickness onto the lateral surfaces should be added to the height (or width) of the range. The worst case assumes that the reflection point of the absorber treatment is at the tips.

7.4.5 Typical Absorber Layout

Recall that the typical layout of a compact range for antenna measurements calls for pyramidal absorber in all the surfaces. The wall behind the reflector may be left partially untreated, although in the case of rolled-edge reflectors it is recommended that it have a full treatment.

Table 7.2 Absorber Treatment Recommendations for a Serrated-Edge Compact Range

Surface	*Absorber Size in* λ_{low}
Wall opposite reflector	$t_e \geq 2.5$
Wall behind reflector	$t_r \geq 1.0$
Floor and ceiling	$t_s \geq 2.0$
Side walls	$t_s \geq 2.0$

Corner and edge treatment can be done with LFB as shown in Figure 5.10. Thus there is no reason for miter cuts on the edges. Behind the QZ, toward the range end wall it may be desirable to switch to wedge absorber.

With the end wall being the critical one, it is desirable not to have doors there; however, often there are doors to aid in the loading and unloading of AUTs onto the positioners.

The lateral wall absorber can be mounted in the standard configuration as shown in Figure 5.9(c). Configurations like the ones shown in Figure 5.9(a, b) have been used, but it is my belief that they do not add as much benefit to the range as may be the case in far-field ranges.

One peculiarity of the indoor compact range is that the feed illuminating the reflector is roughly in the middle of the specular region. Because of this, it is always desirable to have a path for the range users to access the feed positioner to switch feed antennas or to service the positioner itself. The walkway should be minimized. If possible, in fact, it should be avoided, and the absorber should be removed to access the feed positioner. This last recommendation is not very practical. Feeds must be located at the focal point since the phase center should be at the focal point of the reflector; hence typical broadband antennas (LPDAs and ridged horns) are not ideal compact range feed antennas. Typical feed antennas, such as those shown on Figure 7.12, cover a waveguide bandwidth; thus, the need to switch feeds is important, and access to the feed positioner is a must.

There are automatic feed changers, which are typically carrousels with several feeds mounted; the positioner slides or rotates different band feeds to the focal point. These are complex positioning systems with electronic switching between feeds and accurate positioning of different feeds at the focal point. These are recommended for locations switching often between waveguide bands or testing wideband antennas where the downtime switching antennas outweighs the cost of the feed changer.

The walkway absorber to the feed should be shorter than the adjacent absorber to ensure that it is mostly hidden by the surrounding absorber field.

7.4.6 Feed Fences

Continuing with the topic of the feed being in the specular region is the issue of the feed fence. The feed positioner needs to be treated, but the presence of the absorber causes a disruption on the floor absorber treatment. In addition, it is common to have the RF source or the receiver (or both) close to the feed, and behind the positioner is an ideal location that places the RF system close to the feed and the AUT. The fence also hides the AUT positioner base from the reflector. The final use of the fence is to reduce the direct coupling between the feed and the QZ.

Since any disruption of the absorber field can give rise to edge diffraction, the feed fence treatment is arranged to divert the edge diffraction away from the QZ, similar to the way in which the serrations do on the reflector. Figure 7.30 shows the typical treatment of a feed fence. The serration of the absorber treatment at the top of the fence redirects the edge diffracted fields toward the lateral walls and away from the QZ.

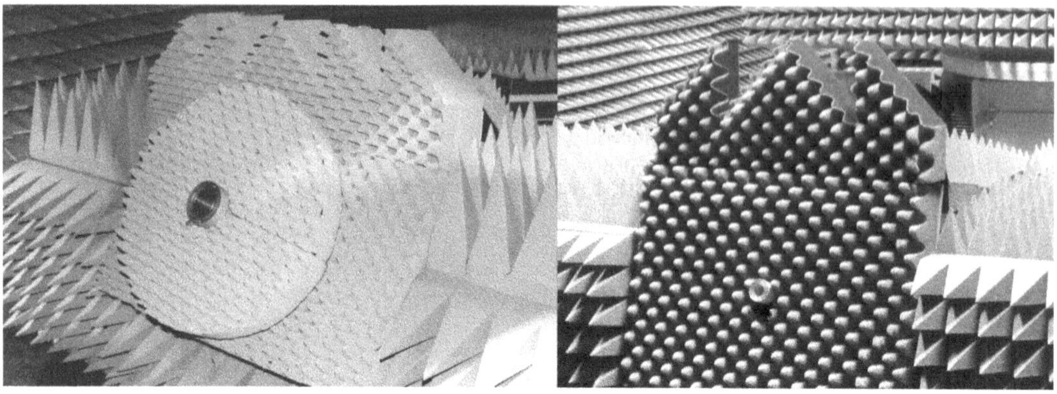

Figure 7.30 Samples of two feed fence implementations.

7.5 High-Power Concerns

The number of active antennas has increased recently. Electronically steerable arrays, or AESA antennas, are becoming the workhorse of many radar systems. Even in 5G technologies it is expected that the base station antennas will be AESA antennas that can switch between beam stages, steering the beam as necessary to communicate with specific devices. For these antennas, reciprocity does not apply. These are nonlinear devices due to their internal electronic circuits and must be tested at their intended power levels of operation. The compact range approach is ideal for these antennas since in many cases they have their own internal sources and digital controls as well as amplifiers. For many of these antennas it is not possible to get a phase reference to measure phase; hence, using near- to far-field systems as those covered in Chapter 6 is not possible. The compact range allows for testing electrically large antennas and obtaining their radiation patterns without measuring the phase. It should be noted that phase is important, and even in compact ranges measuring the phase is required in determining polarization sense or other parameters of the antenna. However in some of these active antennas, it is increasingly more difficult to obtain a transmitted signal that can be used as a reference to measure the phase.

Measuring antennas transmitting high levels of power in a compact range is not a trivial issue. The reflector, after all, is a concave mirror that concentrates the radiated power density at the focus where the feed and the absorber surrounding it are located. Going back to the work by Bickmore and Hansen [22], it can be seen that there is little to no attenuation of the field from an aperture antenna as it travels along the axis normal to the aperture. The results for a tapered circular aperture (like a parabolic dish antenna) were provided by Hansen and shown in [23]. The results for the three different types of apertures studied in [22] are shown in Figure 7.31. The equation for a square uniform aperture, such as it will be for a uniform square array antenna is given by [22]:

$$P = 16.4\left[C^2\left(\frac{1}{2\sqrt{x}}\right) + S^2\left(\frac{1}{2\sqrt{x}}\right)\right] \tag{7.11}$$

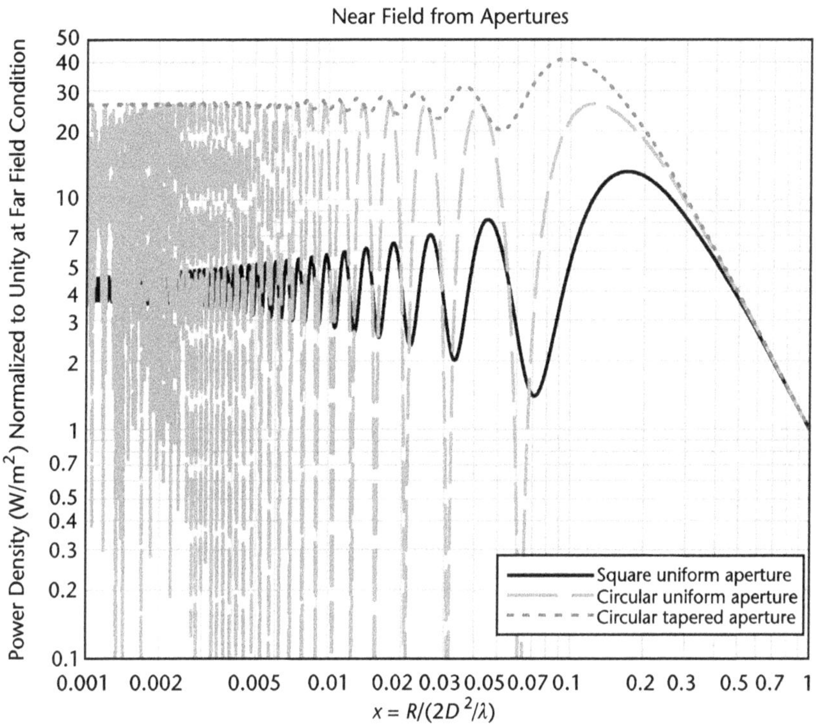

Figure 7.31 Normalized near-field power distribution from some apertures.

where C and S are the cosine, and sine Fresnel integrals and $x = R/(2L^2/\lambda)$, R is the distance from the aperture plane, and L is the square side dimension. The equation for the uniform circular aperture is given by [22]

$$P = 13.14\left(1 - \cos\left(\frac{\pi}{8x}\right)\right) \tag{7.12}$$

For the case of the circular apertures, $x = R/(2D^2/\lambda)$. R is the distance from the aperture plane, and D is the diameter of the aperture. The equation for the circular tapered aperture given in [22] is given as

$$P = 26.1\left[1 - \frac{16x}{\pi}\sin\left(\frac{\pi}{8x}\right) + \frac{128x^2}{\pi^2}\left(1 - \cos\left(\frac{\pi}{8x}\right)\right)\right] \tag{7.13}$$

which is the equation used by Baggett and Hess [24] with the modifications shown in Chapter 4 [see (4.2)].

From Figure 7.31 it can be seen that the power density oscillates around the power density at the aperture of the antenna and peaks at between 0.1 and 0.17 of the far-field condition. If we use 25λ as an approximate value for the focal length, then the reflector is at around 41.5λ from the center of the QZ, where the AUT will be. The far field for a given antenna of size $n\lambda$ is $2n^2\lambda$; hence, antennas that are as small as 4.55λ will illuminate the reflector as if the reflector is in the far field.

Antennas with far fields such that $41.5\lambda \approx 0.27n^2\lambda$ or around 12.4λ in aperture size will illuminate the reflector at the peak of their near-field distribution. That peak can be up to 5.3 dB higher than the aperture power density for the case of square apertures.

Thus, it is possible for the AUTs to illuminate the reflector with a very high-power density. Those fields on the reflector will then be focused at the feed. Hence, there will be a very high-power density at the focal point where the feed is; that needs to be addressed and understood.

Hess [25] developed a transfer function for the compact range to relate the power density at the aperture of the AUT to the power density at the focal plane of the reflector. Hess starts from the idea that the compact range is an angular filter of plane waves. That is, only the plane wave component of an incident wave will be reflected to the focal point. Referring to Figure 7.32, Hess first defines the criterion for the compact range to accept a plane wave as

$$R_o \cdot \left(\frac{1}{2}\Delta\Theta\right) \leq \frac{d}{2} \tag{7.14}$$

The angular area is defined as

$$\Delta\Omega = \int_0^{2\pi}\int_0^{\frac{1}{2}\Delta\Theta} \sin\theta\, d\theta\, d\varphi = 2\pi[-\cos\theta]_0^{\frac{1}{2}\Delta\Theta} = 2\pi\left[\left(-\cos\left(\frac{1}{2}\Delta\Theta\right)\right) - (-1)\right] \tag{7.15}$$

Hess uses the Taylor series expansion of the cosine

$$\cos x = \sum_{n=0}^{\infty}(-1)^n \frac{x^{2n}}{(2n)!} \tag{7.16}$$

then proceeds to drop all the terms where $n > 1$, using the assumption that the argument of the function is small; he then arrives at

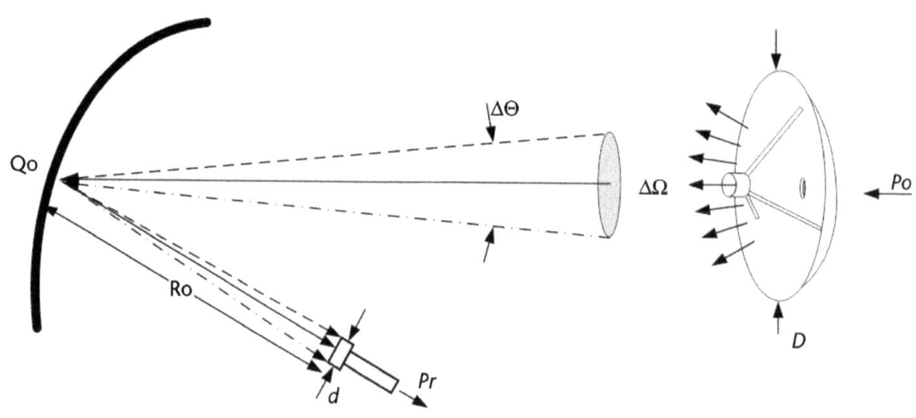

Figure 7.32 Focusing of the power density.

$$\Delta\Omega = \pi\left(\frac{1}{2}\Delta\Theta\right)^2 = \pi\left(\frac{d}{2R_o}\right)^2 \tag{7.17}$$

The radiation intensity of the AUT is given by

$$\Phi^{\text{AUT}} = \frac{dP^{\text{AUT}}}{d\Omega} = G_{\text{AUT}}\frac{P_o}{4\pi} \tag{7.18}$$

Given the angular filtering of the compact range reflector the power at the reflector is

$$P_r = G_{\text{AUT}}\frac{P_o}{4\pi}\Delta\Omega = G_{\text{AUT}}\frac{P_o}{4\pi}\pi\left(\frac{d}{2R_o}\right)^2 \tag{7.19}$$

Approximating the gain of a circular feed aperture by using $G_F = (\pi d/\lambda)^2$, the dimension d can be taken out of (7.19), and dividing both sides by P_o provides the following coupling equation of the compact range:

$$\frac{P_r}{P_o} = \left(\frac{\lambda}{4\pi R_o}\right)^2 G_{\text{AUT}}G_{\text{F}} \tag{7.20}$$

At the feed the power density is the received power divided by the effective aperture of the feed. The effective aperture of the feed is given by $A_e = (\lambda^2 G_F)/4\pi$ [26], so the power density at the feed is given by

$$S_{\text{feed}} = \frac{P_o}{\dfrac{\left(\lambda^2 G_F\right)}{4\pi}} \tag{7.21}$$

Using (7.20) into (7.21) provides the power density at the feed as

$$S_{\text{feed}} = \frac{P_o G_{\text{AUT}}}{4\pi R_o^2} \tag{7.22}$$

Now the power density for the aperture of the AUT is given (using [25]) by

$$S_{\text{AUT}} = \frac{P_o}{\dfrac{\left(\lambda^2 G_{\text{AUT}}\right)}{4\pi}} \tag{7.23}$$

Combining (7.23) and (7.22) we arrive at the equation for the focusing power at the feed of a compact range:

$$S_{\text{feed}} = \frac{\lambda^2 G_{\text{AUT}}^2}{\left(4\pi R_o\right)^2} S_{\text{AUT}} \tag{7.24}$$

The S_{AUT} term is the power density at the reflector. That power density can be computed or obtained from (7.11) to (7.13).

Let us consider a square uniform array of physical size 11λ by 11λ. Assume that this is a very efficient array and that it has an input power of 100W. The gain of this array is 31.82 dB [26]. Let us further assume that the frequency of operation is such that the area of the array is 1m^2 (3.299 GHz). For this uniform square-shaped aperture, the power density at the reflector is given by (7.11). The peak power is at 332 W/m^2 at 0.171 of the far-field distance. The far field is at 242λ so the peak is at 41.4λ, which happens to be the distance to the reflector from the AUT for a focal length of 24.8λ.

Using the 332 W/m^2 into (7.24) gives us 7,903 W/m^2, which is much higher than what standard absorber can handle. [Remember that the gain in (7.24) is linear gain.] While for many years the focusing effect of the compact range was not an important issue, the use of active antennas has changed this. It is important to keep this in mind when designing a range, not only for selecting the proper absorber that can handle these powers, but also to understand the potential power into the receiving RF system to avoid amplifiers and other components being damaged.

7.6 Uncertainty and Effects of the Range

In 2015, Blalock and Fordham attempted to define the terms associated with the uncertainties in compact range measurements [27]. Following the approach used in the IEEE recommended practice for near-field measurements [28], Blalock and Fordham define 16 uncertainty terms or sources of uncertainty; they are listed in Table 7.3.

Of the sources of error listed in Table 7.3, two terms are related directly to range design and the absorber layout of the range (the terms listed third and seventh).

The range antenna or feed-to-QZ coupling will be an important source of error in gain and sidelobe measurements. Minimizing this error can be done by having the proper feed antenna with very low radiation in the direction of the QZ. Indeed, this can be in part accomplished by using feeds such as corrugated apertures. However, as shown in Figure 7.13 the illumination toward the QZ may be about 20 dB lower than the one toward the reflector. This does not mean that the signal is –20dB lower. The propagation losses to the reflector for the main beam direction will be different than the ones from the focal point to the QZ. The main beam of the antenna will have space losses for the focal length. The direct coupling energy will travel roughly $\sqrt{(2fl/3)^2 + (CRh/2)^2}$. If we assume that the losses on the direct reflector path and the direct coupling path are the same, the –20 dB can amount to close to ±0.92 dB of error. This reinforces the importance of the feed absorber treatment and of the serrations shown in Figure 7.30 to ensure that the lateral radiation from the source is attenuated and that any edge effects are directed away from the QZ.

The other term in Table 7.3 related to the range is the seventh, room scattering. Avoiding this error in compact ranges is as important as in the far-field ranges discussed in Chapter 5. The effects of stray signals in a compact range are discussed in [29], which shows the effects of a stray signal of significant value arriving at a given angle on the pattern of a known aperture. The pattern for a perfect plane

Table 7.3 Uncertainty Terms in a Compact Range

No.	*Source of Uncertainty*	*Related to*
1	Range antenna alignment	Installation
2	Polarization mismatch	Positioner tolerances
3	Range antenna to QZ coupling	Absorber layout and fence
4	Reflector edge diffraction	Reflector design
5	Reflector surface roughness	Reflector construction
6	Leakage and cross-talk	RF-system
7	Room scattering	Absorber layout
8	QZ amplitude taper	Feed, alignment
9	Mismatch	Impedance
10	AUT positioning system	Tolerances of positioner
11	Receiver nonlinearity	RF-system
12	Receiver dynamic-range	RF-system
13	RF repeatability	RF-system
14	Multiple reflections (reflector flash)	(Not an issue on antenna measurements)
15	Gain standard	Calibration of the standard antenna
16	Other errors	Potential sources not accounted for in the previous terms

wave illumination is computed and then compared to the pattern of an illumination where a stray signal is introduced as arriving from some angle. That angle of arrival is varied from about 10.8 degrees to 3 degrees to model the rotation of the AUT in the QZ. The results show how the effects on the sidelobe levels of the pattern are critical. In [29], the authors introduced a large amplitude stray signal to show the large effect. In general, the worst effect will be at the lowest frequency of operation. As illustrated in [27], at higher frequencies the stray signal in the range will easily be well below −60 dB. The effect is reduced to no effect on the gain, and for sidelobe levels the error is reduced to 0.05 dB for −30-dB sidelobes. In the example described in [27], the lateral wall absorber is 10λ at the frequency of the analysis.

For compact ranges, the effects of the stray signals are most important at the lowest frequencies of operation. As the frequency increases, the alignment and the reflector constructions become the determining factors on the uncertainties achieved on the range.

References

[1] Matytsine, L., P. Lagoiski, and S. Matitsine, "Antenna Measurement Using Large Size, Lightweight, Broadband Convex RF Lens," in *6th European Conference on Antennas and Propagation (EuCAP 2012)*, Prague, Czech Republic, 2012.

[2] Rodriguez, V., et al., "A Cone Shaped Tapered Chamber for Antenna Measurements Both in Near Field and Far Field in the 200 MHz to 18 GHz Frequency Range and Extension of the Quiet Zone using an RF Lens," *Journal of the Applied Computational Electromagnetics Society*, Vol. 28, No. 12, 2013, pp. 1162–1170.

[3] Koch, W. E., "Metallic Lens Antenna," U.S. Patent 2,733,438, 31 January 1956.

[4] Koch, W. E., "Metal-Lens Antennas," *Proceedings of the IRE*, Vol. 34, No. 11, 1946, pp. 828–836.

[5] Abdelrahman, A. H., A. Z. Elsherbeni, and F. Yang, "Transmission Phase Limit of Multilayer Frequency-Selective Surfaces for Transmitarray Designs," *IEEE Transactions on Antennas and Propagation*, Vol. 62, No. 2, 2014, pp. 690–697.

[6] Courtney, C. C., et al., "The Measured Performance of a Planewave Generator Prototype," in *24th Annual Meeting of the Antenna Measurement Techniques Association (AMTA 2002)*, Cleveland, OH, 2002.

[7] Rowell, C., and A. Tankielun, "Plane Wave Converter for 5G Massive MIMO Basestation Measurements," in *12th European Conference on Antennas and Propagation (EuCAP 2018)*, London, U.K., 2018.

[8] Johnson, R. C., "Antenna Range for Providing a Plane Wave for Antenna Measurements." U.S. Patent 3,302,205, 31 January 1967.

[9] Johnson, R. C., H. A. Ecker, and R. A. Moore, "Compact Range Techniques and Measurements," *IEEE Transactions on Antennas and Propagation*, Vol. AP-17, No. 5, 1969, pp. 568–576.

[10] Johnson, R. C., and R. J. Poinsett, "Compact Antenna Range Techniques," USAF Technical Report No. RADC-TR-66-15, Griffiss Air Force Base, New York, April 1966.

[11] Crysdale, J. H., "An Optical System for Antenna Measurements at Microwave Frequencies," A Thesis submitted to the faculty of Graduate Studies and Research, McGill University, Ottawa, Ontario, April 20, 1953.

[12] Johnson, R. C., H. A. Ecker, and J. S. Hollis, "Determination of Far-Field Antenna Patterns from Near-Field Measurements," *Proceedings of the IEEE*, Vol. 61, No. 12, Dec. 1973, pp. 1668–1694.

[13] Lee, T. H., and W. D. Burnside, "Performance Trade-off Between Serrated Edge and Blended Rolled Edge Compact Range Reflectors," *IEEE Transactions on Antennas and Propagation*, Vol. 44, No. 1, 1996, pp. 87–96.

[14] Pape, J. H., "Evaluation of the Compact Antenna Range for Millimeter Wave Antenna Measurements," in *AMTA 1980*, 1980.

[15] Johnson, R. C., and D. W. Hess, "Performance of a Compact Antenna Range," in *IEEE Antennas and Propagation International Symposium*, June 1975, Urbana, Illinois, pp. 349–352.

[16] Baggett, M., interviewee, private communication, [interview], 10 February 2019.

[17] Liu, K., "Termination of Edges of a Reflector in a Compact Range." U.S. Patent 201000109932A1, 6 May 2010.

[18] Stuzman, W. L., and G. A. Thiele, "Cross-Polarizations and Scanning Properties of Reflector Antennas," in *Antenna Theory and Design* (Second Edition), New York, NY: John Wiley & Sons, Inc., 1998, pp. 338–342.

[19] Joy, E. B., and R. E. Wilson, "Shaped Edge Serrations for Improved Compact Range Performance," in *Proceedings of the Antenna Measurement Techniques Association*, 1987.

[20] Rodriguez, V., "Basic Rules for Indoor Anechoic Chamber Design," *IEEE Antennas and Propagation Magazine*, Vol. 58, No. 6, December 2016, pp. 82–93.

[21] Brumley, S., "A Modeling Technique for Predicting Anechoic Chamber RCS Background Levels," in *Proceedings of the Antenna Measurement Techniques Association*, Seattle, WA, September 1987.

[22] Bickmore, R. W., and R. C. Hansen, "Antenna Power Densities in the Fresnel Region," *Proceedings of the IRE*, Vol. 47, No. 12, December 1959, pp. 2119–2120.

[23] Saad, E. T. S., R. C. Hansen, and G. J. Wheeler, "Power Density in the Near-Field Normalized to unity at 2DD/lambda," in *Microwave Engineers' Handbook, Volume Two*, Norwood, Massachussets: Artech House, 1971, p. 34.

[24] Baggett, M., and D. Hess, "Power Handling Considerations in a Compact Range," in *35th Annual Meeting and Symposium of Antenna Measurement Techniques Association*, Columbus, Ohio, Oct. 2013.

[25] Hess, D., "Note on Power Densities in Point Source Compact Ranges" (internal memorandum), Suwanee, GA: MI Technologies, June 2012.

[26] Stutzman, W. L., and G. A. Thiele, Equation (7.77) in *Antenna Theory and Design* (Second Edition), New York, NY: John Wiley & Sons, Inc., 1998, p. 294.

[27] Blalock, S., and J. Fordham, "Estimating Measurement Uncertainties in Compact Range Antenna Measurements," in *Annual Meeting and Symposium of the Antenna Measurement Techniques Association (AMTA 2015)*, Long Beach, CA, November 2015.

[28] IEEE, "IEEE Std 1720™-2012 Recommended Practice for Near-Field. Antenna Measurements," IEEE, April 2012.

[29] Wayne, D., J. A. Fordham, and J. McKenna, "Effects of a Non-ideal Plane Wave on Compact Range Measurements," in *36th Annual Symposium of the Antenna Measurement Techniques Association (AMTA 2014)*, Tucson, AZ, 2014.

CHAPTER 8

Anechoic Ranges for RCS Measurements

8.1 Introduction

This chapter revisits the RF absorber, studies backscattering, and investigates the RCS per unit area (σ°) of the absorber. After discussing the backscattering of the absorber, the chapter examines the size and absorber treatment of RCS ranges and covers a method of estimating the background RCS of an RCS range. The chapter concentrates on rectangular RCS ranges, illuminated using a parabolic reflector, like the ones discussed in Chapter 7. The chapter concludes by once again examining the tapered range, this time to explore its use as a low-frequency RCS measurement range.

RCS measurement facilities are a totally different type of range. They are different from antenna pattern measurement ranges, in which an object transmits a signal (intentionally or unintentionally) that is received by a range antenna, and from EMC ranges, in which there is a range antenna that transmits and an antenna that receives that signal. Even in immunity EMC ranges, a signal is generated by an antenna, and the EUT receives that signal (the amount of which, one hopes, is small enough that the device continues its normal operation).

In RCS chambers, the range antenna transmits and receives the scattered field radiated by the device under test, commonly known as the target. Anechoic chambers for measuring RCS were under development as early as 1957 during improvements to the U-2 aircraft, according to Lovick [1], who was involved in developing them.

The key to RCS ranges is to eliminate the background RCS of the range, or make sure that the walls of the range do not reflect radar signals back to the range antenna. If too high of a signal is reflected back, it may not be possible to measure the true RCS of the target. This so-called background RCS of the range is a critical parameter when designing RCS ranges.

Chapter 3, which mainly discusses the bistatic reflectivity of the RF absorber, mentions that backscattering only occurs for a pyramidal absorber when the tips are spaced by one wavelength and introduces wedge absorbers as the solution to this problem. Wedge absorbers, however, do have a level of backscattering, which we will discuss in this chapter.

8.2 Absorbers Revisited

The different way that RCS measurements are performed calls for a study of RF absorbers from a different point of view. While reducing bistatic reflectivity (so that it does not affect the illumination of the target) is important, it is more important to reduce what is coming back toward the range antenna that is used to measure the RCS.

8.2.1 RCS and RCS Per Unit Area

RCS is an equivalent area. The definition provided in [2] is that "the radar cross-section of a target is the projected area of a metal sphere that would scatter the same power in the same direction that the target does." In defining RCS and scattering cross-section, the IEEE [3] states the following:

> "For a scattering object and an incident plane wave of a given frequency, polarization, and directions, an area that, when multiplied by the power flux density of the incident wave would yield sufficient power that could produce by isotropic radiation, the same radiation intensity as that in a given direction from the scattering object."

I prefer the IEEE definition. It implies that for a given incident wave with a given power density, multiplying that incident power density by some area yields a given power. If we let that power radiate isotropically, the resulting power density in a given direction is the same as that of the scattered field from the target. This explanation of the definition is also described in [2]. Imagine that we have an incident wave with a given power density in watts/meter^2. Let us call that incident power density ρ_i. Per the definition, if we multiply that power density by some area that we call RCS, and that typically is denoted by the Greek letter σ, we get some power P that if we let radiate isotropically will provide the power density $\rho_s = P/4\pi R^2$. Hence,

$$\rho_s = \frac{\rho_i \sigma}{4\pi R^2} \tag{8.1}$$

or

$$\sigma = 4\pi R^2 \frac{\rho_s}{\rho_i} \tag{8.2}$$

which should be recognized as the typical RCS equation. To make it independent of the distance R, the far field is assumed. Thus, (8.2) is typically shown as

$$\sigma = \lim_{R \to \infty} 4\pi R^2 \frac{\rho_s}{\rho_i} \tag{8.3}$$

It should be noted that the area σ does not necessarily relate to a physical size, and physically large objects can have very small RCS (think stealth aircraft).

Imagine now a large physical area. Let us assume that we illuminate only a portion A of that large physical homogeneous area. We can define a RCS per unit area $\sigma°$ [4], which allows us to obtain the RCS for a given portion of a homogeneous surface S by using the relationship that the RCS is $\sigma° S$, where S is the area of the large homogeneous surface that we are illuminating or interested in finding its RCS contribution.

Hence, we can look at a large wall of absorber as one of these homogeneous surfaces and compute the $\sigma°$ of the absorber treatment. When the treatment is used on one of the surfaces of an indoor RCS range, we can then estimate the area of the surface illuminated by the range antenna or the compact range reflector and obtain the RCS from that absorber surface.

8.2.2 RCS of Absorbers

It is possible to measure the RCS from an absorber. Indeed, in [5], Brumley and Droste present measured results for different types of absorbers. One of the issues when measuring and computing the $\sigma°$ of the absorber is that we want to avoid the effects of the edges of the sample, since those edges are not illuminated on a large absorber-treated surface. As shown in Figure 8.1, for a sample being measured or computed, we have some excitation that illuminates the sample of the geometry in the computational domain. Part of the reflected fields (used to compute the RCS of the absorber) come from the sides of the absorber sample. These returns are not seen on a large treatment, such as a large wall of absorber, so they should not be used to obtain the $\sigma°$. Indeed Brumley and Droste [5] show that postprocessing was used on their measurements, on the inverse synthetic aperture radar (ISAR) images that were obtained to exclude the edges from the RCS measurement. Similarly, when obtaining $\sigma°$ through computational methods the effects of the edges can be avoided by simulating the absorber as a unit cell surrounded by periodic boundaries [6]. The periodic boundaries model the absorber as an infinite wall treatment, and the scattered fields radiating out of the top surface are used to compute the $\sigma°$. This approach with periodic boundaries is available through numerous commercial packages. Figure 8.2 illustrates this approach where a single pyramid can be modeled, thus reducing the overall computational domain. The results, however, represent the reflected field from a portion of a large surface treated with an absorber. The models, as in many cases with absorbers, provide a best-case scenario response. Brumley and Tanakaya demonstrated this by showing potential issues related to the manufacturing of RF absorber [7]. In [7], Brumley and Tanakaya showed that even rubber grabbers' marks can be seen in ISAR images. Rubber grabbers are basically

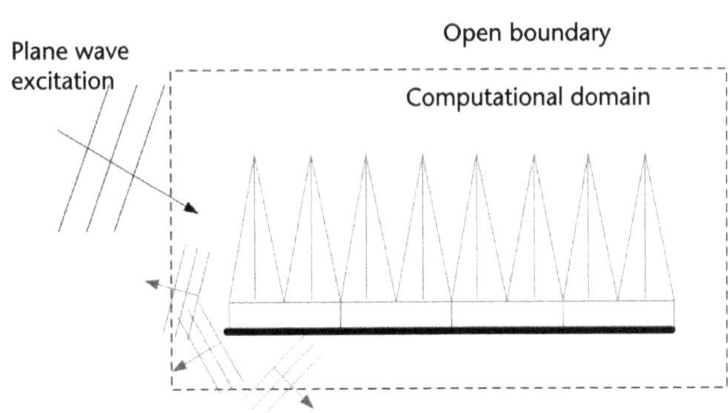

Figure 8.1 Computing and measuring RCS per unit area.

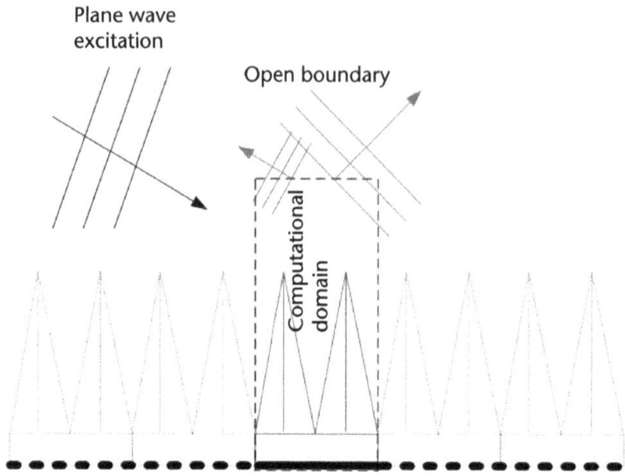

Figure 8.2 Computing the RCS per unit area using periodic boundaries.

treads on the rollers that move the foam through the impregnation tanks where the foam absorbs the carbon black. The rollers are used to move the foam blocks and to squeeze out the extra liquid from the foam. The rubber treads add more pressure to the foam in specific areas pushing out more of the lossy liquid thus creating areas with different permittivity. The ISAR images reported in [7] show the imprints of these roller features. The results also demonstrate the importance of a good installation, where the installers of the RF absorber take care to ensure that there are no gaps between pieces and that the wedges on two adjacent pieces are aligned.

8.2.3 Computed RCS Results for Some Absorber Types

Using a commercial software package with periodic boundary capabilities, I analyzed several absorber geometries. The results should provide readers with additional values to those that can be found in the literature, especially from the work by Brumley [5, 7, 8]. In addition, I use some of the values to show an example of an RCS background estimation for RCS ranges and to compare with measured data.

8.2.3.1 Normal Incidence

Most RCS ranges are rectangular footprint chambers. The first RCS ranges, like the ones described by Lovick, were of this "rectangular chamber" type [1]. There have been reports of taper ranges used for RCS measurement [9], but these are few. (Taper ranges are discussed in Section 8.6.) In either case, the main contributor to the RCS background of the range is the end wall. As discussed previously in Chapter 5 and Chapter 7, it may seem that simply treating the end wall with an absorber of adequate thickness is enough to achieve the desired results.

For example, Chapter 5 shows that it is sufficient for the end wall absorber to have normal incidence reflectivity that is equal or better to the desired QZ reflectivity to meet the design goals. However, RCS is a different animal. We are dealing with the power reflected, and that power is going to be related to the area illuminated and its geometry. That means more of a wall that is illuminated, the higher the RCS of that wall.

Consider a 24-inch RF pyramidal absorber. This type of pyramidal absorber has a base thickness of 4 inches (10.16 cm) and a pyramidal section that is 20 inches (50.8 cm) tall. Each pyramid has a square base that is 8 inches by 8 inches (20.3 cm by 20.3 cm). The manufacturer's typical reflectivity for pyramidal absorbers of this type in the range 2–6 GHz ranges from −45 dB at 2 GHz to −55 dB at 6 GHz. If the $\sigma°$ is computed, the results shown on Figure 8.3 are obtained.

These results do not seem to correspond to the reflectivity of the absorber. The reflectivity appears to increase, like $\sigma°$, with frequency, potentially due to material properties, but the RCS per unit area is more than 10 dB higher than the reflectivity.

Now, consider a plane wave–generating device, such as a lens or a compact range reflector, being used to illuminate the end wall of the chamber treated with the 24-inch absorber. What RCS levels are we measuring, assuming that this ideal device only illuminates a $4m^2$-section of the back wall? This assumption, of course, is not realistic, but let us imagine a scenario in which such a reflector and feed exist. The RCS from the end wall is $\sigma°S$, where S is $4m^2$ or, expressed in decibels, 6 dBsm, so the RCS of that illuminated area of absorber is −40.87 dBsm at 2 GHz, and as high as −28.8 dBsm at 5.75 GHz. That is, it would not be possible to measure targets with RCS levels lower than −28.8 dBsm at 5.75 GHz.

On a side note, some absorber manufacturers have proposed the use of RCS ranges to measure the reflectivity of absorbers as the reduction of RCS for a given conductive flat plate when covered by the absorber being measured. For a PEC plate that is $4m^2$ at 2 GHz (λ_{2GHz}= 0.15m), the RCS is given by: σ (dB) = $10 \cdot \log_{10}(4\pi A^2/\lambda^2) = 39.5$ dBsm. For the same $4m^2$ plate at 5.75 GHz ($\lambda_{5.75GHz}$ = 0.0521m), the RCS is 48.7 dBsm. Thus, the absorber has reduced the RCS of the plate by 80.38 dB at 2 GHz and by 77.5 dB at 5.75 GHz. Clearly, these values are not related to the reflectivity of the absorber, and these absorber measurement methods need to be revised.

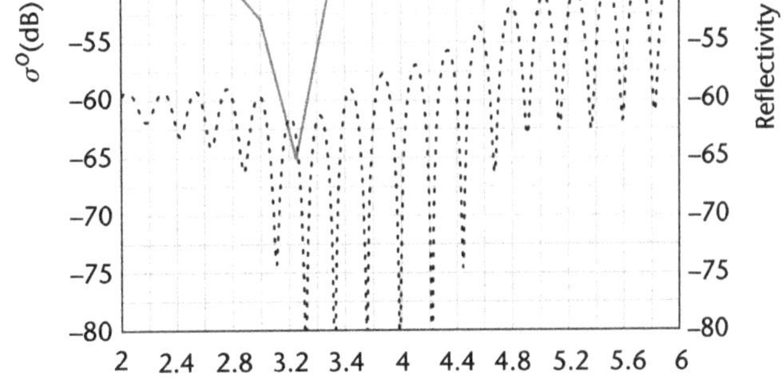

Figure 8.3 RCS per unit area for a wall treated with 24-inch pyramidal absorber.

It could be argued, that by using the periodic boundaries in the numerical models, there is an error on the RCS per unit area of the absorber, but we can use the same method to calculate a single PEC plate. A model is created using a flat plate that is 2 ft. by 2 ft. (0.61m by 0.61m). The lateral walls of the model are periodic boundaries, and the incident wave is normal to the plate. When calculating the scattered field pattern, only the fields on the top surface are used. The lateral surfaces (the periodic boundaries) are ignored, and so is the bottom surface, which is a PEC boundary. The calculated RCS for that 2 ft. by 2 ft. plate is compared to the results from the equation for the RCS of a plate and plotted in Figure 8.4 and validates the numerical approach.

Figure 8.5 shows the RCS per unit area for a few absorber geometries that have been computed. The difference in performance between the standard pyramid and the twisted pyramid (see Figure 8.6) is remarkable. It should be noted that the material properties used on those two computations are the same, and the results agree with what is reported by Brumley [10]. He states that the presence of points, rather than edges to the incident wave on the twisted pyramid geometry, provides a better RCS, although he did not provide measured data.

8.2.3.2 Oblique Incidence Wedge Absorber

Typically, because of the construction of RCS ranges, the incidence onto the lateral surfaces is between 65° and 85°. We can analyze the RCS per unit area for the monostatic return of a large (theoretically infinite) wall of wedge absorber treatment. The results can be surprising if we look at two different sizes of absorber. For example, looking at 36-inch-tall wedge and comparing it to 48-inch wedge, we see that the monostatic RCS per unit area is very similar in both geometries (see Figure 8.7).

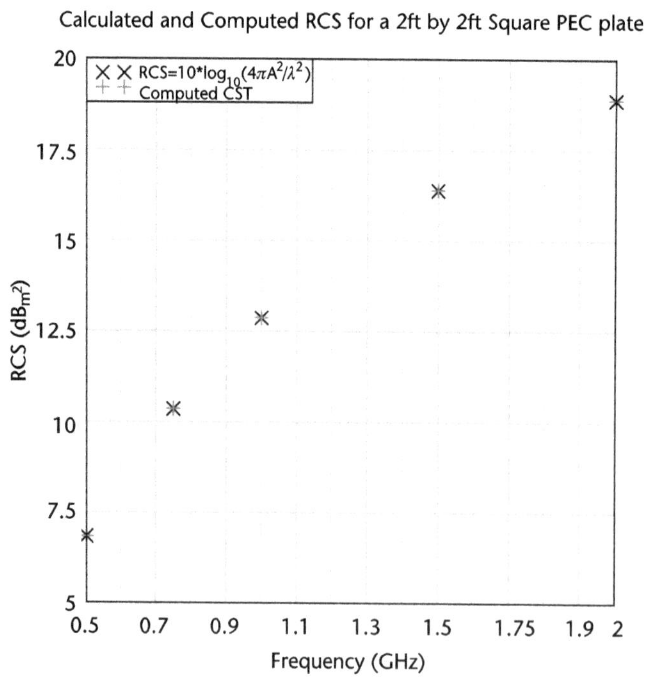

Figure 8.4 Comparison of computed and calculated RCS.

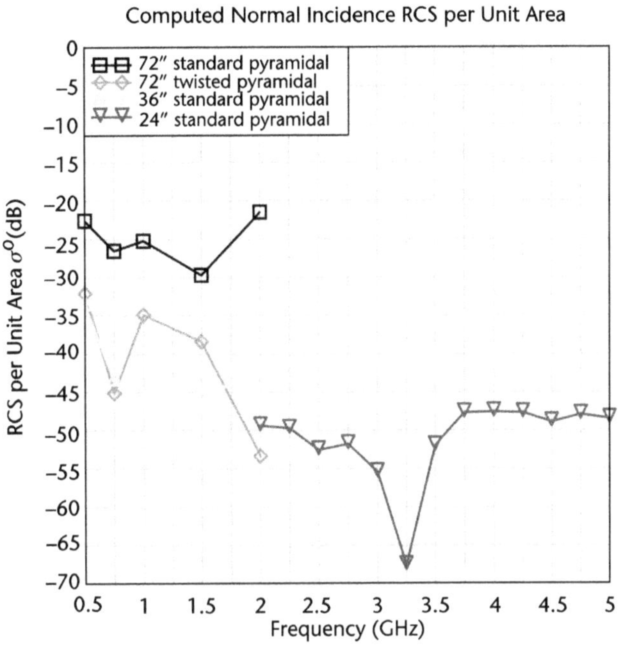

Figure 8.5 Twisted and traditional pyramid geometries.

Figure 8.7 shows the two polarizations of the incident field. The perpendicular polarization indicates that the electric field is perpendicular to the plane that contains the incident field and reflected field propagation direction. The parallel polarization is when the electric field is parallel to the plane that contains the incident and reflected field direction.

Figure 8.8 shows the two principal polarizations of the incident field for two different sizes of wedge absorber. While the similarity in results for the two different absorbers may seem odd, the reason that the RCS for the two types of absorbers is very similar (except as we move toward higher frequencies) is that the geometries are quite similar. Both absorbers have a wedge shape, and both have an angle of wedge that is very similar (17.06° for the 48-inch wedge versus 22.62° for the 36-inch wedge). The interesting conclusion from these results is that the larger absorber, with a 33% increase in height, does not necessarily improve the RCS response of the material, at least at the frequencies analyzed (shown in Figure 8.8). Figure 8.9 shows the results for a 36-inch wedge and an 18-inch wedge absorber from 2 to 5

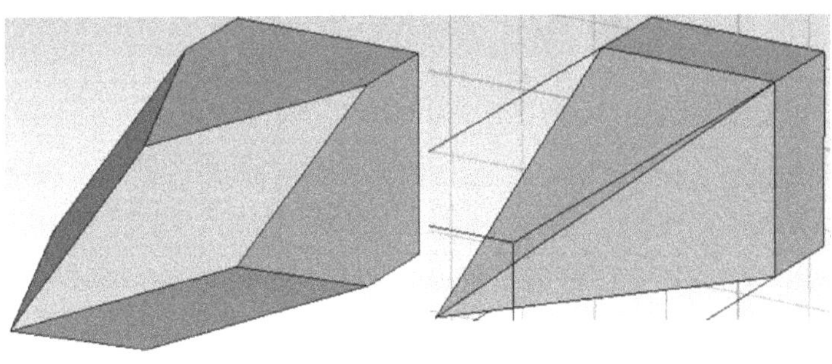

Figure 8.6 RCS per unit area for different pyramidal absorbers.

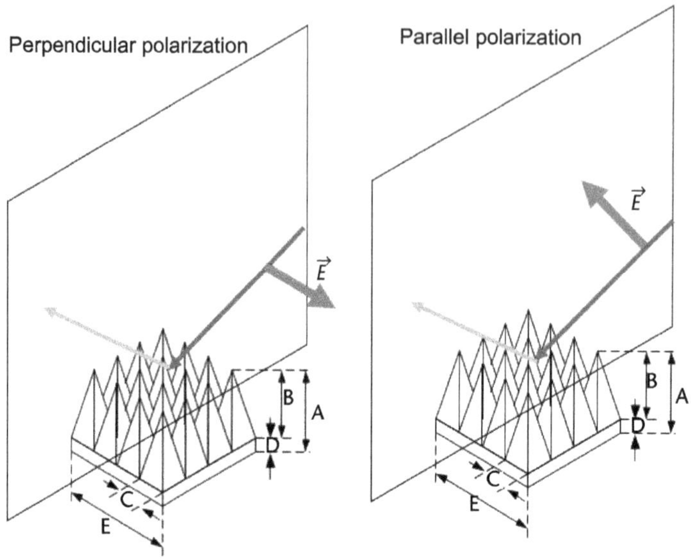

Figure 8.7 Principal polarizations of the incident field.

GHz for different angles of incidence, and it is clear that the longer absorber (50% increase) is better, but not necessarily at the lowest frequency.

As a comparison, the RCS per unit area versus frequency for a 75-degree incidence for different absorbers is plotted in Figure 8.10. According to the data, the trend is for the RCS per unit area to improve with frequency. Notice that the peaks of the plots for the different absorbers improve with frequency. Notice also that potentially for some angles, a shorter absorber size may be better than a longer absorber size at a given frequency.

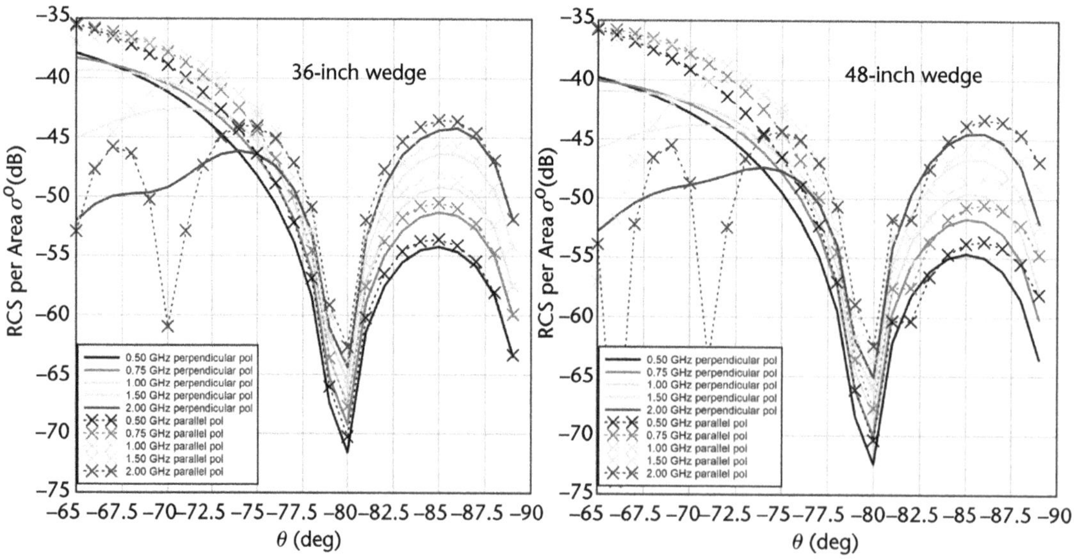

Figure 8.8 Monostatic RCS per unit area for a 36-inch wedge and 48-inch wedge at different frequencies.

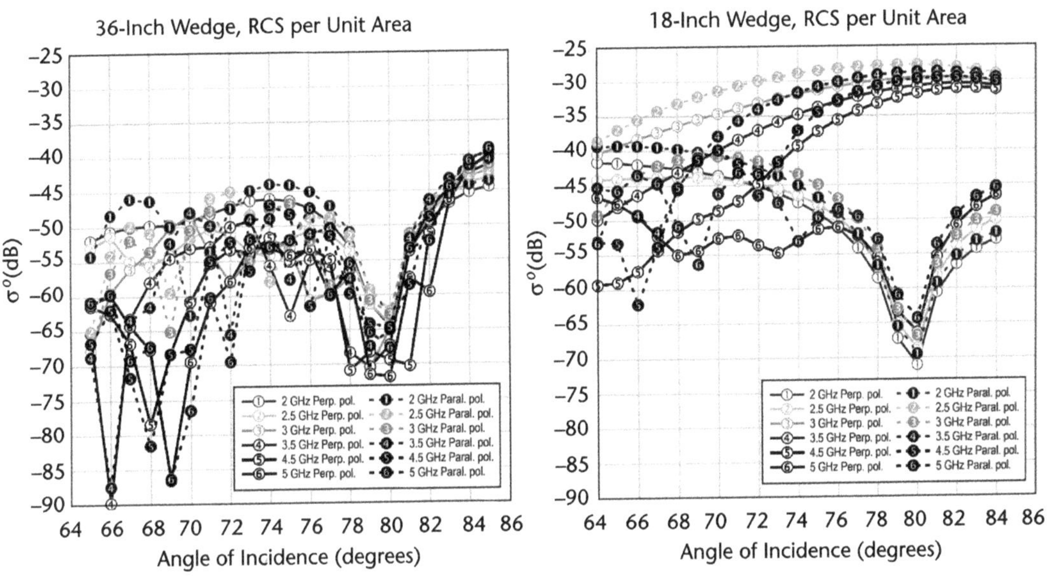

Figure 8.9 Monostatic RCS per unit area for an 18-inch wedge and a 36-inch wedge from 2 to 5 GHz and for angles from 65 degrees to 84 degrees.

Thus, while in general the longer the RF absorber the better (and this is the case for the bistatic reflectivity as described in Chapter 3), sometimes this is not the case for some frequency bands for RCS. Unfortunately, there is no approximation that can be used, and manufacturers do not measure the RCS of absorbers, although approaches to measure it are described in the literature [8]. Computational

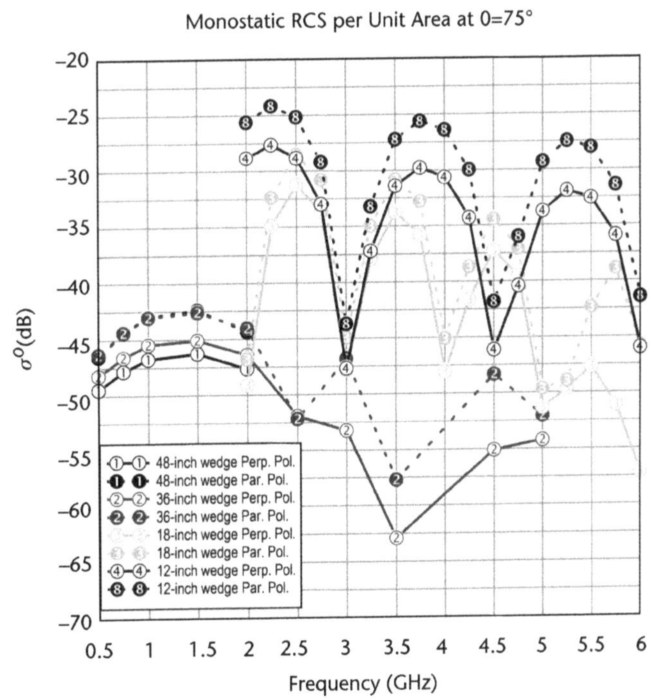

Figure 8.10 Different wedge absorber sizes versus frequency for a wave incident at 75 degrees.

electromagnetic methods can be used, and several commercial computational electromagnetics packages have the capability to perform the computation (i.e., periodic boundaries and oblique incidence), but they still require knowledge of the material properties of the absorber material to allow for a good computation.

8.3 Sizing the Chamber

For RCS chambers, bigger is better. The larger that you can make the room the better off you are going to be. The goal is to minimize the background RCS from the walls, and placing the walls as far away as possible is the best way to reduce their effects. There are postprocessing schemes that can be used to reduce the clutter or chamber background [11, 12], and there are ways to calibrate the ranges and obtain estimates of the uncertainty [13]. However, a larger chamber makes these techniques easier to apply since it is possible to do time-domain gating to exclude returns from those surfaces. It may reduce the clutter to levels that are not an issue since they will be smaller than the expected targets.

RCS measurements require some type of radar system where pulses are transmitted and then the system switches to receive mode. The width of the pulse (PW) drives the chamber size [14]. The QZ will be at least one pulse width; that is, the radar system is specified to provide a pulse narrow enough to encompass the QZ. This rule is a generalization, and if we take a look at the typical pulse, like the one shown in Figure 8.11, we see that radar manufacturers denote the pulse width as the −3-dB point to −3-dB point. The rise time is from the −10-dB point to the −0.46-dB point (10% to 90%), and the fall time is from the −0.46-dB down to the −10-dB level. Pulses may have ringing due to the fast rise time, and it is desirable to avoid using that part of the pulse when measuring the RCS. A desirable approach is for the pulse to be twice the QZ size, hence a 10-ft (3-m) QZ requires a 20-ns pulse It is important to note that typical rise and fall times may be in the range of 1–2 ns. It does not seem like much, but that is about 2 ft. (61 cm) of pulse length. As Figure 8.11 shows, the rise and fall are down to −10 dB, so it may be wise to add as much

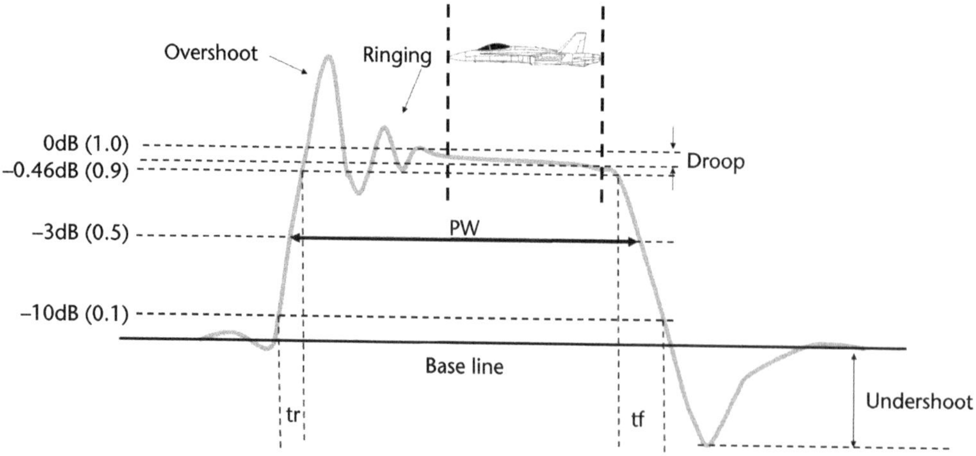

Figure 8.11 Anatomy of a typical pulse envelope.

as 4 ns to the nominal –3-dB to –3-dB PW specified by the radar manufacturer [14] to ensure that no significant energy is present.

8.3.1 RCS Measurement Systems

Before continuing with sizing the range, we need to understand the typical systems used in RCS measurements. The range size and the background levels are linked to the system. The most common architectures are CW measurements, pulsed CW measurements, and short pulse.

8.3.1.1 CW Measurements

The earliest forms of RCS measurements were performed using CW systems, partly because the same equipment used for antenna pattern measurements could be used [14]. These systems can be used with a circulator as an isolating device or with two antennas, a transmit antenna and a receive antenna. The two-antenna architecture should easily remind the reader of an antenna pattern measurement system where we have a source, a receiver, a range antenna, and an AUT. Figure 8.12 shows a CW system with a circulator and CW system using two antennas.

Looking at Figure 8.12, it would appear that the single-antenna system is better since it avoids the spillover and the potential RF leakage related to the cross-talk between cables, and since these two antennas are located as close as possible.

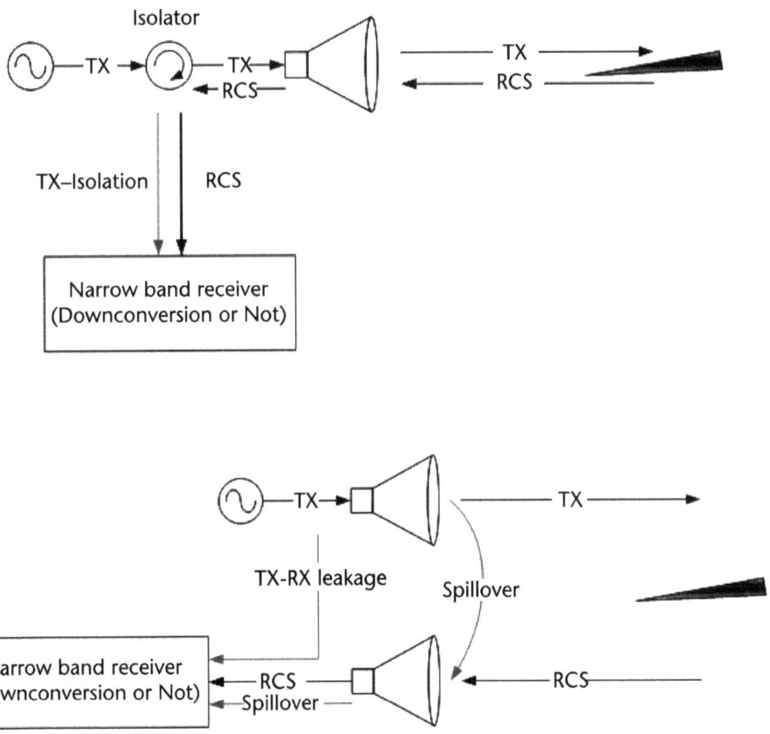

Figure 8.12 Two CW RCS measurement system configurations.

However, circulators or other types of isolators, may only provide 30 dB of isolation. With proper cables and torqueing of connectors, the RF leakage can be reduced to levels as low as −80 dB, which is a significant improvement over isolators. The spillover between the antennas can be a more significant problem. Separating the antennas too much can lead to a bistatic RCS rather than a monostatic RCS measurement. A RF absorber baffle is inserted between the antennas to reduce the spillover to levels that may be better than the 30-dB isolation that the typical isolator can provide. Postprocessing can help in reducing the effects of the spillover. This requires measuring a large number of frequencies, and then, through an inverse fast Fourier transform (FFT) converting the data to time domain. Then in the time domain the spillover and some of the clutter can be gated out. Be aware that some areas of the chamber that fall in the same time window as the target cannot be eliminated; those areas are the ones that we must take into account when estimating the RCS background clutter of the RCS range.

The main limitation of CW RCS systems is the dynamic range; other limitations include windowing effects used with the FFTs in the postprocessing.

8.3.1.2 Gated IF or Short-Pulse Measurements

To overcome some of the limitations of CW systems based on antenna measurement equipment, dedicated RCS systems were developed [14]. These systems have been called "short-pulse" or "gated IF systems." Figure 8.13 shows a typical architecture for such a "short-pulse" system.

The IF frequency of these systems is very high, typically four or five times the matched pulse bandwidth. The matched pulse bandwidth is given by the inverse of the pulse length, so a 5-ns pulse will yield a 200-MHz matched bandwidth. The IF pulse is converted to the desired frequency and then amplified. The "noise-blanking" switch right before the antenna is open for a period of time only slightly

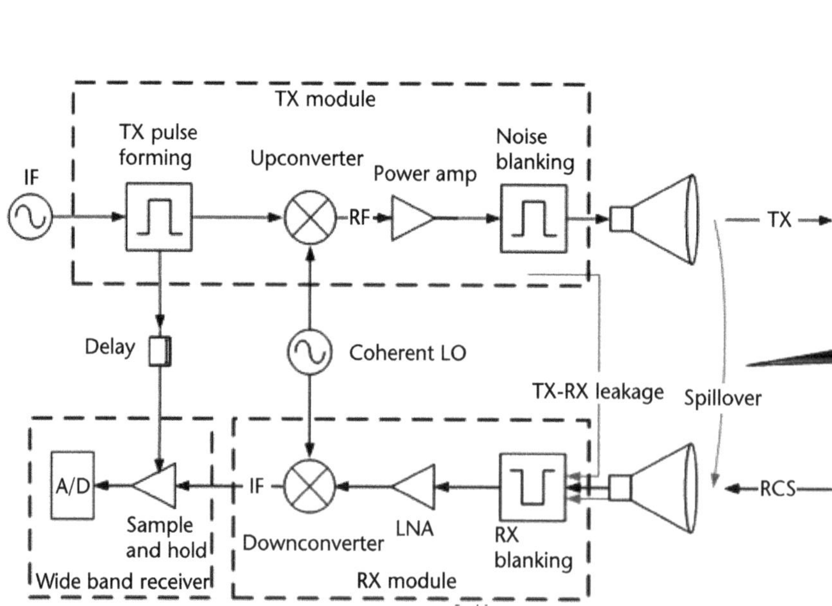

Figure 8.13 Two CW RCS measurement system configurations.

longer than the radar pulse. This switch reduces the spillover coupling when the system is on listening mode.

Analogous to the "noise-blanking" switch is the "RX blanking switch." This one is set to only allow signals to pass for the time that the expected RCS return should be present. This switch also stops the receiver from listening to the transmitted pulse. As shown in Figure 8.13, a "sample and hold amplifier" is triggered to freeze the signal based on the delay time from the TX pulse until the expected arrival of the RCS pulse. Very fast A/D units with conversion rates of 125 MHz or more are used to digitize the IF signal. Note that the system only reads data for a very short time during the arrival of the pulse, effectively "gating out" the clutter returns that are later in time, such as the end wall [14] or earlier in time such as the reflector flash.

8.3.1.3 Gated CW Systems

An in-between approach can be found in the gated CW systems. Figure 8.14 shows this system architecture, which uses a standard narrowband receiver paired with what is called a "gating box." This gating box has a noise-blanking switch and a receive-blanking switch. This is a reduced-cost approach compared to the IF gated systems; however, since the switching is not being done at IF frequencies, the switches in a gating box generally produce longer pulses. Another issue with the gated CW systems is that the receiver is "listening" all the time, even when the receive-blanking switch is not letting signals through. This means that depending on the duty cycle of the pulses, the average signal measured by the receiver will be much lower than the peak. This translates into a smaller dynamic range [14].

From these architectures, we know that we can time-gate or switch-out some of the features of the range. Depending on the system approach, there will be more or less features that can be effectively gated out.

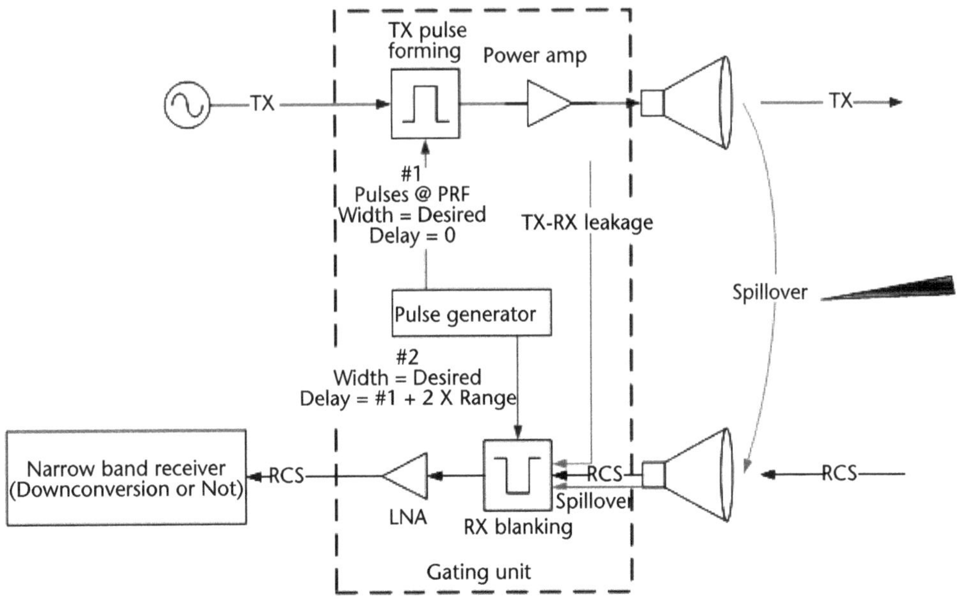

Figure 8.14 A gated CW RCS system architecture.

8.3.2 Range Length

While there are different approaches to measuring RCS, the most common is the use of a compact range reflector; hence we concentrate on sizing a room based on the use of a compact range. Based on that assumption we start with the length of the range, which may be related to the pulse length. Consider a pulse traveling in the range toward a target. The target is inside a QZ, and a certain distance behind is the end wall of the range. Much like the Compact Range (CR) for antenna measurements described in Chapter 7, space is needed for the reflector and its structure. Then, as shown in Figure 8.15, there is the space required for the focal length of the reflector.

Here we start seeing some deviations compared to the antenna measurement compact range. Chapter 7 shows that a distance to the center of the QZ from the reflector of $5/3f_l$; however, in RCS ranges that rule should only be used if the range is physically large, as in the case of ranges operating down to 500 MHz. For other ranges the preferred rule is a full focal length, making it $2f_l$ from the reflector vertex to the center of the QZ. The QZ itself should be large enough to encompass the target.

Behind the QZ is the end wall. We want to minimize the effects of the back wall and, if possible, time-gate it out. A minimum safe distance is two-thirds the PW_{length}, which is defined as the distance traveled by the pulse wave in the time given by the length of the pulse. Given that the propagation is the speed of light, then a 20-ns pulse will have a PW_{length} = 20 ft or 6.09m. That is, $PW_{\text{length}} = PW \cdot c$.

Figure 8.16 shows the reason for the minimum $2/3PW_{\text{length}}$ for the separation to the absorber. As the pulse enters the QZ, we start having echo traveling back, and there are reflections from different areas of the QZ. For simplicity, Figure 8.16 shows the return pulses for the front and the back features of the target, but a series of returns from different areas of the target appear to come back as a continuous pulse. As the pulse continues to travel it will reach the end wall where it will be partially reflected. Choosing the two-thirds minimum ensures that the reflected pulses from the back of the QZ have totally cleared and that the reflection from the wall can be easily gated out from the target's reflected signals.

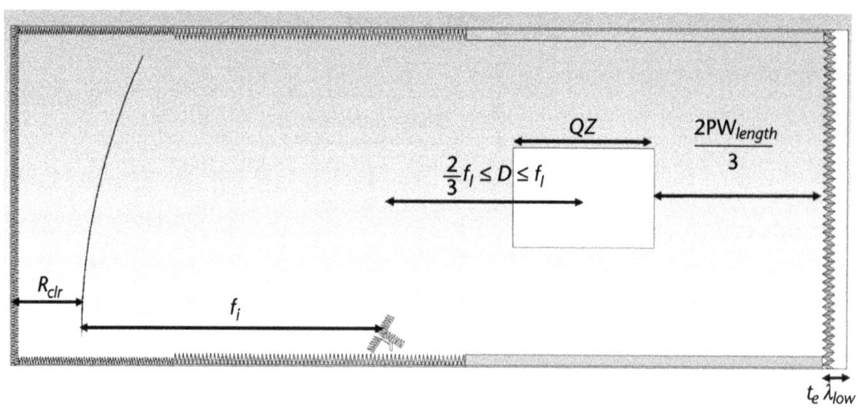

Figure 8.15 RCS range length.

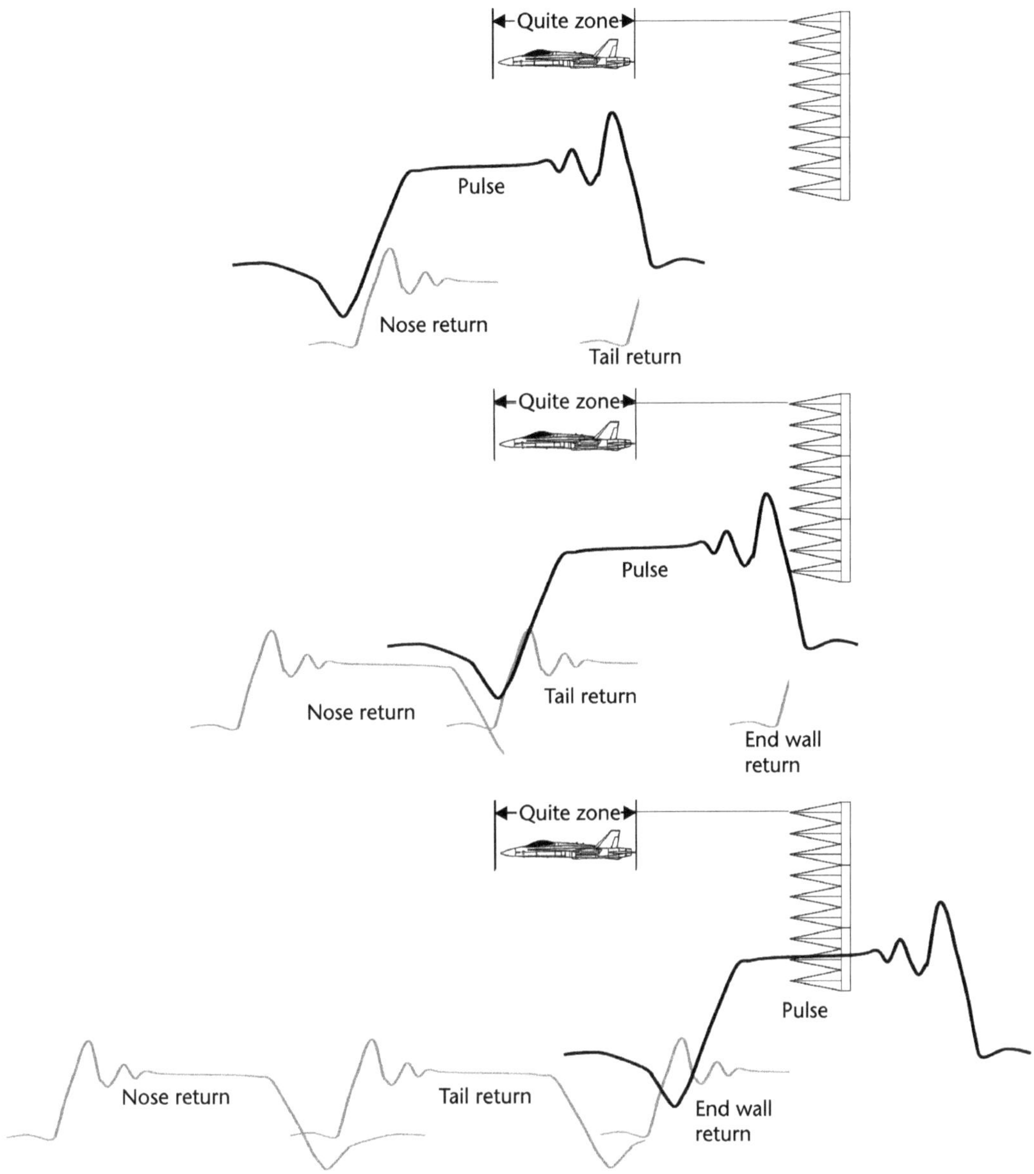

Figure 8.16 Pulse traveling in the RCS range.

Hence, the length is typically given by

$$L = R_{clr} + 2f_l + \frac{QZ}{2} + \frac{2PW_{\text{length}}}{3} t_e \lambda_{\text{low}} \tag{8.4}$$

The end wall absorber is typically 2λ at the lowest frequency, but it does not hurt to go as long as 3λ; hence, $2 \leq t_e \leq 3$.

If we analyze a chamber similar to the one shown in Figure 8.15 using a commercial FDTD simulation (the same one used in the simulation in Chapter 7), we can see the some of the different features on the range. The data of that simulation is shown in Figure 8.17 with a series of comments that show the focal length and the QZ. In this case the distance to the center of the QZ from the reflector is two focal lengths.

We have to remember that we are looking at the time from the excitation port and seeing what is received by the feed antenna. Hence, since the focal length is 75 ft. (22.86m) from the aperture of the feed antenna after 150 ns, we see what is called the "reflector flash." Part of that echo from the reflector will be reradiated by the source, and it will hit the reflector again creating a secondary reflector flash. The direct path to and from the target is approximately six focal lengths from the initial transmit time. The reflector reverberation will mingle with the returned target pulse on the third reverberation, which will be essentially reduced to a negligible value by that time. At 325 ns we seem to see part of this "reverberation" from the reflector and some other signals that are being received and may be arriving from other areas of the range. At six focal lengths in time is the QZ. Finally, 2.5 QZs later in time the end-wall reflection appears. Notice that the levels of the reflected signal from the end wall are significant.

8.3.3 Range Width

As is the case with the length, the width of the range can be related to the pulse width and the distance from the reflector to the target or QZ. Consider the sketch shown in Figure 8.18. The rule is that the reflected path length $A + B$ should be larger than $2f_l + PW_{\text{length}}$.

The reason for this is to separate the target return from the reflected target return by at least a pulse length. Hence the width is given by

$$W \geq 2\sqrt{f_l PW_{\text{length}} + \frac{PW_{\text{length}}^2}{4}} \tag{8.5}$$

The other constraint for the width is that is must be able to house the reflector and keep a distance from the edge treatment to the absorber tips; hence the width must also meet

$$W \geq R_w + (4 + t_s)\lambda_{lf} \tag{8.6}$$

where t_s is the electrical size of the side wall treatment, and λ_{lf} is the wavelength at the lowest frequency of operation of the range.

8.3.4 Range Height

For the range height similar rules apply as those concerning the width, but in the case of the height, we need to consider the feed positioner. Looking at the sketch in Figure 8.19 we see that the height is given by two constraints.

One constraint is the one given by the reflector size, or

$$H \geq R_h + (4 + t_{cf})\lambda_{lf} \tag{8.7}$$

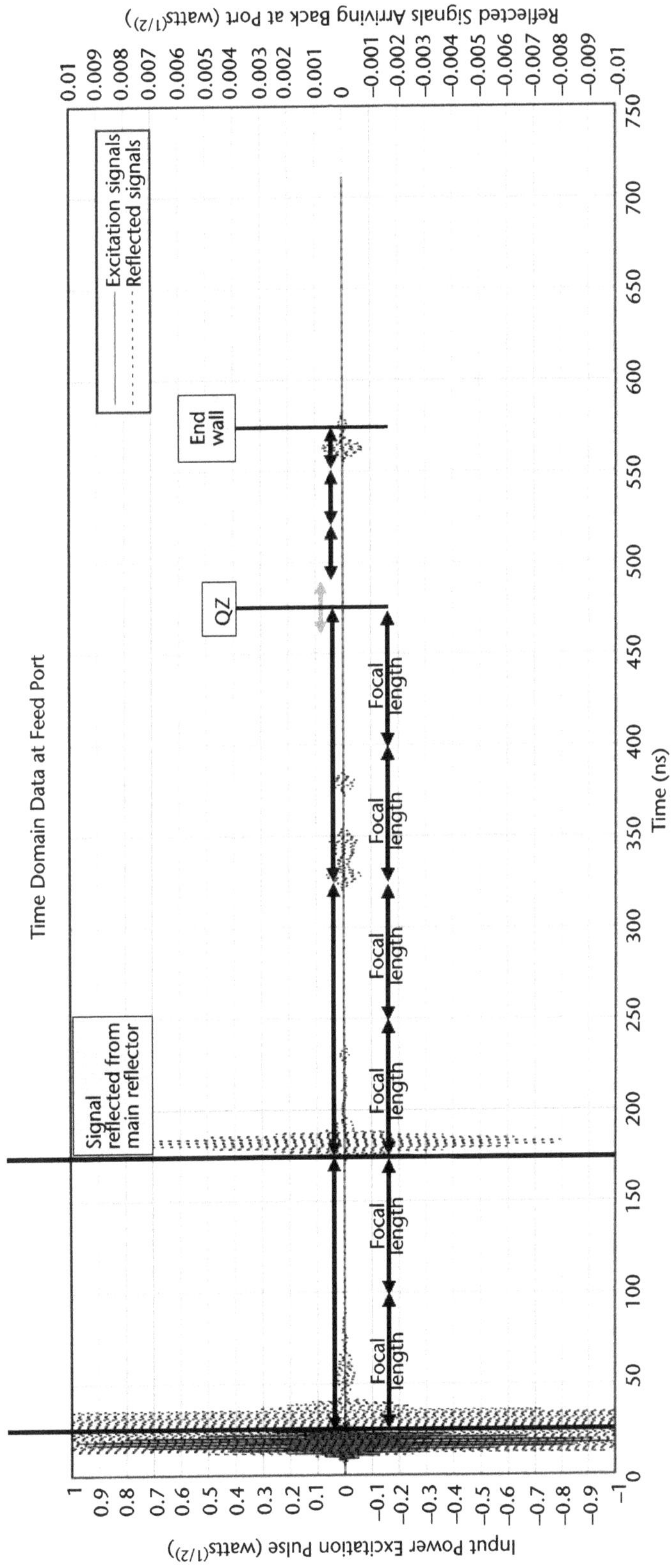

Figure 8.17 Time domain reflected signals for a compact range.

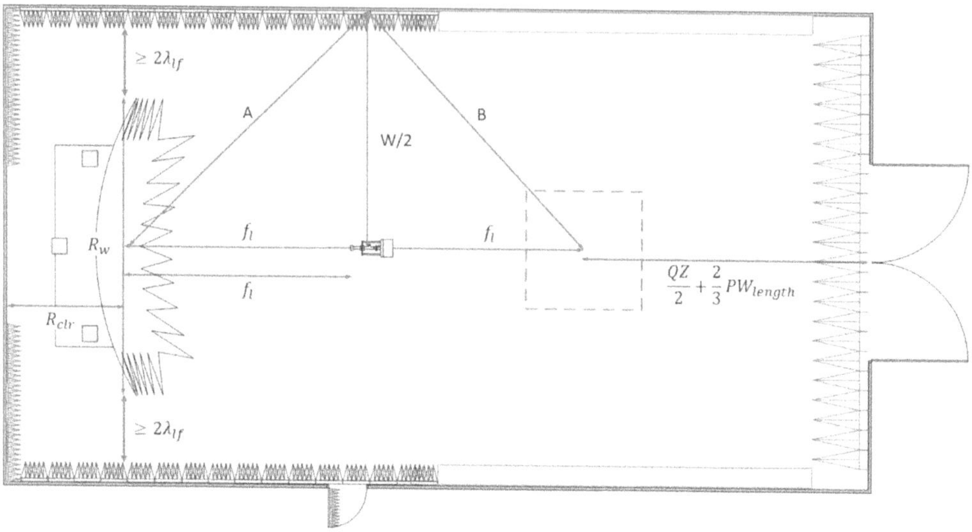

Figure 8.18 Width of a compact range for RCS.

where t_{cf} is the electrical size of the floor and ceiling absorber treatments. Equation (8.7) means that the range must be large enough to house the absorber and the reflector height and keep the absorber at least two wavelengths from the reflector at the lowest frequency.

Using the rules for the width, we can choose the distance above the range elevation (the height of the center of the QZ with respect to the range floor) to be given by

$$H_t \geq \sqrt{f_l PW_{\text{length}} + \frac{PW_{\text{length}}^2}{4}} \tag{8.8}$$

Then we have the feed-positioning system and the distance between the range elevation to be at least $H_B + FP_h \geq H_t$. The distance from the range elevation to the

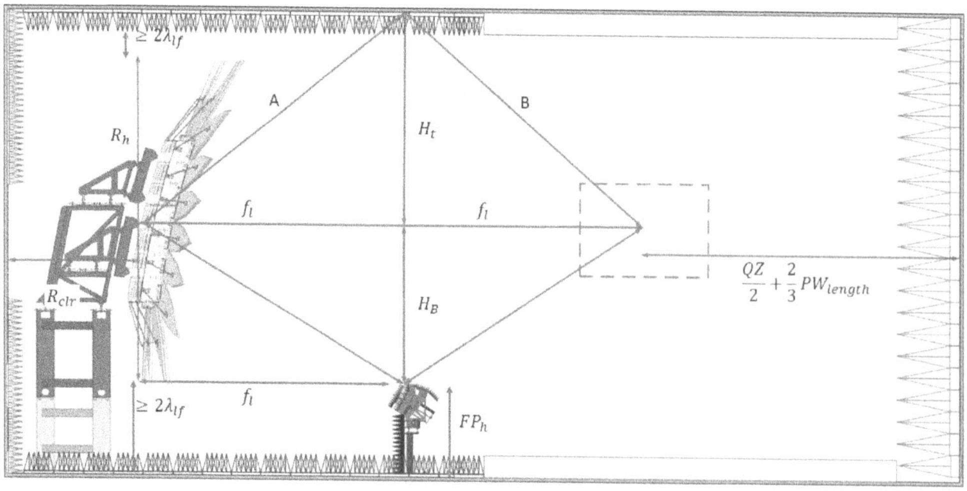

Figure 8.19 Height of a compact range for RCS.

top of the feed positioner is given by the reflector height and the location of the vertex, typically $H_B = R_h/2$. Hence, the height of the feed to the floor is given by

$$FP_h \geq \sqrt{f_l PW_{\text{length}} + \frac{PW_{\text{length}}^2}{4}} - \frac{R_h}{2} \tag{8.9}$$

It is desirable to have H_B as large as possible to minimize the effects of the reflection from the top of the feed positioner. This goal is another reason for the preference for a serrated-edge reflector versus a rolled edge in addition to the reasons specified in Section 7.4.4.

8.4 Absorber Layout

Now that we have roughly a geometry for the range, we can start looking at the absorber requirements. The absorber treatment must meet two goals. First, it must reduce the reflected signals into the QZ to minimize the ripple and second, to provide the best possible plane wave illumination. These are the same requirement that is needed in a compact range for antenna measurements (as discussed in Chapter 7).

8.4.1 Reducing Reflected Energy

Taking care of reflections in the RCS range follows the same approach as described in Chapter 7. Recall that the end wall absorber is typically a pyramidal absorber with an electrical size of at least two wavelengths at the lowest frequency or, if possible, three wavelengths. A treatment of that thickness ensures that the reflected energy is below a certain level.

For the lateral surfaces, the approach used in Chapter 7 for the lateral wall absorber should be followed. This reduces the bistatic reflected energy into the QZ and ensures that the plane-wave illumination is as close to a plane wave as possible.

Let us assume that we have a 6-ns pulse, and a 10-ft focal length. Per (8.5), the angle of incidence for the reflected ray that travels to the center of the QZ is $\theta = \tan^{-1}(f_l/H_t)$, where for the side walls H_t becomes $W/2$. Thus, the angle of incidence for this case is 50°. Typically, since the illumination of the side walls (floor and ceiling) is about −20 dB (the rule of thumb given in Chapter 7, although the actual value will depend on the reflector), bistatic reflectivity better than −20 dB at 50° is desired. Hence, about 1-wavelength-thick pyramidal absorber is a safe size.

As mentioned in Chapter 7, the end wall behind the reflector usually has the shortest absorber, and in some cases part of the end wall is left untreated. Following the rules given in Chapter 7, the electrical thickness of the pyramidal treatment is between three-quarters to one-half of a wavelength at the lowest frequency of interest.

8.4.2 Wedge and Pyramids

In addition for the lateral surfaces—the ceiling, the floor, and the side walls of the range—a combination of wedge absorber and pyramidal absorber is used. As shown

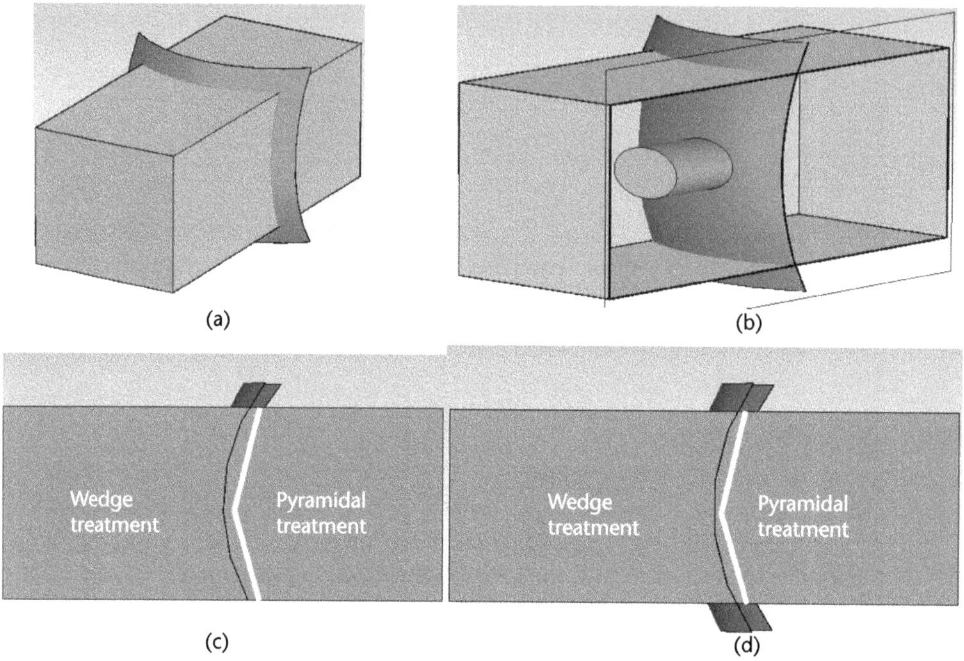

Figure 8.20 Defining the boundary between wedge and pyrimidal treatment on the RCS range. A spherical wavefront in the RCS range centered at the reflector and extending to the QZ: (a) the wavefront intersection with the walls of the range, (b) cut out showing the spherical wavefront touching the elliptical OZ, (c) boundary on the floor and ceiling, and (d) boundary on the lateral walls.

in [6], wedges have less backscattering than pyramids, but pyramids tend to have better bistatic reflectivity than the wedges.

Let us consider a spherical wavefront originating in the center of the reflector. Using that spherical wavefront and extending it to the edge of the QZ, the sphere surface should show the locations where the pulse radiated by the reflector arrives at the same instant in time (see Figure 8.20).

Looking at the intersection between the wavefront and the lateral surfaces, we can select a straight line that marks the boundary between the pyramidal treatment and the wedge treatment. The reflector is surrounded by the pyramidal TET. This way, the bistatic reflection into the QZ is absorbed by the pyramidal treatment.

Because the back scattering is always a concern, the pyramids are rotated 45 degrees to reduce the backscattering by exposing an edge rather than a flat surface to the incoming wave. Figure 8.21 shows the RCS per unit area for a 36-inch pyramidal absorber with the incoming wave incident onto the pyramid sides (ϕ = 90°) and also with the incident wave onto the edges of the pyramid (ϕ = 45°). The results are given for different angles of incidence (θ = 0° to θ = 65°) and for both 1 GHz and for 2 GHz. Notice that for θ > 35°, the incidence on the edge provides a much lower $\sigma°$.

Another interesting conclusion we can draw from Figure 8.21 is that at some frequencies, a slight change in the incidence angle can reduce the $\sigma°$ by more than 10 dB. This raises the question about tilting the end wall and the possible advantages of doing so. Tilting the wall has been used since the very beginning of RCS

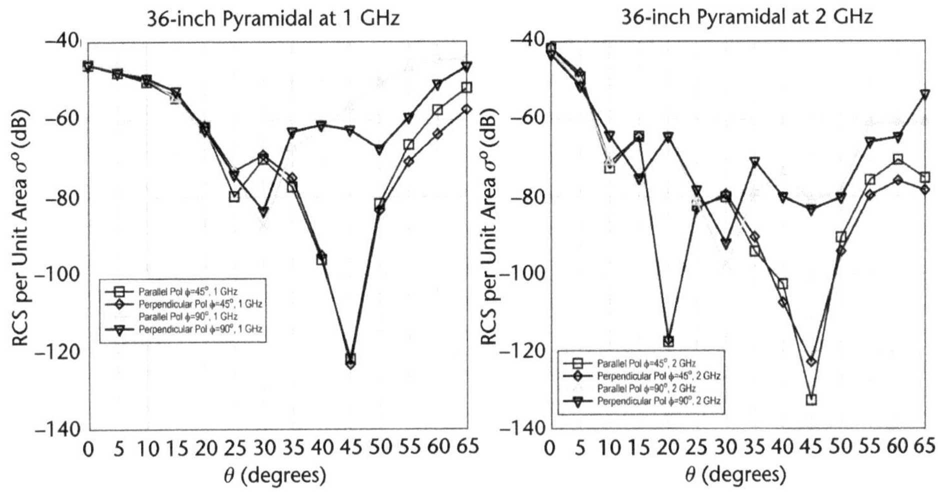

Figure 8.21 RCS per unit area for a 36-inch pyramidal absorber at 1 and 2 GHz for $\phi = 45°$ and $\phi = 90°$.

measurements. Lovick [1] mentions tilting the back wall as much as 45° and having an opening on the ceiling for the main backscattering to exit the range.

In some past research, tilting the wall did not appear to provide an advantage [15]. However, the work presented used a standard gain horn to illuminate a tilted wall of absorber and not a plane-wave illumination. The improvements shown in Figure 8.21 assume a plane wave incident onto an infinite wall of absorber; this is not the case presented in [15]. Thus the tilted wall was exposed to a series of plane waves with different angles of incidence and amplitudes. If a compact range reflector is used to illuminate the tilted end wall, an advantage can be seen.

Figure 8.22 shows the time domain of two identical reflector-feed systems with the same lateral surface treatments. The only difference is that one of the end

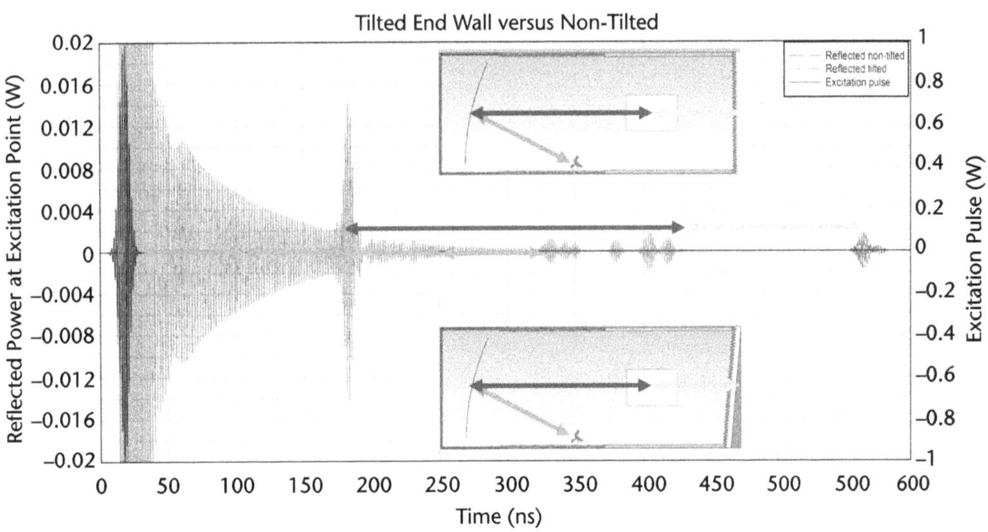

Figure 8.22 Tilted end wall when using a compact range to illuminate the range.

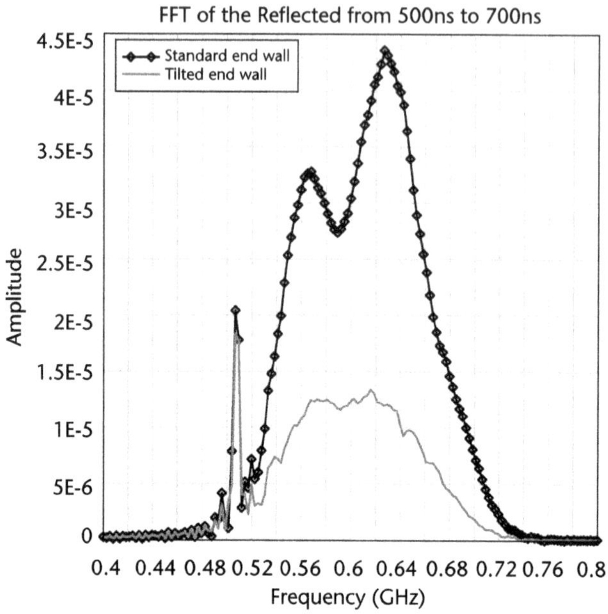

Figure 8.23 Fourier transform of the data from the tilted and standard wall.

walls is tilted 5°, and the other is a standard wall. The reflected signals in time are virtually identical for both ranges until we get to the end wall reflections where we see the tilted wall having a much smaller signal compared to the standard wall. A Fourier transform using FFT of the data is shown in Figure 8.23. Contrary to the results for a standard gain horn illumination [15], the compact range illumination showed an improvement for the tilted wall.

Tilted walls, however, do not completely eliminate the end wall reflection, and the cost of implementation can be significant. Pulsed systems paired with long ranges are a far better choice to eliminate the effects of the end wall.

Going back to the rotated pyramidal absorber installation and the pyramid/wedge boundary, the typical recommended layout for the absorber on a RCS range

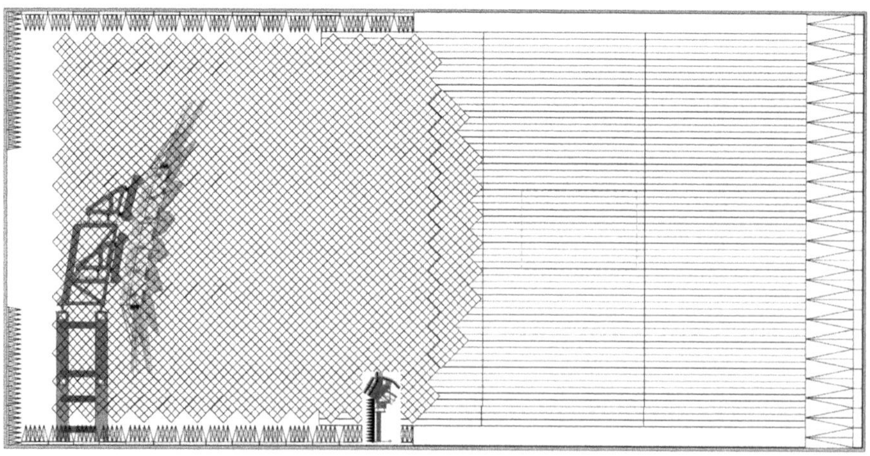

Figure 8.24 Typical absorber layout for a compact range for RCS.

is roughly as follows. The longest absorber is on the end wall opposite the reflector where the absorber thickness ranges from 3λ to 2.5λ at the lowest frequency. The lateral surface will have absorber of about 1.5–1λ. This absorber will be a mixture of wedge and of pyramids where the pyramids are rotated 45° during installation. The boundary between the pyramids and the wedge roughly follows the intersection of a spherical wavefront, as described in Figure 8.20. The wall behind the reflector has the shortest absorber, and it may be devoid of any anechoic treatment in the area right behind the reflector. This layout is illustrated in Figure 8.24. The ceiling and floor will have a similar layout. If possible, lights and other disruptions of the absorber layout should be minimized or moved toward the edges and behind the QZ.

8.5 Background RCS Estimation

In the same manner that the reflectivity on a QZ limits the level of sidelobes or the accuracy of the measurements in an indoor range, for a RCS indoor range, the background levels inside the range limit the level of RCS that can be measured. Brumley provides a general equation for the background RCS level in an indoor range [16]. For a given chamber, the background RCS is given by

$$\sigma_{\mathrm{BG}}\left(\vec{E}^i,\vec{r}'\right) = \left| \iiint_{\text{Chamber Surfaces}} A\left(\vec{E}^i,\hat{x},a(\vec{s})\right) \cdot h(\vec{s})\, dx\, dy\, dz \right|^2 \tag{8.10}$$

Brumley's equation is general, including the expected terms that affect the background RCS. The main term is the absorber backscattering $A(E^i, \hat{x}, a(s))$, which is a function of the incident field and the orientation and type of absorber $a(s)$. The other term in the integral is the effects of the time gating $h(s(x,y,z))$. It should be noted that both the backscatter from the absorber and the absorber type are functions of the vector $\vec{s}$, which defines the locations of the chamber surfaces. The incident field is a function of the illuminating reflector.

From Brumley's approach [16], we can perform an analysis to estimate the background RCS. The first step is to calculate the total surface area that contributes to the chamber background RCS. This basically includes those surfaces that are "seen" by the radar during the time of the target-scattering window and is a function of the radar-gating system as Brumley states in his equation [16]. This excludes the end wall, which is important, as we will show below.

The next item is to calculate the monostatic RCS per unit area of the absorber located in the areas of the walls that contribute to the RCS. Recall that this can be computed using numerical packages, or it can be measured [5].

Another task is to determine the two-way loss to those areas of interest. That is, when the reflector is illuminated, part of the energy arrives at the walls at a level lower than the main collimated beam. Similarly, the monostatic scattering from the absorber will be attenuated by the same amount before it returns to the feed. The reason for this is that the reflector (as discussed in Chapter 7) is an angular filter.

Analyzing the reflector geometry can provide information on the near-field illumination of the contributing surfaces. This can also be done with available numerical electromagnetic packages, such as GRASP, or by using the −16-−20 dB rule of thumb presented in Chapter 7.

From the results of these simulations or measurements we reduce Brumley's equation to a simpler form:

$$\sigma_{BG} = \sigma^{o} \cdot A \cdot FS \cdot W \quad (8.11)$$

where σ^{o} is the RCS/unit area of the candidate absorber, A is the surface area of the chamber walls, FS is the two-way field strength at the absorber walls, and W is a weighting function based on the illumination of the wall. The weight W is not necessary if we know exactly what the illumination of the walls is, but for estimating purposes it is a good tool.

To understand this, let us run through an example. Let us assume a very large RCS range with the following dimensions: 25m tall by 29.25m wide and 60m long. This range has a compact range reflector and a combination of 72-inch twisted absorber on the end wall and 36-inch wedge and pyramidal absorber on the lateral surfaces. The range is expected to operate down to 1 GHz. Let us assume that we must provide an estimate of the RCS background performance at both 1 and 2 GHz. The end wall is 731.25m^2 or 28.64 dBsm, and based on the performance of the 72-inch twisted pyramidal absorber, the RCS of the end wall is a large value of −6.28 dBsm at 1 GHz and −24.72 dBsm at 2 GHz. This, of course, assumes that the entire end wall is uniformly illuminated by a plane wave. However, the illumination from the reflector has a certain taper across.

A physical optics approach or other numerical technique can be used to find the reflector illumination in the near field at the lateral surfaces and at the end wall. Figure 8.25 illustrates the results for the reflector illuminating this example range. The results are obtained on two cuts: a vertical cut along the vertical symmetry plane of the range and a horizontal cut along the horizontal symmetry plane of the range. The results are copolarized in vertical and horizontal polarizations. This provides the one-way illumination from where the parameter FS in (8.11) can be obtained. If the field at a given point on the chamber is −20 dB down from the center part of the illumination, then the monostatic reflectivity from that point is also attenuated an additional 20 dB; hence, FS will be −40 dB for that point or area of the range.

From the plots in Figure 8.25, we can estimate that on average the FS is −30 dB at the lateral surfaces. That is, we average the one-way illumination at the lateral surfaces to be −15 dB.

Let us assume that the QZ or target area in this range is a horizontal elliptical cylinder, with the ellipse having a major axis of about 10m and a minor axis of 7m, with the major axis oriented on the horizontal. The length of this elliptical cylinder is along the axis of the range, and it is also 10-m-long. If the QZ is illuminated with the typical 1-dB amplitude taper described in Chapter 7, then the end wall will have a theoretical illumination that looks like the plot in Figure 8.26. Then an estimate for the weight factor W for the illumination can be obtained based on the average illumination of the entire end wall. The estimated W is −8 dB; hence, the end wall contribution drops to −14.28 dBsm at 1 GHz and −32.72 dBsm at 2GHz. These are still significant levels, and they provide an idea of the importance of the time gating of the end wall.

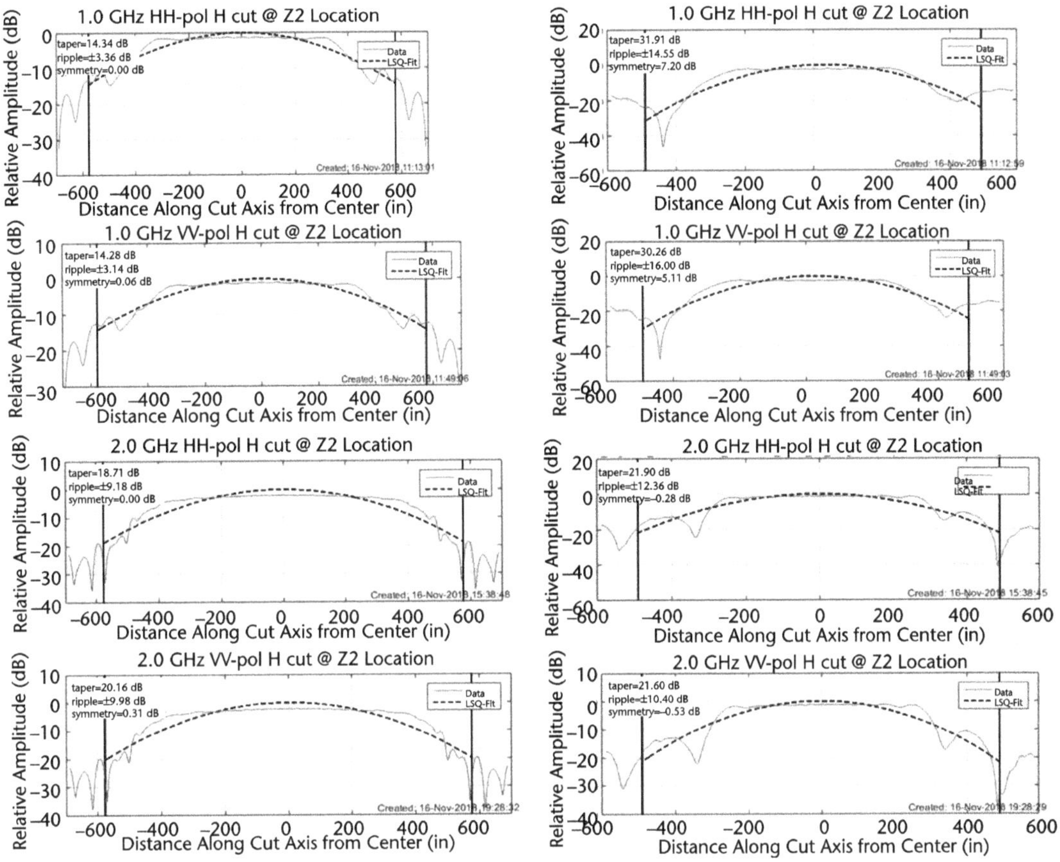

Figure 8.25 Reflector illumination at 1 GHz and 2 GHz. The results are given along a horizontal line (H-cut) and along a vertical line (V-cut) for both polarizations vertical and horizontal per line.

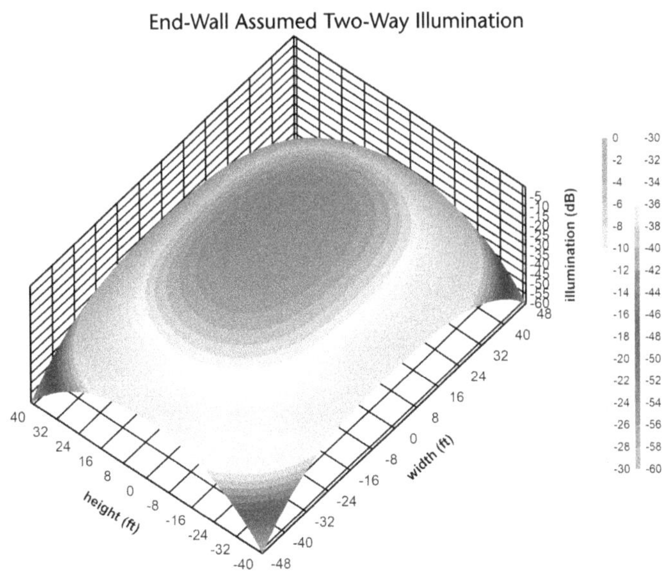

Figure 8.26 Theoretical illumination of the end wall.

Even with time gating there are areas of the lateral surfaces that cannot be effectively time-gated. Let us assume that we have a pulse of about 35 ns that covers the QZ. If that is the case, the end wall can be time-gated out; however, part of the lateral surfaces cannot be gated out.

Consider a chamber with a pulse that encompasses the QZ length. Let us draw circles centered at the reflector to mark the areas of the lateral surfaces where the pulse should arrive at the same time as it arrives at the QZ (and leaves the QZ). This is the same approach used to determine the boundary between pyramids and wedges on the lateral surfaces. After identifying the portions of the wall that cannot be time-gated out, we can do some ray tracing to draw the potential range of angles of incidence onto those sections of the lateral surfaces. Figure 8.27 shows a sketch of this procedure, depicting a "rectangular ring" section of the range where the background RCS cannot be gated out. The areas of the sections of this "rectangular ring" are from absorber tip to absorber tip as follows: The floor is 10-m-long by 27.43m (273.3 m^2); the ceiling is the same; and each of the lateral surfaces are 10-m-long by 23.18m for an area of 231.8m^2 each side.

The FS term is given from the reflector illumination for each polarization. At 1 GHz, the floor and the ceiling we assume $FS = -40$ dB. The value is the same at 2 GHz. For the lateral surface the FS is −25 dB at 1 GHz and −30 dB at 2 GHz. (We are assuming that both polarizations behave similarly.)

The illumination is assumed to be uniform, and that provides a worst-case estimate; hence W can be assumed to be 0 dB. We can get an average for the RCS for the different angles of incidence (65–83 degrees roughly) and we obtain an estimate for the $\sigma°$ that can be multiplied by the areas of the different surfaces. Taking all those factors into account, the estimated background RCSs in this range are listed in Table 8.1.

These are not exact values, and obtaining a true answer is not an easy task. This approach does not take into account the feed positioner or other potential sources of reflections such as doors, lights, and other penetrations. To account for these is why we assume uniform illumination. Consider a similar analysis conducted for a range that was measured. In Figure 8.28 the predicted RCS background from 2 to

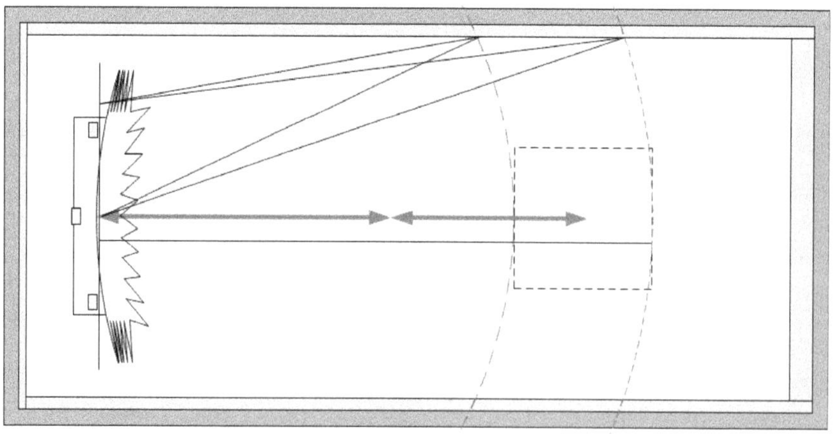

Figure 8.27 A simple ray-tracing approach to estimate the potential angles of incidence from rays departing the reflector and arriving at non-gated-out sections of the lateral wall.

Table 8.1 Estimated Background RCS Level for the Sample Range

Polarization	*Background RCS at 1 GHz*	*Background RCS at 2 GHz*
Vertical	−45.3	−53.2
Horizontal	−43.2	−52.6

6 GHz is compared with measurements. The FS was estimated and given the long pulse; the end wall, which was tilted, was included in the estimate, since it could not be gated out. The approach to arrive to the estimate is the same used in the sample above. The specifics of this range cannot be disclosed, but the example in Figure 8.28 shows the validity of the approach to provide a rough estimate. The big difference at 2 GHz was tracked to the assumed FS value, and the feed used at 2 GHz caused a higher illumination of the lateral surfaces of the range.

8.6 Tapered Chambers for RCS

Reports of tapered ranges used for RCS measurements appeared as early as 1972 [17]. The results presented by Dybdal and Yowell in [17] show deviations between calculated and measured. Their procedure shows that errors of less than 1 dB are achievable for the measured RCS difference of two calibration spheres. One of the spheres was 20 inches in diameter, and the other sphere was 12 inches in diameter. The radar system used by Dybdal and Yowell, shown in [17], is a CW-based system, with no switching. Results are provided for the range of 110–180 MHz. Given the frequencies the results for the sphere RCS are in the upper end of the Rayleigh region where the radius of the sphere (r) approximately follows the constraint given by $2\pi r/\lambda > 1$. At the so-called optical region where $2\pi r/\lambda > 10$, the RCS of a perfect conducting sphere is given by

$$\sigma = \pi r^2 \tag{8.12}$$

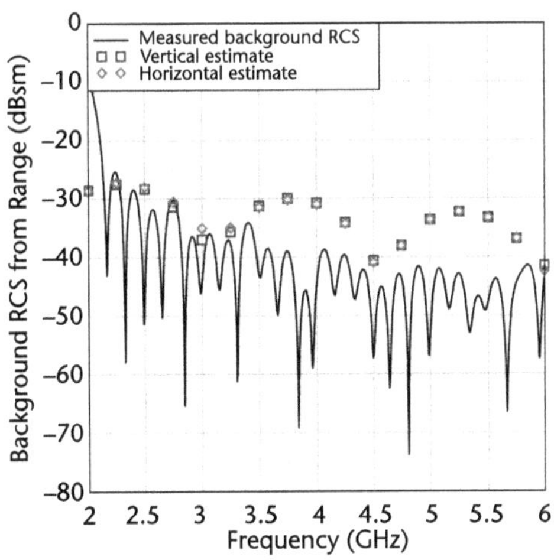

Figure 8.28 Predicted and measured RCS background using the methodology presented in this chapter.

In the Rayleigh region the sphere RCS can be approximated by:

$$\sigma = \left(\pi r^2\right)\left(7.11 \cdot \left(\frac{2\pi}{\lambda} r\right)^4\right) \tag{8.13}$$

For the calibration spheres used and the frequency range, a more accurate approximation (since we are approaching $2\pi r/\lambda = 1$) is to use the Mie series, which can be found in [2] and is given by

$$\sigma = \frac{\lambda^2}{\pi}\left|\sum_{n=1}^{\infty}(-1)^n\left(n+\frac{1}{2}\right)\left(b_n - a_n\right)\right|^2 \tag{8.14}$$

where

$$a_n = \frac{j_n(ka)}{h_n^{(2)}(ka)} \tag{8.15}$$

and

$$b_n = \frac{kaj_{n-1}(ka) - nj_n(ka)}{kah_{n-1}^{(2)}(ka) - nh_n^{(2)}(ka)} \tag{8.16}$$

where $j_n(x)$ is the spherical Bessel function and $h_n^{(2)}(ka)$ is the spherical Hankel function. Using these equations, we obtain an approximate value for the RCS of the 20-inch sphere and for the 12-inch sphere and find the difference between them. Then we can compare these values with the results measured in the taper range by Dybdal and Yowell [17]. From that comparison, shown in Figure 8.29, it appears that the difference between the measurement of the two spheres and the computed RCS difference never exceeds 2dB. These results seem to be an endorsement of the taper range as a potential low-frequency RCS approach. However, [17] mentions that the illuminating antenna, a LPDA, was positioned at 25 ft. (7.62m) from the target. Given that the taper range that they used is described as having a rectangular section of 30-ft-long by 21-ft-wide by 19-ft-tall (9.14-m by 6.4-m by 5.8-m) with a 50-ft- (15.25-m-) long tapered section that terminates in a 10-ft-long by 8-ft-wide by 8-ft-wide (3.05-m by 2.44-m by 2.44-m) source antenna room, it is clear that Dybdal and Yowell are using the taper range as a "rectangular range."

Pellett also has presented data for RCS performed in a taper chamber [9]. The chamber that he used is a much larger 50-ft-tall by 50-ft-wide by 155-ft-long (15.24-m by 15.24-m by 47.24-m). Unlike Dybdal and Yowell, Pellett uses the tapered range as intended, that is as a taper chamber. The system used, described in [18], is a CW system but a pulse box is used to help the isolation by letting it transmit for 250 ns and then switching to receive for 250 ns. Pellett claims sensitivities in the −20 dBsm at 100 MHz that improve to −30 dBsm at 200 MHz and to −40 dBsm for the 250–600 MHz range. In the same paper, Pellett shows results for a compact range located in the same facility. Pellett describes the compact range as having a reflector with a 7.31-m focal length and a reflector projected size of 14m by 8.8m. Using the compact range, Pellett measures a 33-in (84-cm) calibration sphere at

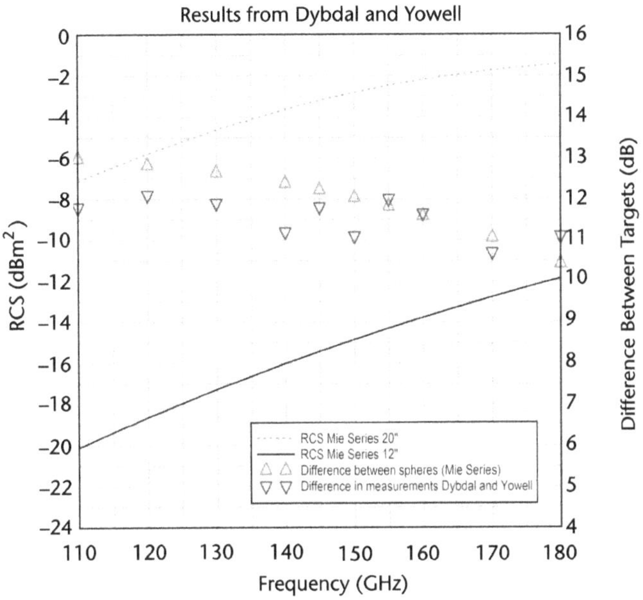

Figure 8.29 RCS of a 20″ calibration sphere at the VHF range.

200–600 MHz and a 2.5-in (6.35-cm) sphere at 200–800 MHz. For the smallest sphere test, we are in the Rayleigh region, and Pellett shows results of about −47.5 to −27.5 dBsm (approximately) since the plots in the paper [9] do not have a very good scale. Per (8.13), we obtain −51.52 dBsm at 200 MHz and −27.43 dBsm at 800 MHz, which are extremely good values. The system appears to provide really good dynamic range if it can measure levels in the −50 dBsm range. These results from the compact range are, according to Pellett, a 15-dB improvement over to the taper range. Postprocessing using FFT and time gating is reportedly used to cut out the back-wall reflections as well as some other chamber reflections.

While taper ranges have been used for RCS, available data suggests that this has been done typically in CW mode. One of the personal concerns I have is that without the proper antenna, tapered chambers are not suitable for broadband measurements and pulse measurements. Additionally, of the main available papers in the literature, only the facility used by Pellett and a similar smaller facility [19] owned by the same company used the taper section to launch the excitation signal. In all the cases that I have seen, the rules shown in Chapter 5 for sizing a taper range were followed when designing these facilities. As mentioned by Pellett [9], the compact range is a much better approach, even when used at really low frequencies. In addition to Pellett, Brumley also agrees, [20] that pushing the compact range reflector to lower frequencies is a suitable approach for VHF measurements of RCS and that it provides better results than far-field illuminated approaches.

References

[1] Lovick, J. E., *Radar Man: a Personal History of Stealth*, New York: iUniverse, Inc., 2010.

[2] Knott, E. F., J. F. Shaeffer, and M. T. Tuley, *Radar Cross Section* (Second Edition), Raleigh, NC: SciTech Publishing, Inc., 2004.

[3] IEEE, IEEE Std 145-2013 IEEE Standard for Definitions of Terms for Antennas, IEEE, 2013.

[4] Long, M. W., *Radar Reflectivity of Land and Sea* (Third Edition), Norwood, MA: Artech House, 2001.

[5] Brumey, S., and D. Droste, "Evaluation of Anehcoic Chamber Absorbers for Improved Chamber Designs and RCS Performance," in *Proceedings of the Antenna Measurement Techniques Asocitaion*, Seattle, WA, September 1987.

[6] Sun, W., K. Liu, and C. A. Balanis, "Analysis of Singly and Doubly Periodic Absorbers by Frequency-Domain Finite-Difference Method," *IEEE Transactions on Antennas and Propagation*, Vol. 44, No. 6, 1996, pp. 798–805.

[7] Brumley, S., and G. Tanakaya, "Diagnostic Evaluation of Wedge Absorbers for RCS Chambers," in *12th Annual Meeting and Symposium of the Antenna Measurement Techniques Association*, Philadelphia, PA, 1990.

[8] Brumey, S. A., *Evaluation of Microwave Anechoic Chamber Materials*, Tempe, AZ: Master of Science Thesis, Arizona State University, May 1988.

[9] Pellett, L., "VHF/UHF Indoor RCS Measurements Using a Tapered or Compact Range," in *13th Annual Meeting and Symposium of the Antenna Measurement Techniques Association*, Boulder, CO, October 7–11, 1991.

[10] Brumley, S. A., "Improving the Performance of Anechoic Absorbers," in *10th Annual Antenna Measurement Techniques Association Proceedings*, Atlanta, GA, September 1988.

[11] Walton, E. K., and L. Beard, "Radar Cross Section Measurements in a Cluttered Environment," in *11th Annual Symposium of the Antenna Measurement Techniques Assocation*, Monterey, CA, October 1989.

[12] Urbanik, E. A., and D. H. Wenzlick, "An Improved Background Subtration Technique," in *Antenna Measurement Techniques Symposium*, Boulder, CO, Ocoober 7–11, 1991.

[13] Muth, L. A., "Radar Cross Section Calibration Errors and Uncertainties," in *21st Annual Meeting and Symposium of the Antenna Measurement Techniques Association*, Monterey Bay, CA, October 4–8, 1999.

[14] Baggett, M., *RCS Chamber Sizing Considerations (Rev C)*, Suwanee, GA: NSI-MI Application Note, 5/02/2018.

[15] Rodriguez, V., "On the Disadvantages of Tilting the Receive End-Wall of a Compact Range for RCS Measurements," in *39th Annual Meeting and Symposium of the Antenna Measurement Techniques Association AMTA 2017*, Atlanta, GA, October 2017.

[16] Brumley, S., "A Modeling Technique for Predicting Anechoic Chamber RCS Background Levels," in *Proceedings of the Antenna Measurement Techniques Association*, Seattle, Washington, September 1987.

[17] Dybdal, R. B., and C. O. Yowell, "VHF to EHF Performance of a 90-Foot Quasi-Tapered Anechoic Chamber," Aerospace Report No TR-0073 (3230-40)-2, December 28, 1972.

[18] Taron, R., and L. Pellett, "Lockheed Advanced Development Company's Electromagnetic Measurement Facility," in *12th annual Conference of the Antenna Measurement Techniques Association*, Philadelphia, PA, 1990.

[19] Urbanik, E. A., and D. G. LaRochelle, "Lockheed Sanders, Inc. Antenna Measurement Facility," in *1996 Proceedings of the Antenna Measurement Techniques Association*, Seattle, WA, November 1996.

[20] Brumley, S., "Low Frequency Operation, Design and Limitations of the Compact Range Reflector," in *16th Antenna Measurement Techniques Association Conference Proceedings*, Long Beach, CA, 3–7 October 1994.

CHAPTER 9

Anechoic Ranges for EMC Measurements

9.1 Introduction

EMC deals with the electromagnetic effects associated with any electric circuit. EMC concerns are the unintended capacitive coupling and inductive coupling that may occur between lines in an electric circuit. Inside this large realm of electrical engineering, there is an area that concentrates on the measurement of the phenomena. EMC measurements are performed, in most cases, to verify that a piece of equipment will operate in certain electromagnetic environments and to check if the device is putting out a large enough quantity of unintended radiation to possibly cause adjacent equipment to fail.

These are not easy tests. A given equipment will not radiate the same when is on a table made out of wood, when on the pavement, or even hanging from a gypsum wall. Any adjacent conductive or dielectric media affects how energy radiates from a piece of equipment and how energy arriving from a given direction couples to the equipment.

Thus, one of the critical parts of EMC testing is to design repeatable test setups that model the reality of how the EUT operates. Tests typically measure the maximum field level emitted by the equipment (emission testing) or determine whether the device operates under certain levels of incident field (immunity testing).

To the antenna engineer, these tests may not make sense given that a lot of the standards call for measuring the fields well in the reactive near-field region of the antenna or EUT; that is, $R \ll \lambda$, where R is the distance from the range antenna to the EUT. The idea is that these tests are standardized, and everybody tests the device in the same way, and in the same environment (within certain tolerance), so the results of the measurements can be compared to some specific goal, or limit.

Since the test environment is guided by the standards, this chapter provides an overview of some of these standards. Some of the standards have test site verification measurements; some other standards just offer specifications regarding the reflectivity of the RF absorber.

This chapter concentrates on those sites for radiated testing. Conductive EMC testing and electrostatic discharge (ESD) testing also have specified test layouts, but I find those tests to fall outside the scope of this book. Thus, this chapter examines the MIL-STD-461G [1] chamber requirements, followed by the RTCA DO-160 [2]. It should be noted that the RTCA DO-160G document is used by other standard associations such as the International Organization for Standardization (ISO) and the European Organisation for Civil Aviation Equipment (EUROCAE). Hence RTCA DO-160G is the same document as ISO 7137 and EUROCAE/ED-14G. It is normal for standards to be copies of each other. However, in some cases there may be slight differences.

The chapter continues with the automotive components testing part of CISPR 25 (as opposed to the full vehicle test section of the same document), SAE J1113/21, SAE J1113/41, and ISO 11452-2 [3]. These standards use very similar chambers, and, like the MIL-STD-461G they perform the measurements with the antenna located at 1m from the EUT or the EUT cable harness.

After defining the facility requirements for automotive components, the chapter describes the requirements for full vehicle testing. These are given, for emissions, by CISPR 12, a section of CISPR 25, and their equivalents SAE J551/2 and SAE J551/5, and for immunity by ISO 11451-2 and its equivalent SAE J551/11 [4]

After describing the standards for full vehicle testing, the chapter continues by discussing the test site requirements for commercial (nonautomotive) product testing. These are the emission test sites defined by IEEE/ANSI C63.4 and by the CISPR 16 standards, and the immunity anechoic room requirements defined by the IEC 61000-4-3.

Recall that the goal of this chapter is not to do an in-depth review of the standards, but to concentrate on the indoor range requirements that these standards mandate.

9.2 The MIL-STD-461 and RTCA DO-160 Chambers

9.2.1 The MIL-STD-461 Series

MIL-STD-461 has always been the author's recommended standard for people entering the EMC testing world. The series dates back to 1967 when the Department of Defense decided to gather into a single document the requirements for military interference control [5]. At the beginning, a separate standard MIL-STD-462 provided the information about the test setup; version D of the MIL-STD-462, which was published in 1993, is the first that shows the absorber loading of the shielded test room and the specifications for the RF absorber. MIL-STD-462D was merged with the MIL-STD-461D when the MIL-STD-461E was published in August of 1999. The later versions of the document, versions F and G (the current version), have not changed the requirements for the test site for the radiated tests.

9.2.2 Sizing the Chamber

Like many other component test standards, MIL-STD-461G [1] calls for a metal-topped test bench where the EUT is placed. The standard covers both conducted and radiated tests. For the radiated tests, there are test, for emissions, and for immunity (which the standard refers to as susceptibility).

The test procedures for magnetic field emissions and susceptibility are done at close proximity, with the sensing coil at 7 cm from the outer surface of the EUT for emissions testing (RE 101), and at 5 cm for the generating coil for the susceptibility test (RS 101). Helmholtz coils are an option for the RS 101 test. These magnetic field tests are performed across the 30-Hz to 100-kHz frequency range. Chapter 3 shows that current absorber technology does not operate at these low frequencies. Hence, an absorber is not required. Shielded rooms are recommended, but not necessary.

The test procedure for electric field emission measurements (Test RE 102) calls for the measuring antennas to be located 1m from the cable harness attached to the component. The standard is very specific on what antennas are to be used and where they are to be located with respect to the EUT. Figure 9.1 shows the test setup as illustrated and described in the standard [1]. The EUT may be floor-standing or a tabletop component. The interesting approach of the MIL-STD-461 test is that the cable harness is the one illuminated rather than the EUT itself. This makes electromagnetic sense since for smaller EUTs most of the energy coupled will be through the cable harness. The latest edition of the standard also calls for the test antenna to be repositioned for large system testing. The number of positions is equal to the

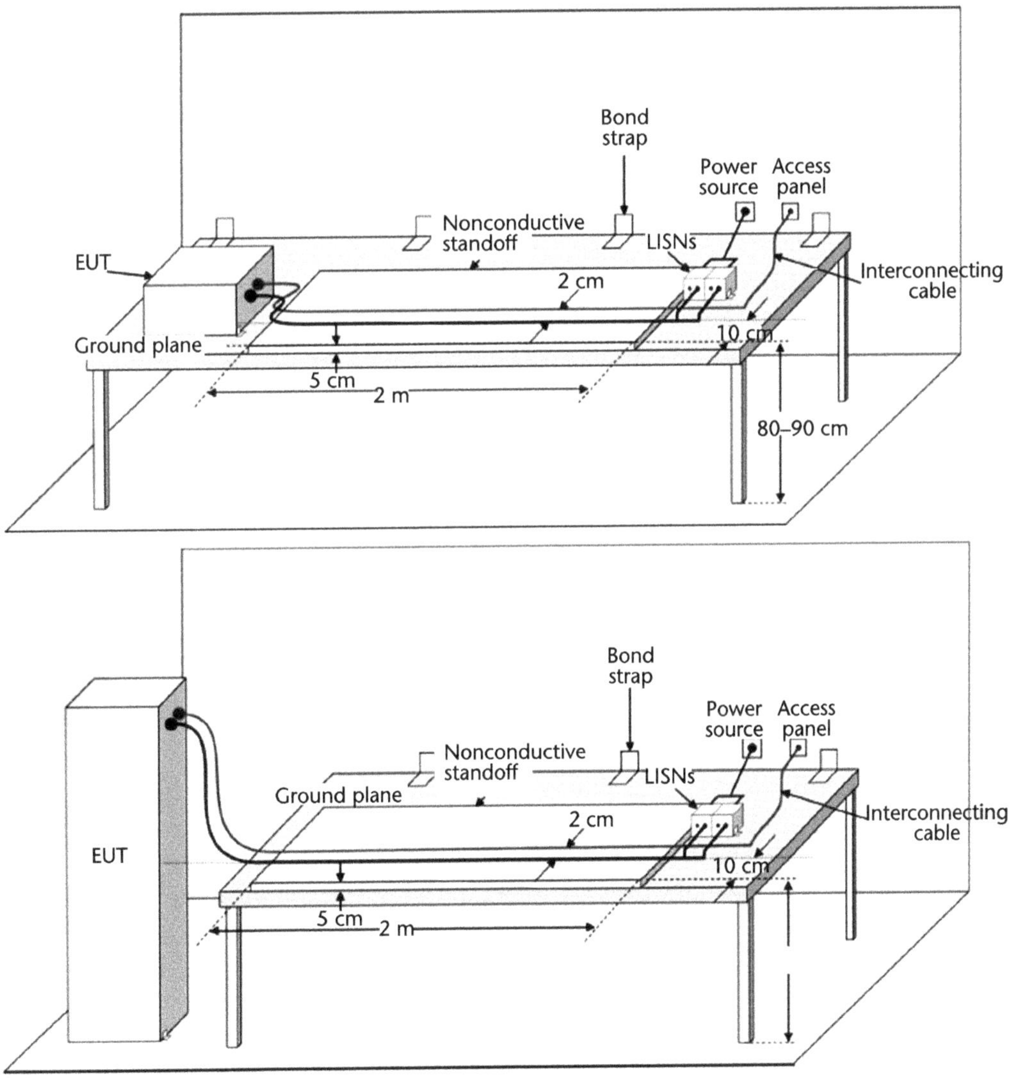

Figure 9.1 Test setup for radiated emission test for an electric field as presented in [1].

size of the test setup divided by three; the result is rounded up to the next integer to obtain the number of antenna positions.

The indoor range for MIL-STD-461 testing must be wide enough house the EUT and the 2m of cable harness. The cable harness is placed on a dielectric plate about 5 cm in thickness above the bench. The bench is between 80 and 90 cm tall and usually has a metallic top. The cables are placed 10 cm from the front to the bench. The power cables that feed the EUT are passed though a line impedance stabilization network (LISN) that also resides on the bench top.

In general, the width of the MIL-STD-461 range is given by

$$W \geq S_{\text{EUT}} + 2\text{m} + S_{\text{anc}} + 60 \text{ cm} + 2 \cdot S_{\text{abs}} \tag{9.1}$$

where S_{EUT} is the size of the EUT, and S_{anc} is the size of any ancillary equipment such as LISNs. S_{abs} is the thickness of the absorber treatment on the lateral walls. The 2m accounts for the cable harness, and the 60 cm accounts for a minimum distance from the test setup to the tips of the absorber. The depth of the indoor range is given by

$$L \geq S_{\text{EUT}} + 1\text{m} + S_{\text{ant}} + 60 \text{ cm} + 2 \cdot S_{\text{abs}} \tag{9.2}$$

where S_{ant} is the size of the largest antenna. Since the antennas used are mandated by the standard, the largest antenna is the dual-ridge horn that is used from 200 MHz to 1 GHz. This antenna is roughly 98 cm (38.5 inches). Figure 9.2 shows the setup for different antennas, demonstrating that, when using the horn, the test distance is measured from the horn aperture. The back of the horn must be at least 30 cm from the absorber behind.

The height of the room has to be high enough to allow for the antennas and largest EUTs to fit. While for RE102, the electric field emissions test, the antennas are specified, for RS103, the electric field susceptibility test, the size of the antenna is not specified. In some cases large TEM field generators like the one shown on Figure 9.3 need to be used to generate the high fields required for susceptibility testing.

Thus, before sizing the height of the room, it is important to estimate the largest EUT that may be tested. For those performing tests on smaller components the height of the room should be at least given by

$$L \geq 1.2\text{m} + \frac{1}{2} S_{\text{bicon}} + 30 \text{ cm} + S_{\text{abs}} \tag{9.3}$$

Per RE102 the biconical antenna is 137 cm from tip to tip, that is, the largest antenna dimension. The center of the antenna is at 120, as shown on Figure 9.2. The height given by (9.3) is the smallest height, and it is recommended that taller chambers should be built to accommodate the larger antennas typically used for immunity, which can be significantly larger like the one shown in Figure 9.4. The antennas used for immunity may be the largest objects that will access the range. The doors and the indoor range size should be sized so that these large radiators are able to be brought in to perform susceptibility testing.

Both the MIL-STD-461 and the RTCA DO-160 allow for the use of reverberation chambers for susceptibility or immunity testing. These chambers allow for the generation of a statistically uniform field over a large volume. This field can be of

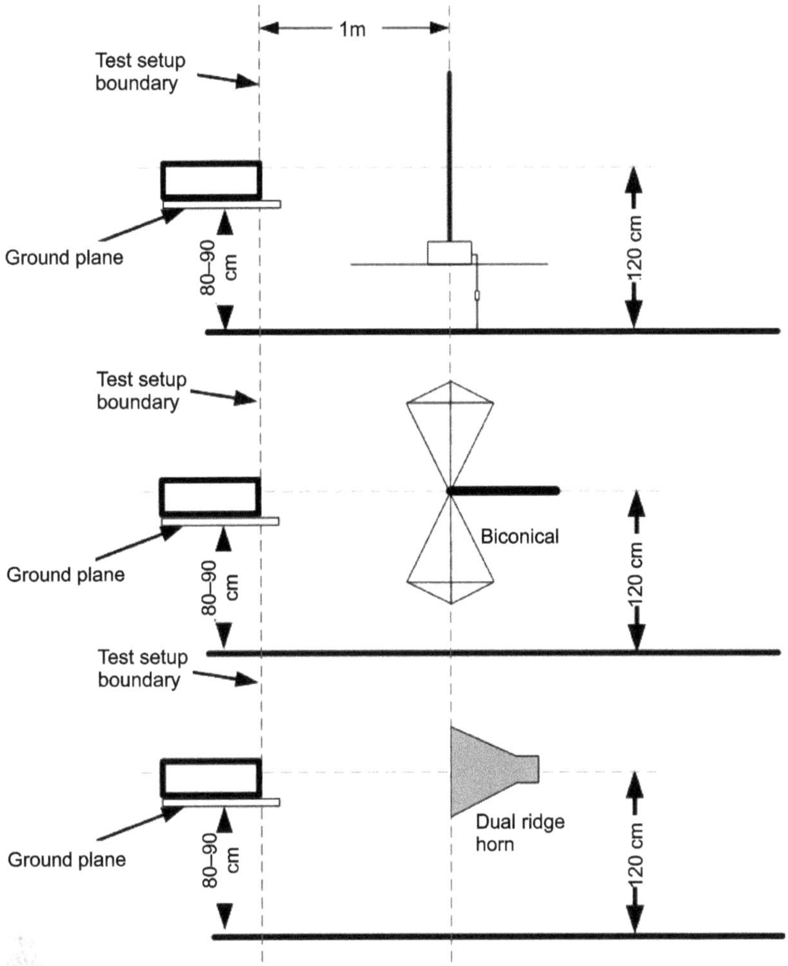

Figure 9.2 Setup for the different antennas used in part RS102 of MIL-STD-461G.

relatively high intensity; some of the testing per these standards calls for the generation of fields in excess of 200 V/m. Indeed RTCA DO-160 calls for CW field levels of up to 490 V/m and pulsed modulated testing with peak field levels of up to 7,200 V/m. Those levels are equivalent to a power density of 137.5 kW/m^2. It is not economically feasible to generate those power densities in an anechoic chamber. Because of the cost of the power amplifiers and the high-power absorber required to line the walls of the room, the reverberation chamber is a far more economical solution.

The last factor is the absorber size. The choice of absorber is also determined by the standard. Section 9.2.3 discusses the absorber layout and the type of absorber.

9.2.3 Absorber Layout

The minimum absorber layout for MIL-STD-462D and subsequent MIL-STD-461E through G is very simple. The absorber covers the entire wall from floor to ceiling on the two end walls that are perpendicular to the test range. The floor is left bare. The ceiling and the side walls are partially covered. The coverage is from

Figure 9.3 A TEM wire E-field generator. The device generates a vertical electric field between the horizontal wires and the floor of the shielded room; the test bench can be placed under the bench to generate the field. (*Source:* Author's private collection.)

Figure 9.4 Immunity testing antennas tend to be physically large to increase the gain and decrease the return loss. Here the author stands next to a 100-MHz to 1-GHz dual-ridge horn antenna. (*Source:* Author's private collection.)

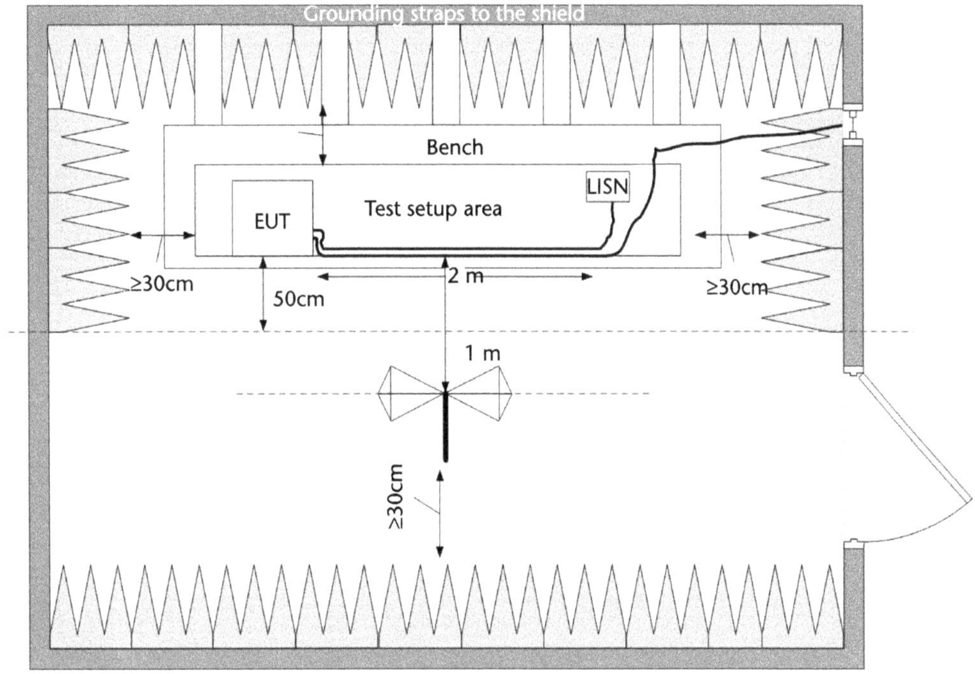

Figure 9.5 Plan view of the MIL-STD-461 typical chamber.

the end wall behind the test setup to 50 cm in front of the test setup. The layout is shown in Figure 9.5. The absorber must be at least 30 cm from the test setup area; this value is specified in the standard and does not change with frequency. The absorber performance is also defined in the standard. The absorber must exceed a given reflectivity at normal incidence for a given frequency range, as stated in the standard. Table 9.1 lists the reflectivity requirements for the RF absorber according to MIL-STD-462D.

Figure 9.6 shows some typical absorbers compared with the limits from the standard. It should be noted that ferrite tile can be used up to about 1.5 GHz, but above that frequency it does not meet the standard limits. Hybrid absorber exceeds the limits as does standard polyurethane that is 24 inches (61 cm) in total height. The 24-inch pyramidal absorber was traditionally the absorber size recommended in prior versions of the MIL-STD-461 document. However, the latest version mentions the use of ferrite tiles and of hybrid absorbers, which can be shorter than the 24-inch polyurethane base pyramidal absorber.

RTCA DO-160G [2] requires RF absorber requirements similar to those of MIL-STD-461G. The only difference is that the DO-160 document frequency range for

Table 9.1 Absorber Reflectivity Requirements for MIL-STD-461 E Through G

Frequency Range (MHz)	*Normal Incidence Reflectivity (dB)*
80–250	6
250 and above	10

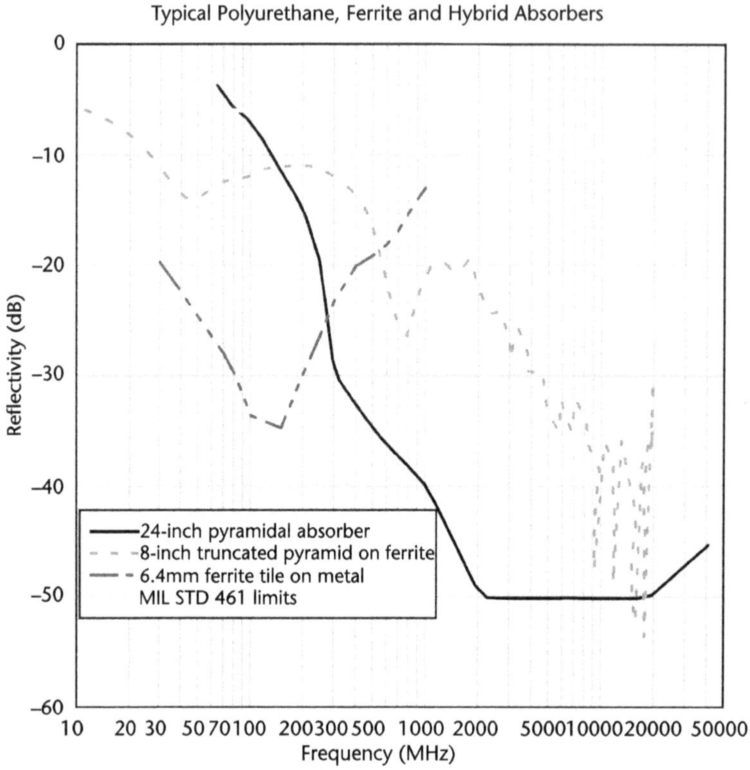

Figure 9.6 Comparison of typical absorber technology with the MIL-STD-461 G limits.

the −6-dB absorption is from 100 to 250 MHz, instead of the 80 to 250 MHz listed in MIL-STD-461G. RTCA DO-160 differs from MIL-STD-461 in the test setup. In the RTCA document test setup, the EUT is in the center with cables stretching out at least 1m on each side of the EUT as shown in Figure 9.7. The power cables going to the LISN are 1m in length with a 10-cm tolerance. The signal cables are at least 1m in length along the front of the test bench. The sketch in Figure 9.7 assumes

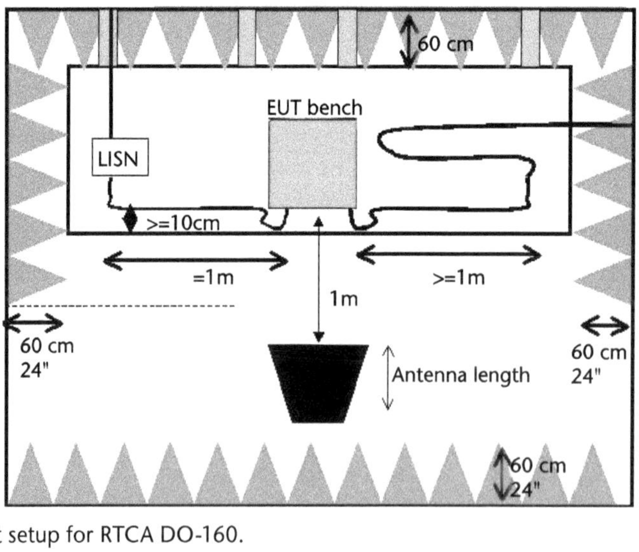

Figure 9.7 Test setup for RTCA DO-160.

that 24-inch pyramidal absorber is used, but hybrid absorbers can also be used, as long as they meet the requirements of absorption.

9.3 The Automotive Component Chambers CISPR 25, ISO 11452, and Equivalents

9.3.1 Introduction

Automotive EMC is not only different from typical commercial equipment or household appliance EMC in the test setup, but also in the application. Minor glitches on the TV screen due to EM noise from a hairdryer or other appliances is a nuisance. On the other hand, a glitch in the antilock braking system or accelerometers for airbag deployment could result in the death of the occupants of the vehicle or pedestrians. Because of the potential liabilities from EMC problems in electronic automotive components, manufacturers of vehicles have developed their internal standards. Any manufacturer of components needs to test their equipment according to these internal standards if they want their product to be part of a vehicle manufactured by a given company.

While this could be a recipe for a chaotic environment, all the original equipment manufacturers (OEMs) tend to follow some specific standard test setup, and the variations between OEMs are small; sometimes different competitors agree to harmonize their requirements to simplify the testing tasks for the component manufacturers.

Section 9.4 discusses the full vehicle; this section considers the main component or electronic subassembly (ESA) test standards. Table 9.2 attempts to list the different automotive EMC standards. It should be noted that these change as new technologies appear. The emergence of electric and hybrid propulsion vehicles is also changing the standards rapidly. Hence, Table 9.2 is not comprehensive but intended to illustrate the large number of standard documents and government regulations.

Most manufacturer standards rely on the ISO or the Society of Automotive Engineers (SAE) standards. Thus, Section 9.3.2 presents the chambers for ISO 11452 and for SAE J1113 and CISPR 25.

9.3.2 Sizing the Chamber

EMC automotive component testing involves conducted as well as radiated tests. For radiated tests there are other methods in addition to the indoor anechoic range. For example, the ISO immunity test standard ISO 11452 has several sections. ISO 11452-1 deals mainly with definitions. ISO 11452-3 presents the test methodology for using a transverse electromagnetic mode (TEM) cell. TEM cells are basically TEM transmission lines where the device under test can be placed in the transmission line to be illuminated by the TEM wave traveling in the transmission line. Other sections of the test look at other types of TEM devices. ISO 11452-5, for example, examines what is called a stripline. The standard that is of interest to us, and that relates to the topic of this book, is ISO 11452-2. Part 2 of ISO 11452, which deals with absorber-lined chambers, is equivalent to SAE J1113/21. Those two standards deal mainly with immunity measurements, while SAE J113-41 and the component section of CISPR 25 deal with emissions. While there are slight differences between these standards, they all share a similar test setup.

The test setup is similar to the MIL-STD-461 and the RTCA DO-160 in that the test is performed at 1m in distance and that the EUT is placed on a bench (with a conductive top, although nonconductive tops are allowed by ISO 11452-2). Figure 9.8 illustrates the setup for the automotive component anechoic chamber test.

The EUT is placed to one side and the cables (power and signals) to the device are extended close to the front of the bench for 1.5m (with a ±7.5-cm tolerance). On the other side of the cable there is ancillary equipment and impedance stabilization networks for the power lines. The cable harness is placed 10 cm from the front edge of the bench. Similarly, the EUT front outer surface is placed 10 cm (±1 cm) behind the cable harness, or about 20 cm (±1 cm) from the front edge of the bench. Like in MIL-STD-461, the cable harness is spaced from the metallic top bench by a dielectric spacer 5 cm in thickness with a requirement for the relative dielectric constant to be $\varepsilon_r \leq 1.4$. The cable harness is placed at least 1m from the absorber tips. The ISO 11452-2, CISPR 25 and the equivalent SAE J113 have similar test setups. It should be noted that CISPR 25 states that the bench cannot be smaller than 2m by 1m.

Figure 9.9 illustrates the test setup showing the test bench from the side, top, and the front views. Notice that the metallic top of the bench is grounded to the chamber shield by a series of grounding straps (as is the case in MIL-STD-461 and RTCA DO-160). Figure 9.10 illustrates a method of grounding the bench by having a copper sheet attached to the shield and sticking out between the absorber. The bench's conductive top has an additional sheet bended, and the grounded sheet attached to the wall is pressure-fitted to the bench to achieve the grounding.

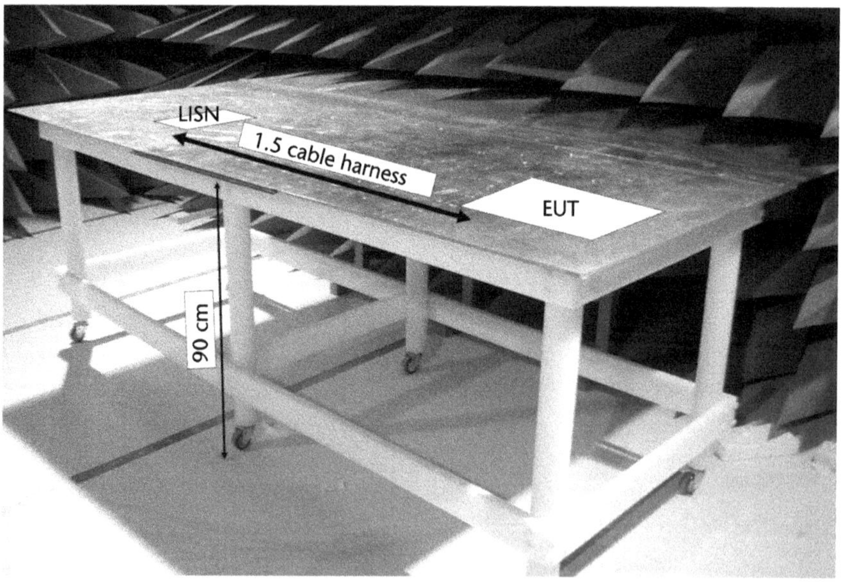

Figure 9.8 A test bench on a CISPR25/ISO11452-2 anechoic range. (*Source:* Author's private collection.)

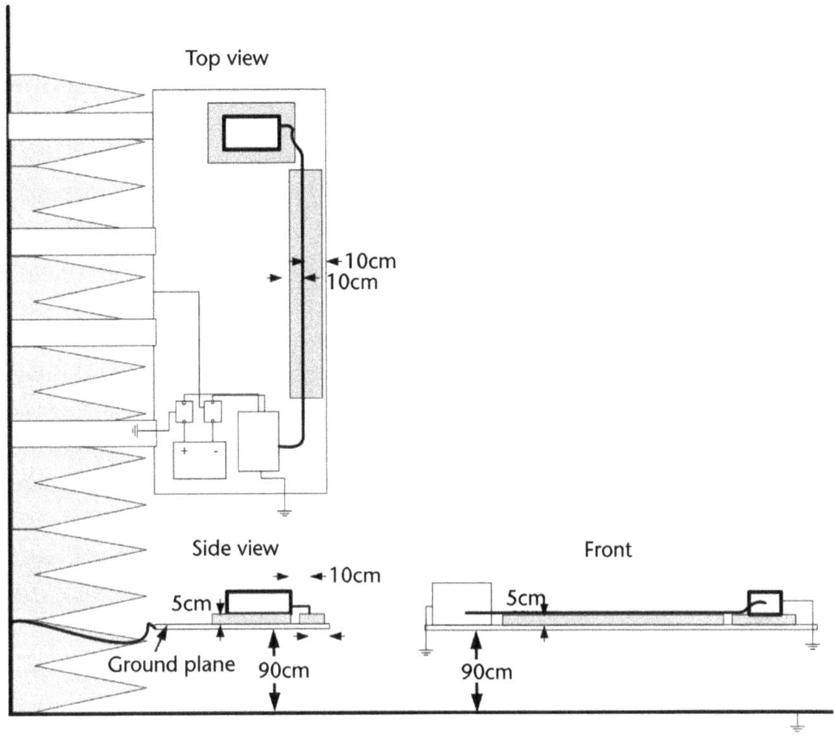

Figure 9.9 Test setup for automotive component testing.

The antenna location for the tests is similar in all these standards. The centerline of the antenna is placed at 1m over the ground, and no part of the antenna can be closer than 25 cm from the ground. This requirement of spacing did not change in the last version of the ISO 11452 document. The latest version of the ISO 11452 standard expanded the applicable frequency range to 80 MHz to 18 GHz from the previous 200 MHz to 18 GHz. This showcases a very common issue with standards. Given that a wavelength at 80 MHz is 3.747m, this means that an efficient radiator that is $\lambda/2$ will not fit, since a dipole operating at 80 MHz will have elements that are 93.68m. Accordingly, with the centerline of the antenna at 1m, when operating on vertical polarizations the element is at less than 7 cm from the floor of the range. This pushes the user to use slightly less efficient antennas like the recommended biconical used in MIL-STD-461, which is typically 1.35m tip to tip.

Figure 9.11 illustrates the positioning of the antenna in front of the bench for different types of antennas used in the testing of automotive components.

Notice that the layout shown in Figure 9.11 is very similar to the ones shown in Figure 9.2. The setup for the monopole antenna is different, and unlike MIL-STD-461E through G, the automotive standards allow for the use of LPDA antennas. (MIL-STD-461 allows LPDA antennas in the reverberation method.)

Now that we have an overall understanding of the antenna positioning and the test layout, we can discuss sizing the chamber for automotive component testing.

Figure 9.10 Grounding of the bench to the chamber shield in a CISPR 25 chamber. (*Source:* Author's private collection.)

The width must be large enough to accommodate the test setup and to maintain at least 1m from the cable harness to the absorber tips as well as from the antenna to the absorber. ISO 11452 allows for the cable harness to be closer to the absorber tips, making the separation a minimum of 50 cm. Thus, designing the room for CISPR 25 will meet the requirements of ISO 11452. Based on this, the recommended size is

$$W \geq S_{EUT} + 1.5m + S_{anc} + 2m + 2 \cdot S_{abs} \tag{9.4}$$

This is a recommendation; neither standard mentions a specific minimum distance from the test setup to the absorber. However, for example, both ISO 11452-2 [6] and Ford Motor Company's FMC 1278 move the antenna laterally to be in front of the EUT at frequencies above 1 GHz. (This makes sense since, as frequency increases, it is more likely for emissions to be from the currents flowing on the devices outer shell.) Additionally, some of those standards require the power supply to be off the bench; hence additional space on both sides of the bench is desirable. Section 9.3.3 discusses the recommended size for the absorber S_{abs}. It should

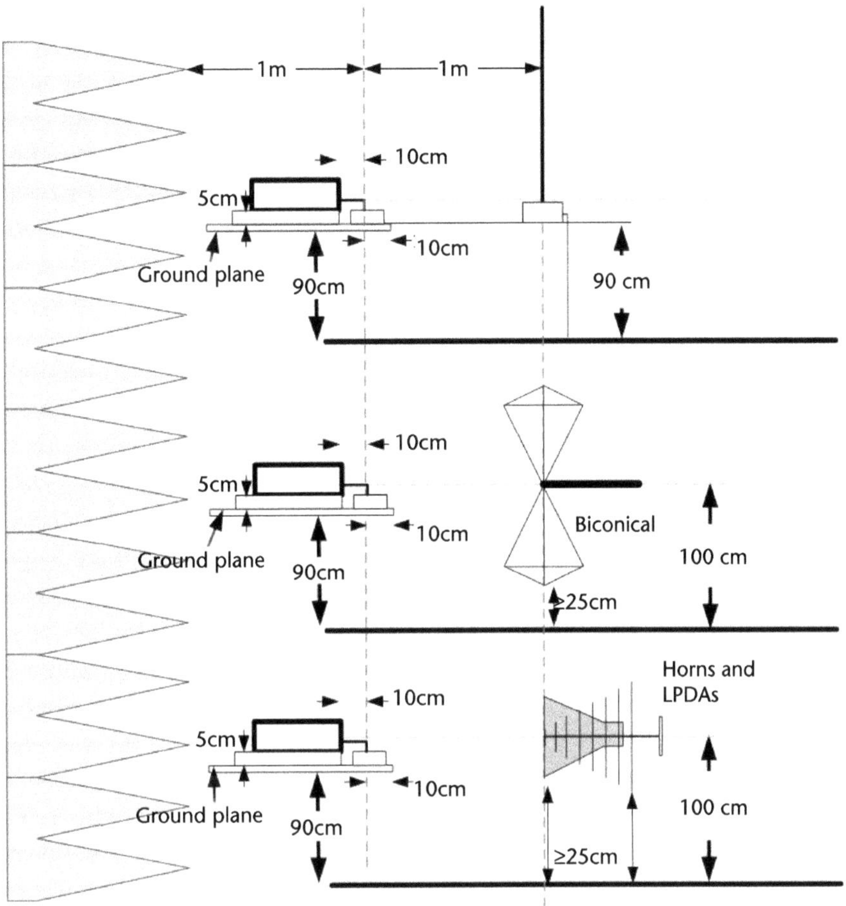

Figure 9.11 Side view of the layout for CISPR 25 and ISO 11452-2 and equivalents.

be noted that both ISO 11452-2 [6] standard and CISPR 25 [7] mention that $S_{abs} \geq 1\text{m}$, or at least that the distance from the harness and the antenna to the shield, should be larger than 2m.

The length of the range is given by

$$L \geq S_{ant} + 3\text{m} + 2 \cdot S_{abs} \tag{9.5}$$

This size allows for the harness to be 1m from the absorber tips and for the antenna to be 1m from the harness and 1m from the absorber tips.

The height is driven typically by the monopole used at the lower frequencies, which is typically 1m in length, making the height of the chamber to be large enough to accommodate that monopole. Hence, the height should be at least

$$H \geq H_{bench} + 2\text{m} + S_{abs} \tag{9.6}$$

This provides space for the bench, a 1-m monopole, and an additional meter from the monopole tip to the absorber. From (9.6) we can estimate that the space from the floor to the tips of the absorber should be at least 3m or that the minimum shield-to-shield dimensions (assuming a minimum 2m by 1m bench) should be 6-m-wide by 6-m-long by 4-m-tall. The ISO chamber could be 5m by 5m by 3.5m. The absorber size could allow for smaller chambers, and typically a lot of chambers for automotive component EMC are smaller than these minimum sizes. Do they meet the standard? To answer this question, let us look at what the standards state about the absorber.

9.3.3 Absorber Layout

The requirements for chamber reflectivity are not clear in the automotive component standards. ISO 11452 used to state that the goal was to reduce the reflectivity in the test area to –10 dB [6], compared to the direct path from the antenna to the EUT. In the meantime, CISPR 25 stated that "error caused by the energy reflected from the walls and ceiling is less than 6 dB in the frequency range of 70 MHz to 1000 MHz." The CISPR 25-2002 version and earlier versions call for the shielded walls to be at least 2m from the EUT or the antenna, in addition to the absorber tips being at a minimum of 1m away from the EUT and the antenna. (See Sections 4.4.1 and 4.4.2 of [7].) Thus, the RF absorber can be as long as 1m in length

The statement in the CISPR 25 standard has been assumed to mean that the reflectivity of the RF absorber should be –6 dB; the reason for that assumption is that typical polyurethane pyramids of 36 inches (91.44 cm) provide that –6-dB level at 70 MHz. Figure 5.38 shows that a reflected level of –6 dB means an error +3.53 when in phase to –6.04 when the reflected energy is out of phase. The maximum error in decibels is indeed only 0.4 dB higher than the 6-dB level when the reflected signal is out of phase. It is not clear if this is what the standard intended.

As a matter of fact, not until the second edition of CISPR 25 [7] was there a methodology for validating or characterizing the range. In that edition, the procedure was included in an informative annex, although it was mandated as a means of checking the absorber-lined shielded room (ALSE). The procedure tests a noise source attached to a load-terminated wire that represents the cable harness. The test is to be done in an open area test site (basically a ground plane on an open field) and then compared to the measurements done in the chamber. The difference between the results should be within 6 dB.

ISO 11452-2 does not have a method for verifying the reflectivity of the room, except for a simplified ray tracing that can show that the reflectivity is close to normal incidence, as shown in Figure 9.12.

Figure 9.6 shows that the standard 24-inch polyurethane used in MIL STD chambers does not meet the requirements of –10 dB reflectivity required by the ISO 11452-2 from 80 MHz to 18 GHz. Hybrid absorbers meet those requirements. Traditionally, polyurethane 36-inch pyramidal absorber has been used to meet those requirements. Figure 9.13 shows the typical reflectivity of the 36-inch pyramidal absorber. This example shows that the material does not quite meet the –6-dB level at 70 MHz as potentially required by CIPSR 25. The modeled 36-inch pyramid has a slightly higher reflectivity at 70 MHz. For the ISO 11452-2 requirements, the 36-inch pyramid model met the requirements at 200 MHz and above as the prior

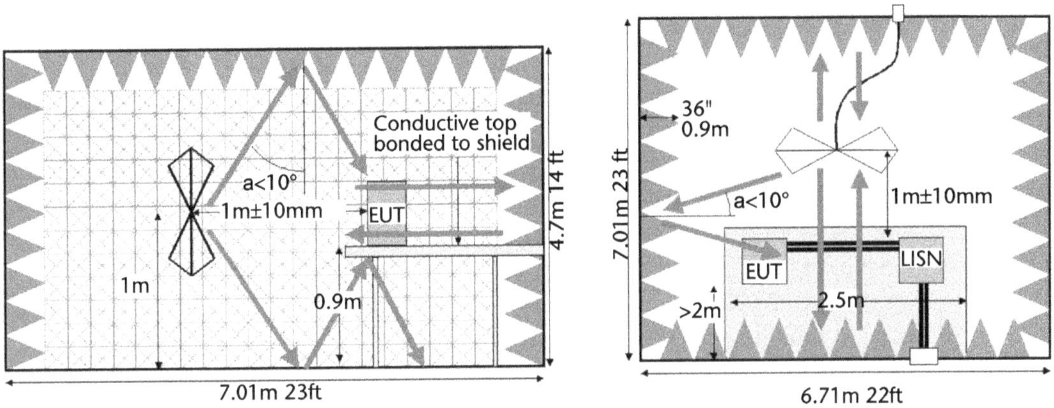

Figure 9.12 Simple ray tracing that shows that given the test geometry, the specular ray incidence is close to normal incidence.

version (1995) of the standard required, but for the newer requirements of 80 MHz, the modeled absorber has reflectivity levels at 80 MHz that are about 1.5 dB higher.

Figure 9.6 shows data for an 8-inch (20-cm) hybrid absorber. That material exceeded −10 dB for the entire range of interest of both ISO-11452-2 and CISPR 25. Indeed for CISPR 25 the better than −10-dB reflectivity is equivalent to errors of 2.38 dB to −3.3 dB (as shown in Figure 5.38), far smaller than the 6-dB error quoted in CISPR 25.

Recall that the 2002 version of the CISPR 25 document tries to clarify the issue of the validity of the ranges by proposing a validation method. The methodology fails to state the range of frequencies at which it is valid. The absorber requirements in CISPR 25 are for frequencies above 70 MHz, yet the standard provides methods

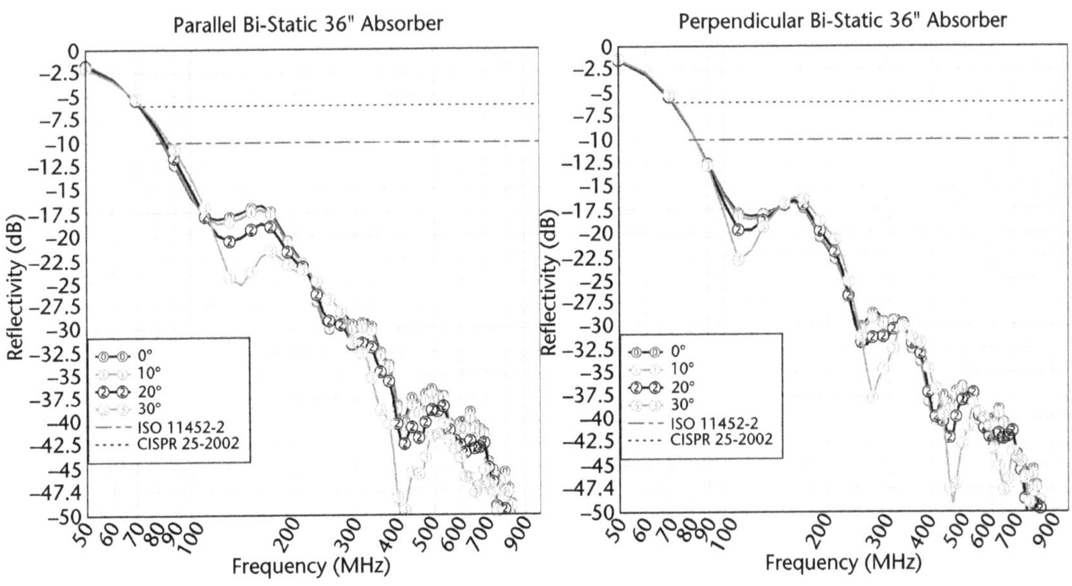

Figure 9.13 Computed reflectivity for a typical 36-inch pyramidal absorber at different angles of incidence and both polarizations.

for testing down to 150 kHz. Is the chamber validation procedure to be used down to 150 kHz or only down to 70 MHz?

One of the issues with the validation approach is the reference measurement. That measurement is to be made in an open area test site. The issue there is the grounding of the bench. In CISPR 25 (and MIL-STD-461), the bench is grounded to the shield walls, rather than the floor (see Figure 9.10). However, in the open area test, the bench cannot be grounded to the floor since there are no walls to ground the bench. That difference in the grounding can make a chamber with a suitable absorber noncompliant. These tests are conducted in the reactive near field of the antenna, so inductive and capacitive coupling to the test setup is extremely critical. It is recommended that the bench in the chamber be grounded in the same fashion as it is in the reference measurement outdoors.

In addition to testing components and subassemblies, the full vehicle must be tested. Section 9.4 explores the testing of full vehicles.

9.4 The Automotive Chamber CISPR 12, SAE J551, and ISO 11451

9.4.1 Full-Vehicle Standards

Full-vehicle testing is also divided on emissions and immunity testing. There is an additional part that can be defined as self-immunity. This type of testing checks the effects of electronic systems on the vehicle on the receivers installed in the vehicle itself. (Anyone who has heard the humming of the engine on the car radio understands this.) The main standards that deal with emissions are CISPR 12 and SAE J551/2. Immunity of full vehicles is mainly addressed in ISO 11451-2 and SAE J551/11, which deal with anechoic chamber testing or open area test site (OATS). There are other test options for immunity that deal with lower frequencies. SAE J551/17 deals with immunity to power-line magnetic fields. ISO 11451-4, and its equivalent SAE J551/13, use a current probe to simulate radiated fields being induced into the cables of the vehicle's electrical system; a method applicable to lower frequencies in the 1–400 MHz range. This book refers to off-vehicle radiating sources where indoor anechoic ranges are required. The self-immunity testing of full vehicles is defined by the full-vehicle section of CISPR 25, and its equivalent SAE J551/4. As is the case for all EMC testing, the size of the chamber is directed by the test setup described in the standard. Section 9.4.2 describes the test setup of these standards.

9.4.2 Test Setup and Sizing the Chamber

9.4.2.1 CISPR 25 Full Vehicle

Like the component testing part, CISPR 25 is specific on the test setup for the full-vehicle test. The absorber requirements are the same as the required ones for component testing. However, while the component testing portion does provide a validation test, that test does not really apply to the full-vehicle test. Figure 9.14 illustrates the setup. CISPR 25 calls for a 1-m-long monopole antenna to be placed on top of the vehicle as the receive antenna; ideally, however, actual vehicle antennas are to be used. Assuming that a monopole is to be used as opposed to other vehicle-mounted antennas, the height of the room is given by

$$H \geq 3\text{m} + V_{\text{Height}} \tag{9.7}$$

where V_{Height} is the height of the tallest vehicle to be tested in the room. The width of the room is related to the width of the vehicle and given by

$$W \geq 4\text{m} + V_{\text{Width}} \tag{9.8}$$

where V_{Width} is the width of the widest vehicle, and finally the length is

$$L \geq 4\text{m} + V_{\text{length}} \tag{9.9}$$

where V_{length} is the length of the longest vehicle to be tested. Possibly shorter rooms can be used if the actual vehicle-mounted antennas are used, especially since nowadays a lot of antennas are integrated into the glass windows of the vehicle, or they are more aerodynamic structures shorter than a 1-m-long monopole.

9.4.2.2 CISPR 12 Test Setup

This section explores the test setup for full-vehicle emissions testing. Both the SAE J551/2 and the CISPR 12 document define a similar setup. The test is defined to be performed on an outdoor range [8]. Like all EMC standards, these documents are living documents that are continuously evolving, as mentioned by Andersen [9]. However, no major changes have been incorporated from the originals except for electric-propulsion vehicles and adding other types of equipment that operate with internal combustion engines, such as water pumps and chainsaws.

The setup is similar to commercial equipment EMC testing, as shown in Section 9.5, but with a few differences. There is no designated QZ or test zone, as this is defined in the commercial EMC test. Additionally, as in the other standards, the measuring antenna is fixed at one position as opposed to the height scanning performed according to commercial standards.

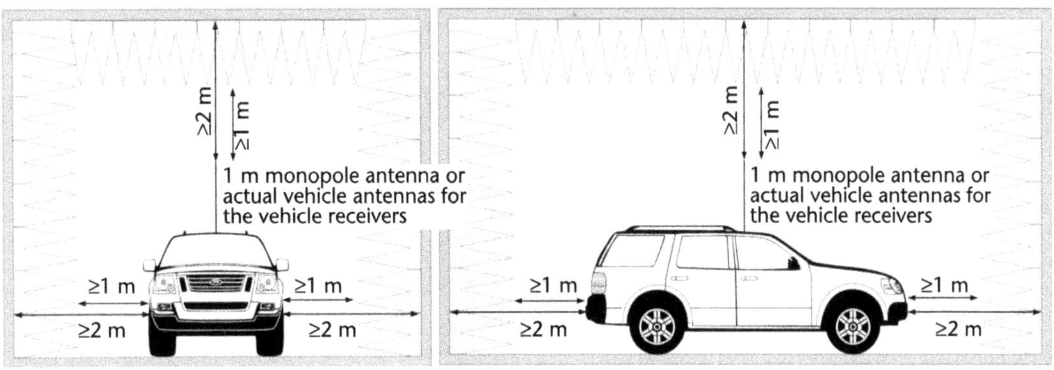

Figure 9.14 Test setup for a full vehicle per the CISPR 25.

Figure 9.15 illustrates the test setup. The test site is an open-area test site (OATS) that consists of a metallic ground plane on a large area free of reflecting structures. The test requires a 30-m-radius area centered between the vehicle and the test antenna. In addition, no reflecting structure or equipment can be located within 15m of the vehicle under test. The measurement can be conducted at 10m from the outer surfaces of the vehicle (or boat or lawnmower, as those also fall under the CISPR 12 standard [8]). If the test is being performed at a 10-m distance, the measuring antennas are placed such that the centerline or phase center is at 3m over the ground onto which the vehicle rests. The distance to the different antenna types is measured as is done in CISPR 25; hence, the distance to an LPDA is measured from the tip, and the distance to a biconical is measured from the axis of the elements (see Figure 9.11).

The measurement can also be taken at a 3-m distance, in which case, the centerline or phase center of the antenna is placed at 1.8m over the ground onto which the vehicle rests. The standard calls for both sides of the vehicle to be tested with the antenna in line with the center of the engine. For frequencies between 150 kHz and 30 MHz, measurements are only taken at vertical polarization using a vertical monopole, and the monopole and its ground plane or counterpoise is placed as close as possible to the metallic ground and physically connected to the ground.

Why perform the measurements at 10m if a 3-m measurement is allowed? The reason is that the entire vehicle must be within the 3-dB beamwidth of the antenna. If measurements at 3m are performed, and the entire vehicle is not inside the 3-dB beamwidth of the antenna, multiple antenna positions are required to ensure that the entire vehicle has been tested.

Clearly, the test setup calls for an extremely large area that is open to the elements and to electromagnetic noise sources. The standard recognizes this issue and states that absorber-lined shielded enclosures (ALSEs) can be used "provided that results obtained can be correlated with those obtained using the OATS described" [8]. The meaning of this correlation is not clear. Does measuring a noise source in the described OATS and in an ALSE and comparing the results suffice? The industry approach is to perform the site validation test that is performed for OATS and chambers for commercial EMC and use those results to validate the site. Thus, the

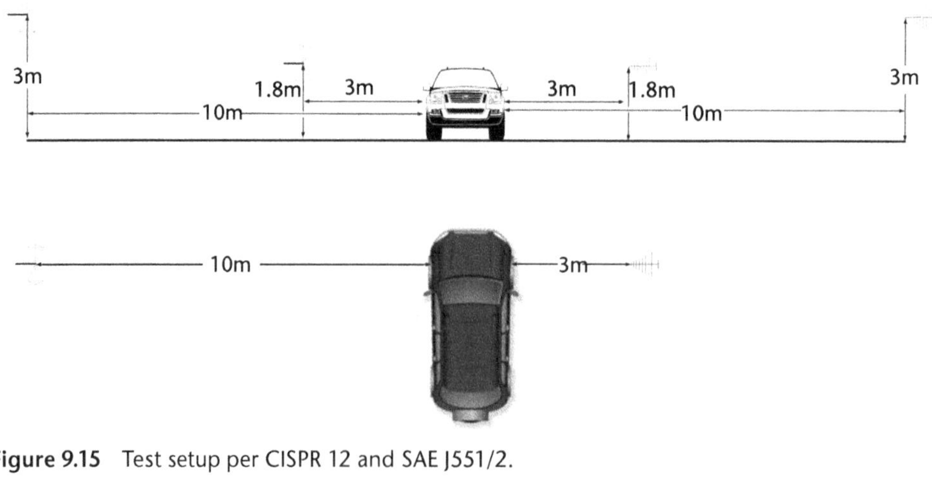

Figure 9.15 Test setup per CISPR 12 and SAE J551/2.

automotive full vehicle test anechoic chamber becomes a specialized version of the commercial EMC chamber where the QZ is large enough to encompass the vehicle. Other special considerations are a method to exhaust the combustion fumes out of the shield since the test calls for the internal combustion engine to be idling at a specific level of RPMs. The enclosure should be large enough to include two ranges to test both sides of the vehicle. This can be accomplished by placing the vehicle on either an azimuth positioner, or rotating platform, or turntable, and rotating it to test both sides of the vehicle following the test geometry described here and illustrated in Figure 9.15. The sketches in Figure 9.16 illustrate the use of the turntable to rotate the vehicle to test both sides with two separate ranges.

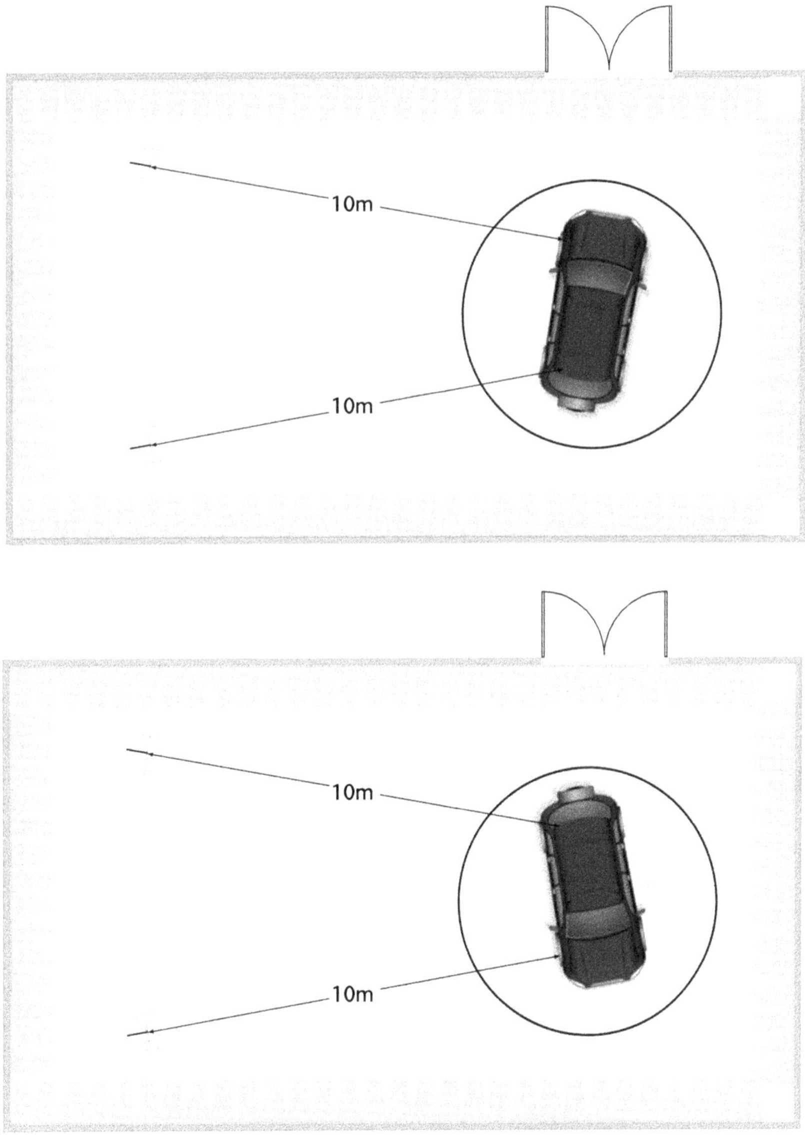

Figure 9.16 Using a turntable to rotate the vehicle inside an indoor range to test both sides of the vehicle as required by CISPR 12.

When using two measurement ranges in a single indoor facility, some precautions should be taken like cross-polarizing the testing antennas to reduce the effects of one antenna on the antenna at the other location. Thus, one antenna is set to vertical polarization, while the other antenna is set to horizontal and vice versa.

Recall that the indoor range for CISPR 12 is basically a specialized commercial EMC range. Section 9.5 will describes the rules to size these ranges and how to specify the RF absorber. Let us now proceed to look at the test setup for full vehicle immunity, where we learn that the same indoor facility used to test the emissions of a full vehicle can be used for immunity testing.

9.4.2.3 ISO 11451 Test Setup

The ISO 11451-2 is the international standard for immunity testing of full vehicles inside an anechoic chamber where the vehicle is illuminated with a field-generating antenna [10]. The document states that the test should be carried out inside an absorber-lined chamber. The document also specifies the reflectivity on the test region to be −10 dB. The level is not applicable to the transmission line systems (like the one shown in Figure 9.3), which are used below 30 MHz to generate the field. Hence, the −10-dB reflectivity is for 30 MHz and above. The total applicable frequency range for the standard is 10 kHz–18 GHz.

The 2015 version of the standard keeps the test setup (illustrated in Figure 9.17) from the previous versions. The reference point at which the field is to be generated is either 20 cm behind the front axel or 1m behind the point where the windshield meets the hood, depending which is closer to the front of the vehicle. The antenna is placed 2m from that reference point, and no part of the antenna can be closer than 0.25m from the floor or 50 cm from the outer surfaces of the vehicle. If the vehicle is taller than 3m, the reference point is moved to 2m above the ground.

Immunity antennas can be physically large to be efficient to radiate high fields with the least possible input power [4].

It should be noted that the standard [10] does provide some information on the shielded anechoic room. The standard mentions that “the radiating elements of

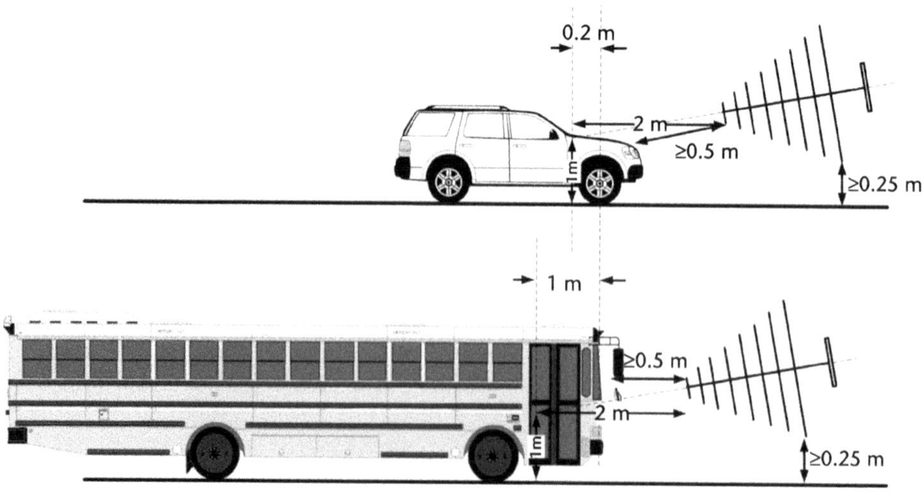

Figure 9.17 Test setup for ISO 11451-2 2015.

the field-generating device shall be no closer than 0,5 m to any absorbing material and no closer than 1,5 m to the wall of the shielded enclosure." This statement is similar to the one used in ISO 11452-2 for components.

The 2015 version of the standard does take into account the findings reported in [11], where the field generated by the transmission line system couples to the ceiling of the enclosure rather than the floor. The user is warned about these effects in the standard. Another inclusion in the document deals with electric vehicles, and test procedures include setups when charging electric vehicles. The minimum size for the chamber can be estimated based on the largest vehicle to be tested and the setup, but there is no methodology given for chamber validation to ensure that the recommended −10-dB reflectivity is met. The minimum length of the room is given by

$$L \geq 3.5\text{m} + V_{\text{length}} + A_{\text{length}} \tag{9.10}$$

where V_{length} is the vehicle length, and A_{length} is the length of the largest antenna. The height is in general given by the largest antenna rather than the vehicle except in the case of buses that may exceed the 3-m height. For typical sedan type vehicles, the height of the room can be given by

$$H \geq 1.5\text{m} + A_{\text{height}} + 0.25\text{m} \tag{9.11}$$

For those cases where the vehicle is very tall, such as a bus, and to take into account the transmission line system, a better approximate equation is given by

$$H \geq 1.5\text{m} + V_{\text{height}} + 0.25\text{m} \tag{9.12}$$

The width of the chamber should be wide enough to allow for a minimum of 2m of cable harness for the charging testing of electric vehicles. Additionally, the room should be wide enough to allow for the longest antenna elements when the radiating antenna is in horizontal polarization:

$$W \geq 1.5\text{m} + A_{\text{largest}} + 3.5\text{m} \tag{9.13}$$

This minimum-size room is illustrated in Figure 9.18.

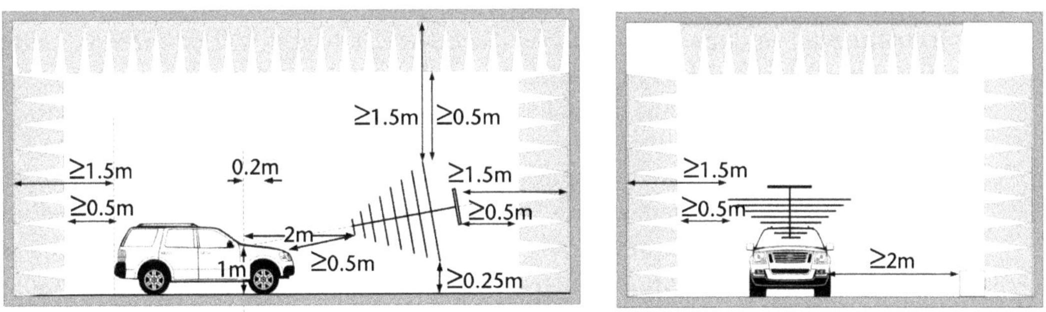

Figure 9.18 Sketch of an immunity setup in an indoor range following the guidelines of ISO 11451-2.

Equations (9.10) to (9.13) are coarse approximations. In general, for vehicles, both the immunity and the emissions testing is conducted within the same indoor facility; hence, as shown in Section 9.5, the size of the room needed for automotive emissions is always large enough to house the setup required for immunity testing per the ISO 11451-2.

9.5 Commercial EMC Measurements CISPR 16 and IEEE/ANSI C63.4

9.5.1 Introduction

Almost every piece of electronic equipment needs to be tested for EMC emissions. This is typically mandated by government agencies to qualify the product for appropriate marking to be sold or operated in a particular jurisdiction. Government agencies typically do not define the method of measurement, but refer to standardized methods. For emissions, typically the methodologies are the ones defined in the IEEE/ANSI C63.4 standard and the CISPR 16 standard. This "alphabet soup" of standards can be extremely complicated. For example, the CIPSR 16 has more than 15 parts. CISPR 16-1-6 deals with EMC test antenna calibration, but CISPR 16-1-5 deals with the sites where the antenna calibrations can take place and with test site validation.

In addition, there is a series of standards that follow CIPSR 16. While CIPSR 16 deals with the methods, calibrations, and uncertainty calculation, CISPR 14, for example, deals with household appliances and electrical tools. CISPR 15 is the device-specific standard for electrical lighting. This book does not detail all these standards. In fact, such an effort is in part futile, as the standards evolve to meet new challenges. For example, the CISPR 22 for office information equipment (e.g., computers and fax machines) was merged with the CISPR 13 (broadcast equipment) to create the new CISPR 32, which deals with emissions from multimedia equipment. This was done because today's TV receivers are also computers. Previously, TV receivers were tested under CISPR 13, while computers had to be tested per CISPR 22. This section concentrates on the test site as defined by both CISPR 16 and by the IEEE/ANSI C63.4 standard.

9.5.2 Test Site

The OATS is the standard defined and preferred site for performing EMC emissions on commercial equipment (and full vehicle testing). Even the indoor anechoic range validation site for the CISPR 25 ALSE calls for correlating to a measurement done on an OATS. If a vehicle is to be tested for emissions in an anechoic chamber, CISPR 12 tells us to correlate those measurements to measurements performed on an OATS.

The OATS is a metallic ground plane, free of reflective obstructions, where the device is tested for emissions. The test consists of placing the EUT on an azimuth positioner or turntable and rotating it while the measuring antenna is scanned from 1m to 4m over the metallic ground plane while measuring and looking for the highest field level radiated. The measurement was originally set to be done at 30m, but given the substantial cost of such a large area, measurements at 10m were selected. Later, measurements at a 3-m distance were also allowed for some equipment.

This was the approach used from 30 MHz to as high as required depending on the EUT. Although the standard was for the frequency range from 30 MHz to 1 GHz, the document stated that the equipment had to be tested to a given frequency related to one of the harmonics of the internal clock of the device. As computers got faster, users of the standard used their OATS to frequencies beyond 1 GHz.

The method for validating whether a site was suitable to be used as an OATS was the normalized site attenuation (NSA) measurement. This method applies both to OATS and to alternate sites. For example, earlier versions of the standard, like the 1992 version of ANSI C63.4-1992, mentioned that alternate test sites were allowed, provided that the alternate site met the attenuation requirements over the QZ and that the ground plane met the planarity requirements that a traditional OATS was required to meet. Anechoic chambers for EMC have been around since the early 1980s, and all of them are designed to meet the NSA criteria for the OATS. Before we look at the NSA, it should be mentioned that during the late 1990s and the first decade of the new millennium, the development of a test site for measurements above 1 GHz took place. The CISPR committee arrived at a method where the ground was now covered in absorber, and the reflectivity of the site was measured using a method different than the NSA approach. Anechoic chambers for EMC measurements spanning from 30 MHz to beyond 1 GHz, must meet the requirements of these two methodologies.

9.5.3 Normalized Site Attenuation

The site attenuation was defined by the IEC as the available input power to a short dipole for a field strength of 100 μV/m at a distance of 3m over a conductive ground plane or screen [12]. Smith et al. provided a derivation for the site attenuation in their paper [12]. The equations used in that paper provide the basis for the theoretical normalized site attenuation used by the ANSI C63.4 document. Barron provided a simple representation of the equations for normalized site attenuation [13], providing the following for the theoretical NSA:

$$NSA_{\text{Th}} = -20\log_{10}\left(f(\text{MHz})\right) + 48.9 + E_D^{\text{Max}} \tag{9.14}$$

where f is the frequency in MHz, and E_D^{Max} is the maximum received field. The test geometry is shown in Figure 9.19.

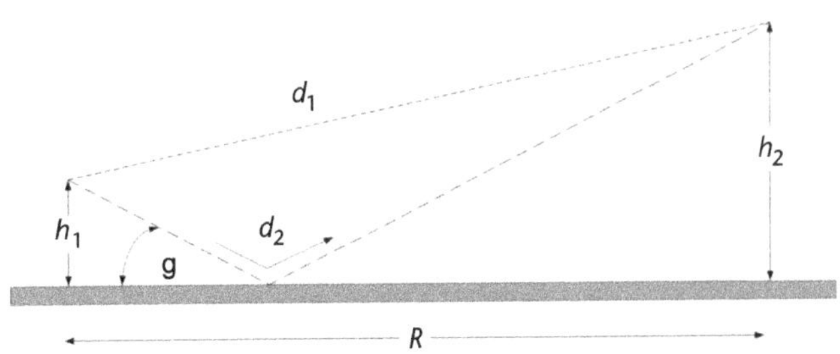

Figure 9.19 The NSA test geometry.

The transmit antenna is fixed at some height h_1. The height of the transmit is set per the standard to 1m, 1.5m, and 2m over the ground. Vertical polarization is done for 1m and 1.5m, while horizontal polarization is done at 1m and 2m over the ground. The receive antenna height (h_2) is varied from 1 to 4m over the ground. The test distance is usually 10m but the NSA test can be also done at a 3-m distance. The dimensions d_1 and d_2 are given by the following:

$$d_1 = \sqrt{R^2 + (h_2 - h_1)^2} \tag{9.15}$$

and

$$d_2 = \sqrt{R^2 + (h_2 + h_1)^2} \tag{9.16}$$

and the angle γ is obtained from $\gamma = \tan^{-1}((h_2 + h_1)/R)$. Barron provides the equations for the maximum field for the vertical and horizontal polarizations. For the horizontal polarization

$$E_{\text{DH}}^{\text{Max}} = \frac{\sqrt{49.2} \cdot \sqrt{d_2^2 + d_1^2 |\rho_{\text{H}}|^2 + 2d_1 d_2 |\rho_{\text{H}}| \cos\left(\varphi_{\text{H}} - \frac{2\pi}{\lambda}(d_2 - d_1)\right)}}{d_1 d_2} \tag{9.17}$$

and for the vertical polarization

$$E_{\text{DV}}^{\text{Max}} = \frac{\sqrt{49.2} \cdot \sqrt{d_2^6 + d_1^6 |\rho_{\text{H}}|^2 + 2d_1^3 d_2^3 |\rho_{\text{V}}| \cos\left(\varphi_{\text{V}} - \frac{2\pi}{\lambda}(d_2 - d_1)\right)}}{d_1^3 d_2^3} \tag{9.18}$$

where ρ_H and ρ_V is the reflection coefficient of the ground plane, and φ_H and φ_V is the phase of the reflection coefficient, for horizontal and vertical polarizations. The reflection coefficient can be set to 1 with the appropriate phase if a perfect electric conductor is assumed as the ground surface. If the conductivity of the ground material is known, then the reflection coefficients are given by

$$\rho_{\text{H}} = \frac{\sin\gamma - \sqrt{(\varepsilon_r - j60\lambda\sigma - \cos^2\gamma)}}{\sin\gamma + \sqrt{(\varepsilon_r - j60\lambda\sigma - \cos^2\gamma)}} \tag{9.19}$$

and

$$\rho_{\text{V}} = \frac{(\varepsilon_r - j60\lambda\sigma)\sin\gamma - \sqrt{(\varepsilon_r - j60\lambda\sigma - \cos^2\gamma)}}{(\varepsilon_r - j60\lambda\sigma)\sin\gamma + \sqrt{(\varepsilon_r - j60\lambda\sigma - \cos^2\gamma)}} \tag{9.20}$$

Both CISPR 16 and ANSI C63.4 required for a site to be usable for EMC emission measurements to deviate no more than ±4 dB from the theoretical. The 4-dB number comes from assigning 1 dB of error to those parts that may deviate from the theoretical; hence, 1 dB per antenna, 1 dB for the site itself, and 1 dB for the instrumentation. However, the antennas may have a smaller contribution as does the instrumentation. Furthermore, if the site is properly made, the contribution is also smaller. If the reflectivity of the absorber on the lateral walls provides an error that is within ±3 dB, the anechoic chamber typically meets the requirements of an OATS. Per Figure 5.38, absorptions between −7 and −10 dB are more than adequate to meet this requirement.

To validate a range, the NSA is measured at five locations and two different heights above the ground for each polarization (vertical and horizontal). The five locations are the center, and the front, back, and left and right points of a cylindrical QZ. The cylinder in the case of EMC chambers is vertical, with the normal axis to the circles being a vertical line. Figure 9.20 shows the test geometry in plan view. At each of the locations, the distance between the antennas remains unchanged (either 10m or 3m). The transmit antennas are on the QZ, while the receive antenna is scanned on a dielectric mast from 1 to 4m. The test is performed in horizontal polarization with the transmit antenna at 1m and 2m over the ground and for vertical polarization with the antennas at 1m and 1.5m over the ground.

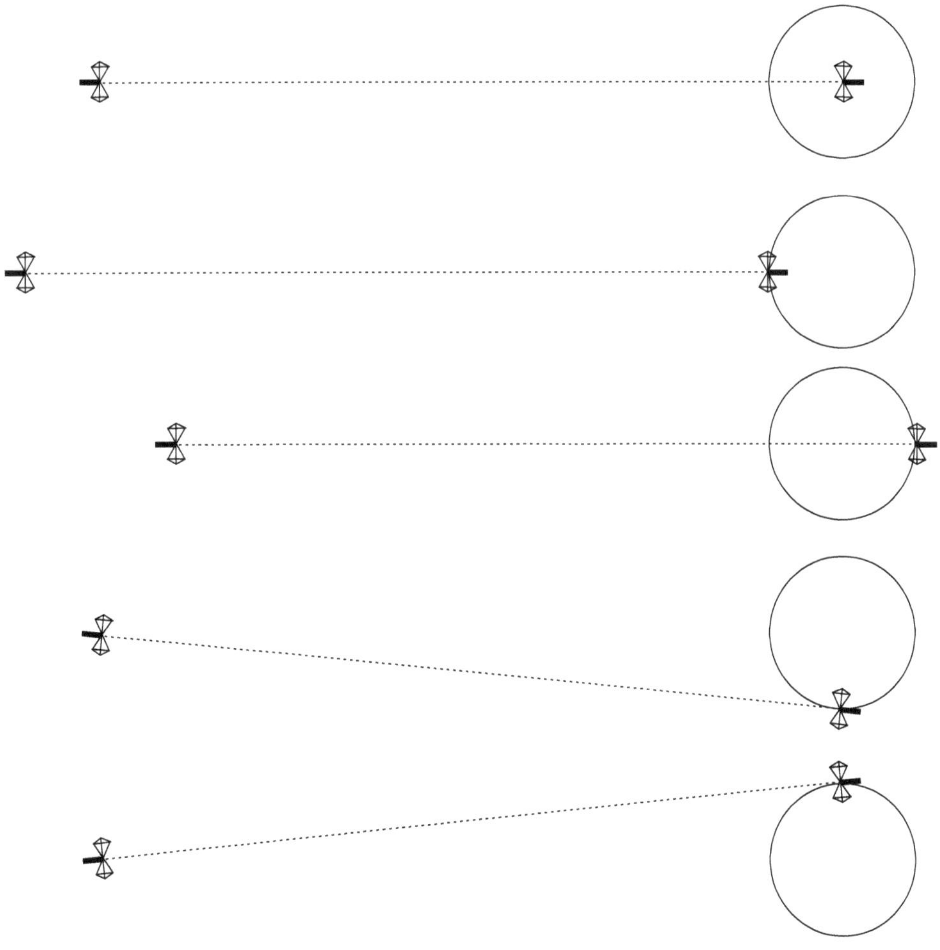

Figure 9.20 Geometry of the NSA test in plan view.

Most hybrid absorbers meet this level at 30 MHz as shown in Chapter 3. Indeed, if the room is large enough there is no need for hybrid absorbers, and a large pyramidal material can be used. An example of such an EMC chamber is the large anechoic range at the Laboratory of Integration and Testing (LIT) at the Brazilian Institute of Space Research (INPE). The room is suitable for doing compliant EMC testing as well as having reflectivity levels lower than –50 dB above 1 GHz to perform antenna testing. The chamber is large to accommodate a 7-m-diameter QZ for automotive EMC. The range is about 28-m-long and 15-m-wide with a 11.5-m height. Figure 9.21 shows pictures of the range. All the absorber was 1.82m in length, with some of it optimized with a curvilinear cut. The absorber had about –9 dB of normal incidence reflectivity at 30 MHz. The room was wide enough to have a good angle of incidence at the lateral walls. Those angles were roughly between 41 and 33.7 degrees, providing almost normal incidence performance. The ceiling was high enough that the angles of incidence were small. Additionally, path loss helps slightly in these large anechoic rooms to be able to meet the deviation from the theoretical NSA.

9.5.4 Site VSWR Testing

The NSA measurement is the accepted approach for validating sites up to 1 GHz. Above 1 GHz, CISPR developed a different approach where a fully anechoic range is required versus the traditional semi-anechoic approach shown to this point. The test requires a transmit antenna to be located at the QZ at discrete locations. Figure 9.22 shows the test procedure. The idea is that the sampling at those discrete points along the line allows for the peaks and valleys of the standing wave to be captured and for the reflectivity of the chamber in the QZ to be measured. The approach is not ideal, but it is currently the only accepted method. The committee that defined the test procedure had to make some compromises to ensure that the test could be done with simple equipment without requiring expensive positioners or additional receivers than the ones used by the laboratory.

Figure 9.21 A large semi-anechoic room for EMC using no ferrite or hybrid absorber. On the left the author stands with Benjamin Galvao, who was at the time director of the laboratory. (*Source:* Author's private collection.)

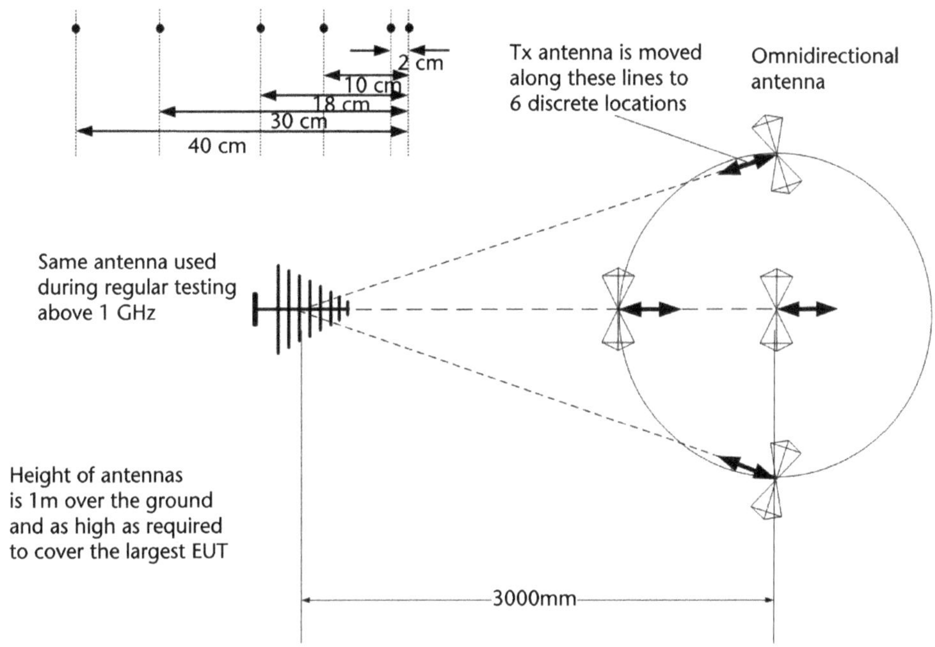

Figure 9.22 The site VSWR test for validating indoor ranges for testing above 1 GHz.

The test procedure has been criticized by, among others, Windler [14] and Chen [15], with both suggesting a time-domain reflectometry approach.

Semi-anechoic rooms used for EMC testing below 1 GHz can be used for testing above 1 GHz; the test distance is reduced to 3m. While some chamber manufacturers have shown chambers that met the VSWR test at 10m, there are no standards that set limit levels for measurements done at 10m. Also, the room has to be fully anechoic. This can be accomplished by placing some absorber on the floor, as shown in Figure 9.23, but there is a big difference between the absorber required to make a 3-m measurement and the amount required for a 10-m test setup.

Figure 9.23 shows a site VSWR test. A horn antenna is used as the receive antenna, as it will be during the normal test. A small omnidirectional biconical antenna is the transmit antenna. The antenna is mounted on a low dielectric Styrofoam support.

9.5.5 Sizing the Indoor Range

What rules of thumb can be used to size a typical EMC emissions chamber for CISPR 16 and ANSI C63.4? The room must be large enough to enclose the test setup, and the rules used here can be applied to estimate the size of CISPR 12 automotive rooms.

The starting point is the test distance and the QZ size. The room must be large enough to accommodate the antenna, its positioned, the test distance, and the QZ diameter. In general, the rule of thumb is that the chamber test distance and the QZ diameter are on average 72% of the overall length. For a 3-m test chamber, the percentage can be as small as 58%. Using this rule leaves enough space for the antenna and its positioner as well as absorber coverage; thus,

$$L = \kappa\left(R + QZ_{D}\right) \tag{9.21}$$

Figure 9.23 The site VSWR test in a semi-anechoic chamber; notice the floor absorber needed to make the range a fully anechoic range above 1 GHz. (*Source:* Author's private collection.)

where $\kappa = 1.39$ for 10-m range chambers and $\kappa = 1.7$. The reason for a different κ is that while the test distance is smaller, the antenna and the positioner do not reduce in size (although smaller absorbers are typically used). This typically allows for about 1–2m of space between the back of the QZ and the absorber, and allows for the absorber to be about 1-m hybrid material.

For the width, the goal is to have an angle of incidence that is less than 45° in the worst case. Thus, using the test distance R and QZ_D as the QZ diameter, the width of the room is given by

$$W_{10m} = R + \frac{QZ_D}{2} \tag{9.22}$$

For a 3-m chamber, it is desirable to have a better angle of incidence and make the room slightly wider to accommodate the antennas (which do not shrink); hence, (9.22) becomes:

$$W_{3m} = \frac{R}{\tan 30^\circ} + \frac{QZ_D}{2} \tag{9.23}$$

It should be noted that the QZ size on a 3-m range chamber cannot be more than 3m in diameter or the test antenna will be in the QZ when measuring the back point of the QZ.

The height of the room should also allow for a good angle of incidence. Ideally, angles smaller than 45 are desirable. The worst angle occurs when the scan antenna is at 4m, and the transmit antenna is at 2m. Because of that the height is given by

$$H_{10m} = \frac{R + 6}{2} \tag{9.24}$$

For a 3-m chamber it is desirable to have the angle at 30°. Thus, (9.24) becomes

$$H_{3m} = \frac{\left(\frac{R}{\tan 30^\circ}\right) + 6}{2} \tag{9.25}$$

Using the equations above, a 10-m range chamber for a 4-m QZ is approximately 19.46m in length, 12m in width, and 8m in height. These are approximate dimensions as the absorber plays an important part. For some manufacturers, these approximations may suffice, while for others they may require additional space to obtain a compliant chamber.

These are approximations, and the size of the rooms can be reduced. To avoid the reflections being in-phase or out-of-phase, the range may be at an angle of a few degrees from the longitudinal plane of symmetry of the room. Figure 9.24 shows a chamber with the configurations of an oblique range.

The equations (9.22) and (9.24) given for a 10-m range chamber can be used to estimate the size of a CISPR 12 chamber, where, in addition, ISO 11451-2 can be performed.

9.5.6 RF Absorber

For 10-m chambers, hybrid absorbers are the best solution. Typically for these chambers, RF absorber that is 1 to 1.5m in size is ideal. If absorber with no ferrite is to be used, it has to be long enough to provide the required reflectivity at 45 degrees of incidence. Since for 3-m chambers the angles of incidence are smaller, a shorter hybrid absorber can be used, typically one that is 60 cm in height.

9.6 The Commercial Immunity Chamber IEC 61000-4-3

9.6.1 Introduction

The other side of commercial EMC testing is radiated immunity. (Remember that only radiated testing falls within the scope of this book.) The main standard covering radiated immunity for commercial applications is IEC 61000-4-3. Anechoic

Figure 9.24 A 10-m range EMC chamber with the QZ (defined by the location of the turntable) offset from the center to avoid in-phase and out-of-phase reflections from the walls. Notice the floor absorber (moved to the side) for the above 1-GHz testing per CISPR 16. (*Source:* Author's private collection.)

chamber testing per this standard applies from 80 MHz to 6 GHz; the frequency was increased in the 2006 version, which was amended in 2010 [16]. There were other changes, such as providing tests for linearity and harmonic levels of the power amplifiers. However, there were no changes for the test setup, except for allowing a smaller uniformity area for tests above 1 GHz and making changes to the requirements of the table where the EUT sits, with wood tables no longer allowed.

The test setup places the antenna at 3m from the EUT. The room is fully anechoic or at least must have a floor absorber between the antenna and the EUT. The test setup in the IEC 61000-4-3 is validated by performing a field uniformity measurement. The intent of this measurement is to ensure that the field is uniform over the EUT or, to be more precise, over a 1.5-m by 1.5-m grid of 16 points spaced 50 cm apart. The grid is placed 80 cm from the shielded floor of the chamber. The field is considered uniform if the largest difference between the 16 points in the grid is 6 dB. Figure 9.25 illustrates the side view and front view of the test setup. Of the 16 points four points can be dropped if they are out of the 6-dB range. Originally the intent was to drop the bottom row of points, since they where the closest to the ground and potentially affected by the absorber on the floor. However, the standard does not specify which points, so any four points out of specification can be thrown out per the standard.

The latest version [16] allows for the use of subgrids if the EUT is smaller than the 1.5-m by 1.5-m grid. If the EUT is small enough that a 0.5-m by 0.5-m grid of four corner points can be used, then no points can be dropped. The 6-dB difference

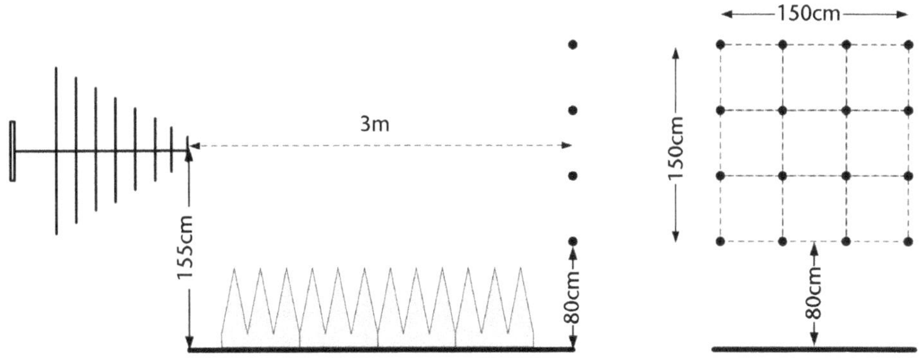

Figure 9.25 The test setup for measuring the field uniformity per IEC 61000-4-3.

is equivalent to a reflected level of –9.6 dB. This is an oversimplification since we have reflected waves approaching from many directions, but typically absorbers with a normal incidence performance of better than 10 dB are suitable to treat the chamber walls and floor.

9.6.2 Sizing the Chamber

These chambers are easy to design and can be as small as 8m by 4m by 3.5m. The size allows for hybrid absorber around 60 cm in height and for space for the radiating

Figure 9.26 The source antenna inside an IEC 61000-4-3 chamber, the source antenna can be fairly large as the lowest frequency is 80 MHz. (*Source:* Author's private collection.)

antenna and the table for the EUT. Figure 9.26 shows one of these small chambers. The room is fully lined with ferrite tile, which is sufficient to meet the requirements of the standard. To achieve the desired performance at frequencies above 1 GHz, a hybrid absorber is placed on the walls and ceiling between the test antenna and the EUT or uniformity plane. The wall opposite the antenna is also covered with hybrid absorber. The floor is bare except for ferrite and polyurethane foam to cover the area of the floor between the radiating antenna and the uniformity plane where the EUT will be located.

It is more common for the IEC test to be conducted in a larger emissions chamber. The larger chamber meets the uniformity requirements, and the only change is to add an area of absorber between the antenna and the uniformity plane, as shown in Figure 9.27. The turntable used for emissions measurements can be seen on the floor in Figure 9.27. There is no requirement for the EUT to be rotated, but the uniformity plane is set at the QZ of the large chamber purely for convenience, since there are power outlets under the turntable to power the EUT.

9.7 Conclusion

EMC measurements constitute a wide subject. Here we briefly cover the main radiated immunity and emission standards and the three main areas of EMC radiated measurements; namely, defense, automotive, and commercial. In addition, the

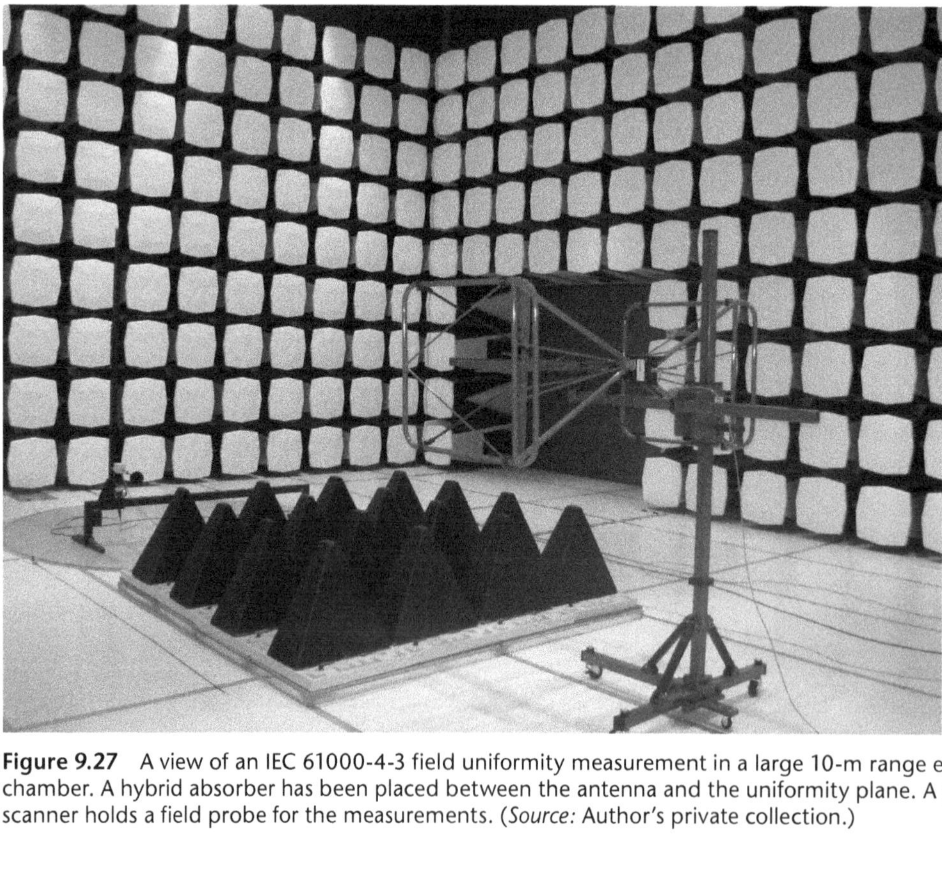

Figure 9.27 A view of an IEC 61000-4-3 field uniformity measurement in a large 10-m range emissions chamber. A hybrid absorber has been placed between the antenna and the uniformity plane. A planar scanner holds a field probe for the measurements. (*Source:* Author's private collection.)

chapter offers guidance on the range size and absorber specifications as well as explanations for some of the test geometries and how they affect the absorber and size for the range.

References

[1] U.S. Department of Defense, "Requirements for the Control of Electromagnetic Interference Characteristics of Subsystems and Equipment," Department of Defense, 11 December 2015.

[2] RTCA, "DO-160G Environmental Conditions and Test Procedures for Airborne Equipment," RTCA, Washington, D.C., December 8, 2010.

[3] Rodriguez, V., "Automotive Component EMC Testing: CISPR 25, ISO 11452–2 and Equivalent Standards," *IEEE Electromagnetic Compatibility Magazine*, Vol. 1, No. 1, 2012, pp. 83–90.

[4] Rodriguez, V., "Antenna Selection for Automotive EMC Emissions and Immunity Applications (100 kHz to 18 GHz)," *Interference Technology, ITEM 2005*, pp. available at https://interferencetechnology.com/antenna-selection-for-automotive-emc-emissions-and-immunity-applications-100-khz-to-18-ghz/, May 2005.

[5] U.S. Department of Defense, "MIL-STD-461: Electromagnetic Characteristics Requirements for Equipment," Department of Defense, Washington, D.C., 31 July 1967.

[6] ISO, "ISO 11452-2: Road Vehicles—Component Test Methods for Electrical Disturbances from Narrowband Radiated Electromagnetic Energy—Part 2: Absorber-Lined Shielded Enclosure," ISO, Geneva, Second Edition, 2004-11-01.

[7] CISPR, "CISPR 25: Radio Disturbance Characteristics for the Protection of Receivers Used on Board Vehicles, Boats, and on Devices—Limits and Methods of Measurement," International Electrotechnical Commission, Geneva, Switzerland, Second Edition, 2002-08.

[8] CISPR, "Vehicles, Boats, and Internal Combustion Engine Driven Devices—Radio Disturbance Characteristics—Limits and Methods of Measurement for the Protection of Receivers Except Those Installed in the Vehicle/Boat/Device Itself or In Adjacent Vehicles/Boats/Dev," International Electrotechnical Commission, Geneva, Switzerland, September 2001.

[9] Anderson, P., "The Present Status of the International Automotive EMC Standards," in *2009 IEEE International Symposium on Electromagnetic Compatibility*, Austin, TX, 2009.

[10] ISO, "ISO 11451-2: Road Vehicles—Vehicle Test Methods for Electrical Disturbances from Narrowband Radiated Electromagnetic Energy—Part 2: Off-Vehicle Radiation Sources," in *ISO*, Geneva, Switzerland, Fourth Edition, 2015-06-01.

[11] Rodriguez, V., "Analysis of Large E Field Generators in Semi-Anechoic Chambers Used for Full Vehicle Immunity Testing: Numerical and Measured Results," in *2014 International Symposium on Electromagnetic Compatibility (EMC'14/Tokyo)*, Tokyo, Japan, 12–16 May 2014.

[12] Smith, A. A., Jr., R. F. German, and J. B. Pate, "Calculation of Site Attenuation from Antenna Factors," *IEEE Transactions on Electromagnetic Compatibility*, Vols. EMC-24, No. 3, August 1982, pp. 301–316.

[13] Barron, M., "Normalized Site Attenuation Calculations Using Standard Spreadsheet Analysis," in *1999 IEEE International Symposium on Electromagnetic Compatibility*, Seattle, WA, 1999.

[14] Windler, M., "Site Qualification Above 1 GHz and sVSWR Systemic Errors," in *2010 Asia Pacific International EMC Symposium*, Beijing, China, 2010.

[15] "Uncertainties in sVSWR and A Proposal for Improvement Using Vector Response Measurements," in *2014 International Symposium on Electromagnetic Compatibility (EMC'14/Tokyo)*, Tokyo, Japan, 12–16 May 2014.

[16] IEC, "IEC 61000-4-3-2006+AMD1:2007+AMD2:2010 CVS: Electromagnetic Compatibility (EMC)—Part 4-3: Testing and Measurement Techniques—Radiated, Radio-Frequency, Electromagnetic Field Immunity Test," International Electrotechnical Commission, Geneva, Switzerland, 2010.

CHAPTER 10

Specialty Ranges

10.1 Introduction

RF anechoic chambers or indoor ranges are not only limited to the measurement of antennas, EMC, and RCS. RF anechoic chambers have a place in every test where a free space or a controlled electromagnetic environment is desired. There is a series of ranges that are specialized for a type of testing where the RF environment must be known and controlled. Among these special tests that require controlled EM environments are the over-the-air (OTA) testing of cell phones per the CTIA standards, MIMO systems throughput testing, measurement of airborne radomes, and target or scene simulators.

Technically, some immunity EMC testing can be considered as hardware-in-loop testing; this is the case for some automotive testing where the vehicle systems are monitored as the vehicles are illuminated by a known high EM field. The vehicle is accelerated, and the brakes are applied, while different field levels illuminate the vehicle. Those immunity tests are performed to ensure that the vehicle operates as intended when affected by external sources. These external sources can be extreme with field levels around 200 V/m or higher.

In HWiL testing, the EUT is tested in a normal environment, not extreme. Is the phone going to make a call as the individual is walking down the street? Is the radome going to affect the weather radar system and provide the wrong direction for the storm or wind shear to the pilot? Is the missile going to follow the intended target and not be confused by decoys? In all these cases, the system under test operates using EM-radiated energy, and the functionality of the device is tied to the EM test. Answering these types of questions is the reason for HWiL testing, and to perform these tests, a specific RF or EM environment is required; this will be provided by the anechoic indoor range.

10.2 CTIA OTA Testing

The Cellular Telecommunications Industry Association (CTIA) is a trade association created by the cellular phone industry for self-regulation. Around 2000, some members of CTIA approached companies involved in RF testing and with experience on standard development to develop a test for testing the OTA performance of cell phones. A standard test procedure was developed during the early part of the first decade of the millennium. Among the parameters of importance to the industry were total radiated power (TRP) and total isotropic sensitivity (TIS). The TRP, as the name implies, is the integration of the power density radiated by the device over the sphere, although the method eliminated the contributions to the total radiated power pointing to the ground and to the sky. The sensitivity test required a modulated signal, or call, to be made to the phone being tested and the attenuation of the signal until the phone failed to respond.

Originally, the CTIA OTA test procedure applied to handsets. The QZ size was set to a 30-cm-diameter sphere [1]. Handsets were mainly physically and electrically small at their frequencies of operation; for example, the 3GPP Band 12 (699–746 MHz) [1] had the longest wavelength at 0.43m. This was clearly larger than a handset, and it also meant that the QZ size required can be smaller than a wavelength. The electrically small QZ means that it is not possible to use the free-space VSWR typically used for far-field chambers to validate the performance of the range. Even at the highest band, the 3GPP Band 41, where the highest frequency, 2,690 MHz, has a wavelength of 0.11m, the QZ is about three wavelengths. However, at that frequency, the typical high-gain probe antenna used for free space VSWR testing is almost as large as the QZ itself.

The CTIA document provided a new range evaluation methodology, which varies a lot from the free-space VSWR test. The probing antennas are omnidirectional instead of the preferred directional probes of the free-space VSWR test (see Chapter 5), and the probes must have a highly axisymmetric pattern on the azimuth plane of the antenna. The pattern, per the standard requirements, cannot deviate more than ±0.1 dB as it is rotated around its axis. The probe antennas of choice are sleeve dipoles [2] and center-fed loops (see Figure 10.1). I refer, as does the standard, to the plane as azimuth, rather than the E- or H-plane For the dipole, the azimuth plane is the H-plane, and for the loop antennas, the azimuth plane is the E-plane. The choice of this antenna is, in part, because in both of these types, the feeding network is located outside of the azimuth plane of the antenna. Quarter-wavelength

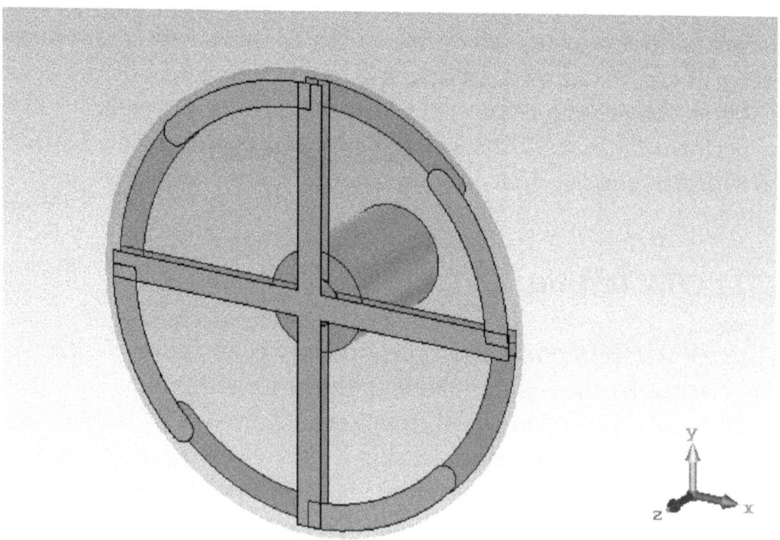

Figure 10.1 A center-fed printed-loop antenna similar to the type used in the range validation test per CTIA.

chokes (or ferrite beads at lower frequencies) are used to minimize cable effects on the measurement.

The test is designed to characterize not only the range, but the entire system, including the positioning systems used in the range to hold the device. The test procedure is to measure the azimuth pattern of the dipoles and loops at different locations within the QZ. The deviations from the circle pattern, what the standard calls the *ripple*, is then used to do a statistical analysis that is converted into an uncertainty for the system. The entire test is based on these reference probe antennas having as perfect of a pattern on the azimuth plane. While achieving ±0.1 dB on a sleeve dipole is not that difficult, as shown in [2], accomplishing it on loops is not necessarily an easy task. Calibrating the dipoles and loops is critical, and users should request pattern data for the actual probe antenna rather than typical data. This will reduce potential error during the test.

The minimum test distance is calculated using different approximations for the far-field boundary and choosing the longest of the different results. Thus, the typical $2D^2/\lambda$ approximation to the far field is only used for frequencies above GNSS (1,574 MHz), and at that frequency, the 30-cm QZ places the far field at 0.96m. The highest frequency of 2,690 MHz places the far field at 1.61m away from the center of the QZ. Below 1,574 the far-field approximation is loosely based on $R \gg \lambda$, and to be more precise, the standard uses $R > 3\lambda$.

Since this document is a standard, the anechoic chamber size is determined by the document [1]. The document states that the anechoic chamber must be large enough to accommodate the largest test distance (i.e., 1.61m), but the document also states that a shorter 1.2-m measurement distance can be used, provided the additional uncertainty from the shorter distance at some frequencies is indicated [1]. In addition to the test distance, the chamber must contain the QZ and the range antenna. As is the case with the rectangular chamber, it is desirable to have some space between the QZ and the absorber tips. The absorber layout is important since the reflections from the absorber have to be minimized; the reason is that there will be contributions from the positioners and the communications antennas in the range. (These communication antennas are there to establish calls with the device being tested.)

The short test distance helps because, it causes the angles of incidence to be more acute at the lateral surfaces of the chamber. This happens because the room should be large enough to accommodate a human entering the room and setting up the test. These setups include, among others, placing the phantom head on the positioner and setting the phone in the phantom hand. These are tasks that require a person to stand next to the positioner to work on the setup, thus requiring a large working area.

One of the first CTIA-Approved Test Laboratories (CATLs) was a tapered anechoic chamber at the ETS-Lindgren manufacturing plant. The reason for the tapered chamber approach is that, at the time of the range implementation, some of the cellular bands operated in the 450-MHz range, and as discussed in Chapter 5, a tapered anechoic room was a better choice to achieve the required reflectivity at those low frequencies. The design, which was successfully used in subsequent implementations, used a 36-inch absorber on the end wall; thus the reflectivity levels were approximately −30dB at the 450-MHz range and improved to about −40 dB at about 700 MHz. These estimates are based on the absorption of the rear end wall materials. Recall that the QZ sizes are about 30-cm cylinders or spheres, per the CTIA document [1]. Hence at the lowest frequency of 450, the QZ is slightly smaller than $\lambda/2$. At 700 MHz, the QZ is about 0.7λ. At that frequency and QZ size, the interference ripple may not be measurable.

Analysis of different CTIA-approved chambers shows that reflectivity levels better than −20 dB at the lowest frequency of operation (to be more precise −22 to −25 dB) appear to be sufficient for the required measurement uncertainty. Most of these ranges operate down to 700 MHz (rather than the 450 MHz mentioned above), so typically 24-inch absorber is used to achieve the −20-dB level at the lowest frequency using range antennas with about 10 dB of directivity. Remember that the purpose of the room is not to measure the pattern. While pattern measurements are performed, these are omnidirectional patterns with no sidelobes or a main lobe so −20 dB provides less than ±1 dB of error, and −25 dB provides ±0.5dB, but this assumes that a full cycle of the wave appears in the QZ, which is not the case in these small QZs; accordingly, the contribution from the range may be smaller.

The typical room is evolving, and as wireless technologies change, so will the test environments. Section 10.3 discusses the testing of MIMO systems. Different environments are being suggested for 5G system testing as mentioned in Chapter 2.

10.3 MIMO OTA Testing

MIMO technology and anechoic chambers do not seem to go together. The concept for MIMO-based systems is that advantage can be taken of reflected signals from buildings, vehicles, and other structures to be able to get a complete message reconstructed from the pieces received by different antennas in the system. Thus, a MIMO system testing should have a way of using potential multipath to increase the throughput of data.

Conversely, anechoic chambers are typically designed to minimize the multipath by having an absorber on the internal surfaces of the range. Because of this drawback of the anechoic range for MIMO testing, the industry looked at loaded reverberation chambers, where a discrete absorber was placed inside the room to lower the resonant behavior of the enclosure and simulate a MIMO channel model [3]. One of the criticisms of the reverberation approach is that the channel model it simulates tends to produce the best MIMO performance environment [4].

Foegelle brought the anechoic environment back into the picture [4] by creating what he called a "boundary array" method. This approach places the MIMO device under test (DUT) inside an anechoic room and uses a spatial channel simulator to excite a series of range antennas. These antennas present the DUT with signals arriving with different characteristics such as directions, polarizations, and fading. The flexibility of this anechoic range with a spatial channel emulator method is that it can create different environments to test the performance of the system.

The MIMO room is typically a cubical room where the sources are placed around the room as shown in Figure 10.2. The design of this room from an anechoic perspective is not different than the design of a spherical near-field system with a movable range such as a gantry or an arch. However, unlike the SNF systems, these

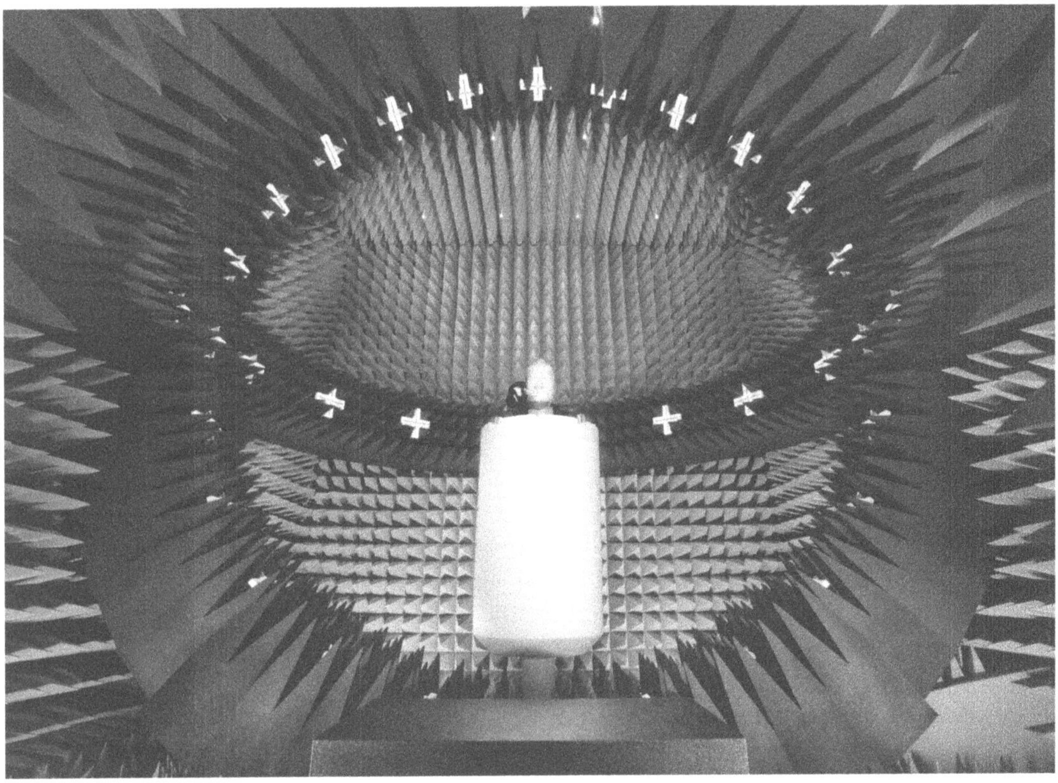

Figure 10.2 A MIMO test chamber. The horizontal ring of switched dual-polarized antennas is used for MIMO testing. The vertical ring is used for pattern measurements and for CTIA type testing. (Photo courtesy of ETS-Lindgren, Inc.)

anechoic rooms are not intended to measure radiation patterns; hence having a certain level of reflectivity to allow for the measurement of radiation pattern features is not required. The goal of MIMO chambers is to measure the system performance over the air, typically throughput (data rates). This does not mean that anechoic performance is not important. Reflections from the lateral surfaces may aid in the throughput of the system. For example if we are trying to measure the throughput in a case where all the reflections come from a single direction, a nonideal MIMO environment [4], with some signals reflected from the range surfaces, may distort the measurement.

10.4 RTCA-DO-213: Commercial Radome Testing

Testing of the radome (a portmanteau of radar and dome) is performed to ensure that the radome does not have detrimental effects on the radiation pattern of the antenna it is protecting or hiding. Radome testing is basically a set of antenna measurements with and without the radome covering the antenna, so antenna measurement ranges are designed to test radomes. The complexity of radome ranges is related to how the antenna may be positioned within the radome, and a special positioning system is required that moves the antenna and the radome precisely with respect to each other to have the main beam of the antenna look through the radome at specific angles.

Military radome testing is the most complicated radome testing, as well as the most precise. Because of the classified nature of a lot of the test procedures, military radome testing is not discussed in this book. However, almost every commercial aircraft has a nose radome that houses a radar system that checks the weather ahead of the aircraft. RTCA has recently developed a new version of the standard for the testing of these nose radomes on commercial airliners [5]. Prior to this latest version, measurements of radomes were allowed at shorter distances if performed indoors. This was probably a compromise to allow for smaller indoor facilities and the standard allowed for the test distance to be as short as $D^2/2\lambda$, which is one-quarter the far-field distance; thus the phase taper is as large as 90 degrees. Furthermore, as shown in [6], testing even at the standard far-field rule causes the first null to fill up and the first sidelobe to merge with the main beam. The new document made a change, not to set the distance to the more standard far-field boundary of $2D^2/\lambda$ but instead, to restrict the phase taper to 22.5°. Clearly, these two are technically the same as shown in Chapter 1, but by setting the phase taper as the parameter instead of as a physical distance, the standard opened the door to the use of compact ranges. Another important change has been the definition of D. Per the previous version, D was the diameter of the weather radar antenna, but now D also includes any portion of the radome in front of the system antenna as reported in [7]. Because the standard looked at blessing the compact range approach, it also entered the typical compact range variation for the wavefront into the standard; therefore, the amplitude variation or amplitude ripple is set to be $\leq \pm 0.5$ dB, and the phase ripple to $\pm 5°$. The reason for those values is that they are the typical specifications offered by compact range manufacturers.

The new standard allows the use of near-field to far-field ranges. It should be noted that the near-field approach can be extremely time-consuming as full acquisitions are required with the radar antenna pointing at specific directions inside the radome. Each position requires a full semispherical acquisition; hence the time for evaluating a radome can be significant. If a near-field to far-field system is to be used, the pattern of a radar antenna must be measured and compared to the pattern measured in a far-field range, that is, a range where the phase taper meets the better-than-22.5° requirement.

The weather radars operate at 9.345 GHz, which is in the X-band. Because of this high frequency, the absorber can be physically short and still provide sufficient absorption to measure small sidelobe levels that are typically found on weather radar antennas.

The operating frequency of these ranges whether they are far-field or near-field, is very narrow. The wavelength is only 3.2 cm. The required absorber is typically less than 8 inches (20 cm). Let us analyze a typical compact range for RTCA DO-160A measurements, like the one shown in Figure 10.3. The overall internal dimensions of the room are 10.65-m-long by 6.5-m-wide and by 5-m-tall. The focal length of the reflector is 144 inches or 3.65m. Using the standard 5/3 focal length (f_l) for the distance to the QZ places, the QZ at 20 ft. or 6.1m. The height of the focal point or vertex of the reflector is 71.8 inches (1.82m) over the floor. The QZ generated by the reflector is an elliptical cylinder with the major axis being 1.82m, the minor axis being 1.22m, and the length of the QZ cylinder 1.82m. The range is not symmetrical in elevation, with the center of the range being at 2.929m or 9.6 ft.

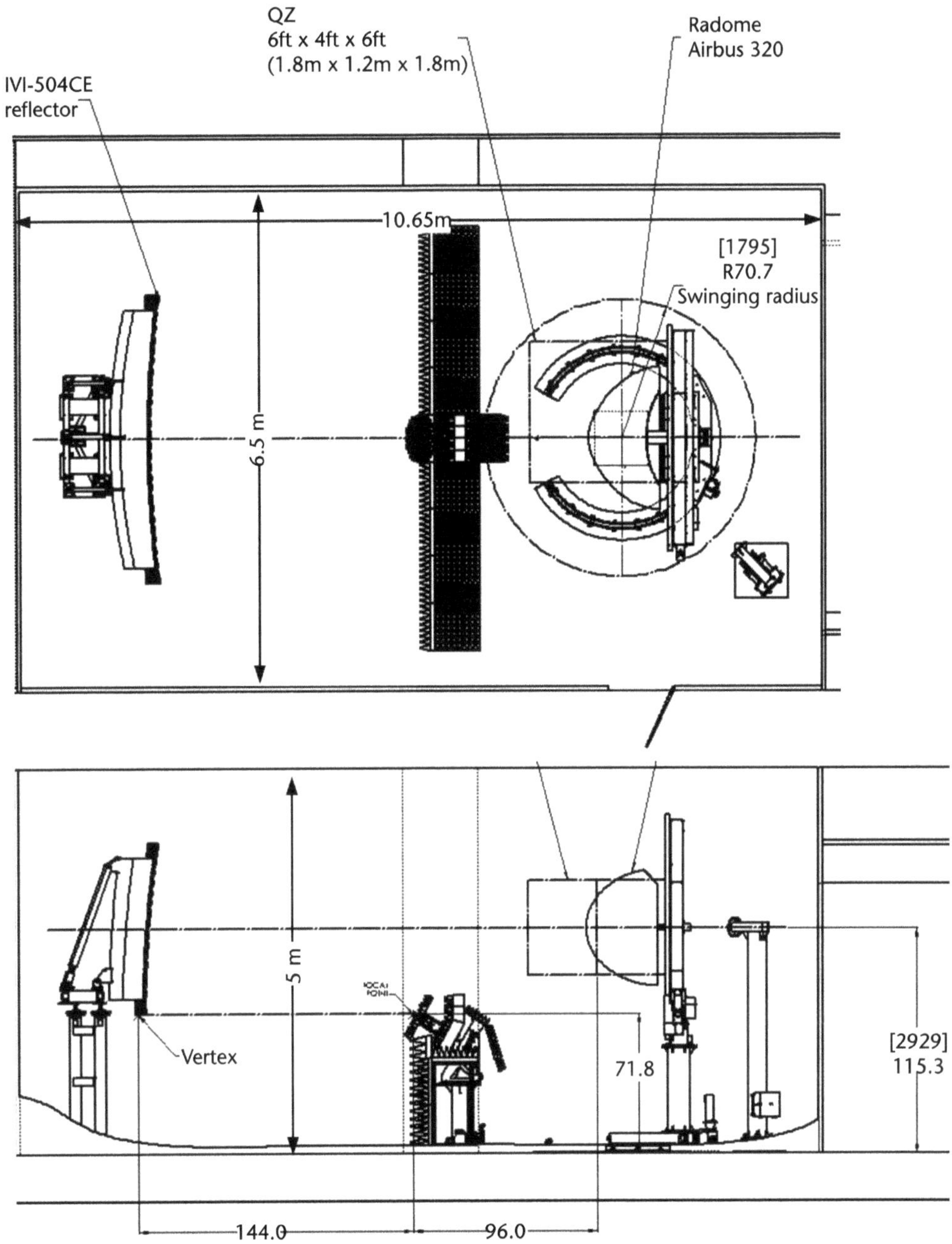

Figure 10.3 Plan and elevation view of a RTCA DO-213A range concept.

This geometry makes the specular reflection at the ceiling 51.6 degrees, while the angle of incidence on the lateral walls is 42.3 degrees. The specular bounce of the floor is hidden by the feed positioner and a fence. A traditional pyramidal absorber of 8 inches in height (20 cm) can provide more than 50 dB of reflectivity. Thus, selecting that 8-inch absorber for the end wall provides reflectivity levels that add an equivalent ripple of ±0.02 dB, well below the ±0.5 required per the standard.

Let us assume that the compact range reflector illuminates the lateral surfaces with field levels that are about 20 dB lower than the QZ illumination levels. Placing a 4-inch (10 cm) thick pyramidal absorber or 3.16λ, provides us with a bistatic reflectivity of −39.7 dB at 51.6 degrees or −48 dB at 42.3 degrees. (See Appendix 3A.) Adding that to the reflector illumination of −20 places the reflectivity of the lateral walls at −68 dB and the reflected ray from the ceiling at −56 dB. Thus, the ripple level from the room scattering is much smaller than the level required by the standard.

10.4.1 Positioning System

The most complex part of a radome testing range is the positioning system. The standard itself states, "Radome performance measurements require a test range that simulates the actual operation of the radome as installed over the antenna system on the aircraft." The standard continues, "...[a] radome/antenna positioner system shall be used that locates the test antenna within the radome at the same location as used in the aircraft installation." Thus the positioning system is not as simple as a roll-over azimuth positioner. There are actually two separate positioning systems working together, one that positions the antenna and one that positions the radome. Figure 10.4 shows one of these positioners. We notice a large structure that supports the radome. Some of these radomes can be very large—think Boeing 747 or Airbus 380—while other radomes can be small, like the ones used in regional jets. The radome rotates on azimuth to present different sides to the reflector, while a roll positioner can rotate the entire radome. This combination allows for the radome to

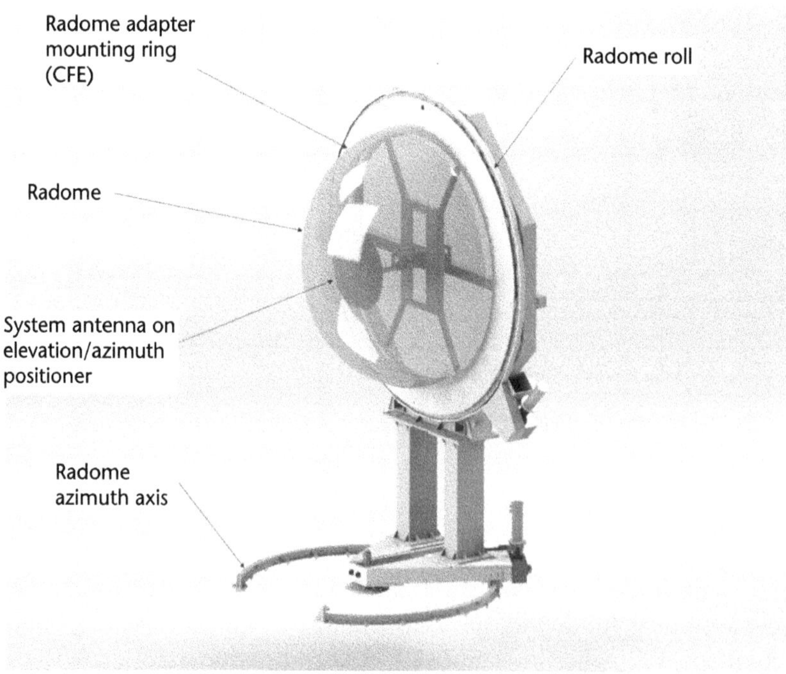

Figure 10.4 A radome measurement system positioner. (3D rendering courtesy of NSI-MI technologies).

be positioned so that every part of its surface can be at some point directed toward the reflector. Supporting the radar antenna inside the radome, there is a small positioner (smaller that the antenna it supports) that can position the antenna around two intersecting axes.

The two tests that are performed on radomes are the transparency efficiency (TE) test and pattern test. For the TE test, the antenna must remain fixed and pointed at the compact range reflector while the radome is rotated, as needed, to put a particular set of angles, elevation, and azimuth in front of the antenna main beam. The polarization of the range antenna (the feed) must match the radar antenna polarization.

The TE test is the most common, and it should be performed after any repair to the radome to ensure that the repair did not affect the transparency of the radome.

A more complex test is the pattern test, which has to be done for qualifying the radome and, in some cases, for repaired radomes as is the case for radomes with forward-looking wind-shear capabilities or any other "sidelobe depending radar functions" [5]. The pattern test, or antenna radiation pattern cuts, calls for the radar antenna to remain pointed at given radome elevation and azimuth angle (EL, AZ) direction or test node or point. While the radome and antenna combination remains fixed (in terms of the components' relationship to each other), the entire assembly is scanned +/−90 degrees in one axis to obtain a pattern cut. Again, the polarization of the range antenna must match the test antenna polarization throughout the test scan. Thus, the pattern is measured for the radar antenna pointing at different nodes on the radome (the number of nodes to be tested is given by the standard).

10.5 Target and Scene Simulators

Target and scene simulators are systems that present a series of RF signals to an EUT or unit under test (UUT) at different angles of arrival and at specific times. Wayne [8] describes how these signals may represent targets, decoys, jammers, or even other environmental conditions. These signals must appear as plane waves to the UUT sensor. Thus, the target simulator is a real-time programmable plane-wave generator.

Technically the MIMO testing systems are a specialized version of a target or scene generator, as they try to generate an environment in which the MIMO device appears to be surrounded by reflecting objects, direct paths, and even reflections from moving objects.

It is the application that makes the target simulator much more complicated than the MIMO testing system. The speed at which the scene changes is several orders of magnitude above what a MIMO system may need to be tested. The AIM-120 AMRAAM top speed is classified [9]; some sources place its speed at Mach 4 [10]. Clearly, these are extremely fast systems, and the system inside the UUT must guide itself as the target potentially changes its direction of travel.

Wayne [8] describes different systems to generate the scene. Some of them are mechanical systems in which antennas are placed on a positioner where they are moved in front of the UUT [11]. Figure 10.5 illustrates one of these types of mechanical systems. This approach limits the locations for the target and the decoy, as they

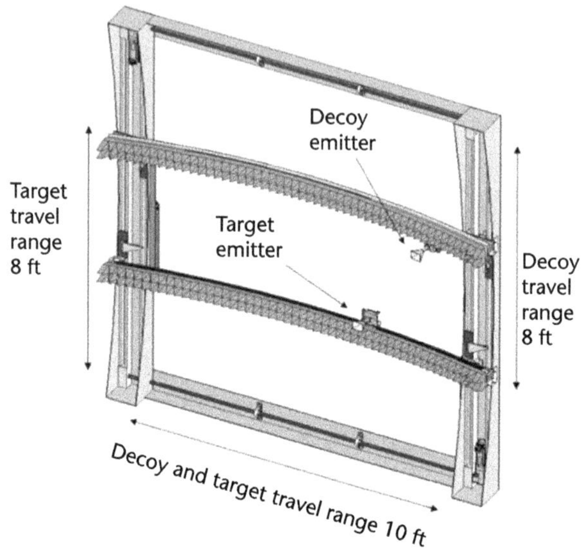

Figure 10.5 A mechanically positioned target simulator with a single target and decoy. (*Source:* NSI-MI Technologies. Reprinted with permission.)

cannot cross each other. The UUT is placed far away in the far field of the antennas mounted on the positioner. Other systems use a compact range reflector that is positioned to change the angle of arrival of the plane wave at the location of the UUT sensors. Finally, there is the large wall of emitter systems, where a large array is positioned away from and opposite the UUT. Portions of the array are excited, creating plane waves that can be generated as if they arrive from different angles. Several targets can be offered to the UUT using this approach, and these targets can cross each other as they simulate objects flying around (see Figure 10.6). This last approach, the wall of emitters is the most complex of these target simulator systems. Every element of the array can be controlled in magnitude and phase; a small subgrid of elements, a subarray, can be activated to present a plane wave at

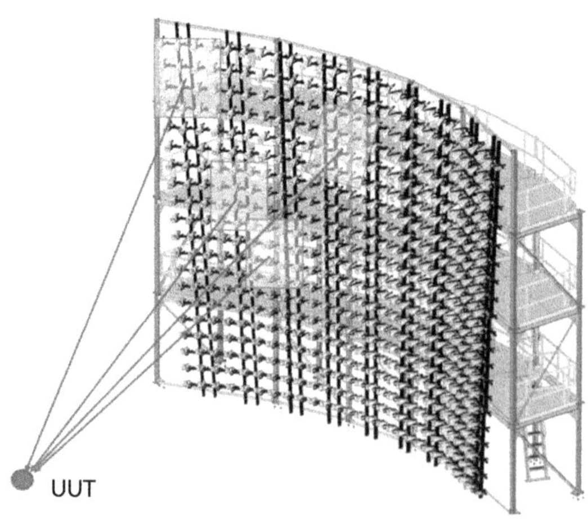

Figure 10.6 A large wall of emitters scene generator. Portions of the large array are activated to create waves arriving at different angles. The phase and magnitude of every element can be controlled. (*Source:* NSI-MI Technologies. Reprinted with permission.)

the UUT. Elements in the array can be turned off and adjacent elements turn on to make the target or decoy shift in position.

Regardless of the approach to generate the scene, thought must be put into the chamber. The target antenna and the decoy antenna become range antennas, and the antennas on the UUT become the AUT in the QZ of the range. This means that for some locations of the target antenna, the angle of incidence onto the lateral wall can be very poor. Consider the target simulator in Figure 10.7. As one of the antennas moves to the side to illuminate the UUT with a different angle of arrival, the antenna gets closer to the absorber wall and changes the angle of incidence of the field onto the absorber—in this case, from about 54 degrees to 66 degrees of incidence. As shown in Figure 3.9, if the absorber treatment is 5λ in thickness (that is 15 cm or 6 inches at the middle of the X-band), the reflectivity changes from lower than −50 dB to about −37 dB. Potentially an additional "target" will appear in the test in the form of the reflection from the wall. This may not be an issue, but it may provide incorrect results in the test and may affect, in some cases, the creation of a plane wave at the UUT.

Ideally, using the design rules introduced in Chapter 5, the rectangular range can be designed to ensure that good angles of incidence are present to all potential locations of the target. If possible, an absorber that is long enough must be used to ensure that for the range of angles, the reflectivity stays below certain level.

It is very often that existing rooms are being retrofitted for installing target simulators, and chamber geometries are being selected based on how well they performed in the previous use. However, a range that was designed for operation with the range antenna and the QZ at fixed locations (typically center-line in the range) is not the ideal geometry for a target simulator where there are multiple range antennas that illuminate the QZ. Setting target simulators in nonideal rooms results potentially in angles of incidence onto the lateral surfaces of the range that exceed 75 degrees. Figure 3.9 shows that $15\text{-}\lambda$ absorber can still provide 50 dB or reflectivity; hence, 18-inch absorber should be a good choice, but it may be that such a

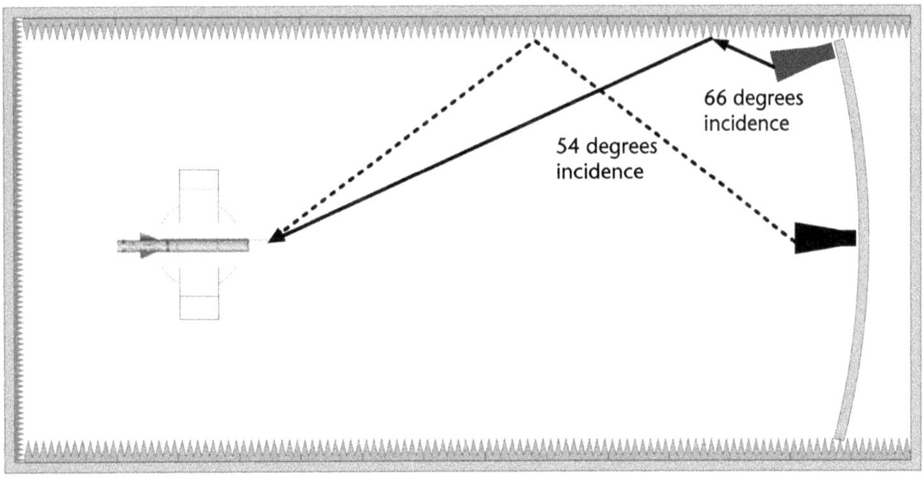

Figure 10.7 A target simulator in an anechoic range. The angles of incidence on the walls can vary significantly.

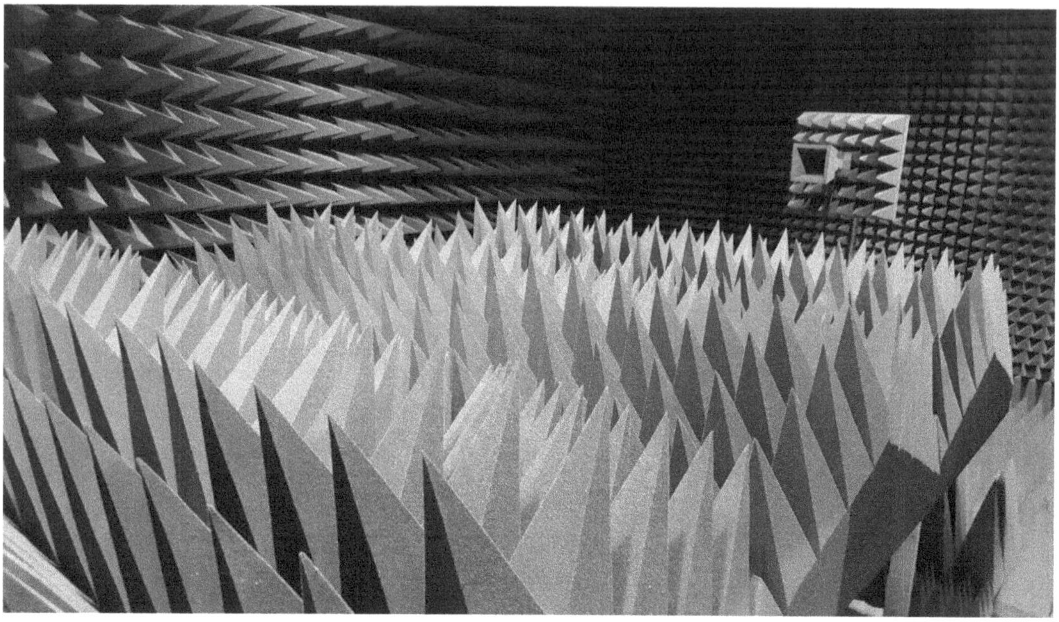

Figure 10.8 Specially cut absorber for high angles of incidence. The absorber cut was designed for treating the lateral surfaces of a target simulator range. (*Source:* Author's private collection.)

choice of absorber is too long, and it may block the direct view from the emitters to the UUT. Thus, in some cases longer absorber is not an option.

In some cases, special absorber may need to be designed to provide a better reflectivity at those angles of incidence. In [12], a method is explored for making measurements of the absorber at high angles of incidence. The purpose is to check the numerical model results for the reflectivity of special cuts of absorber for high angles of incidence. Figure 10.8 shows this type of absorber in the test setup described in [12].

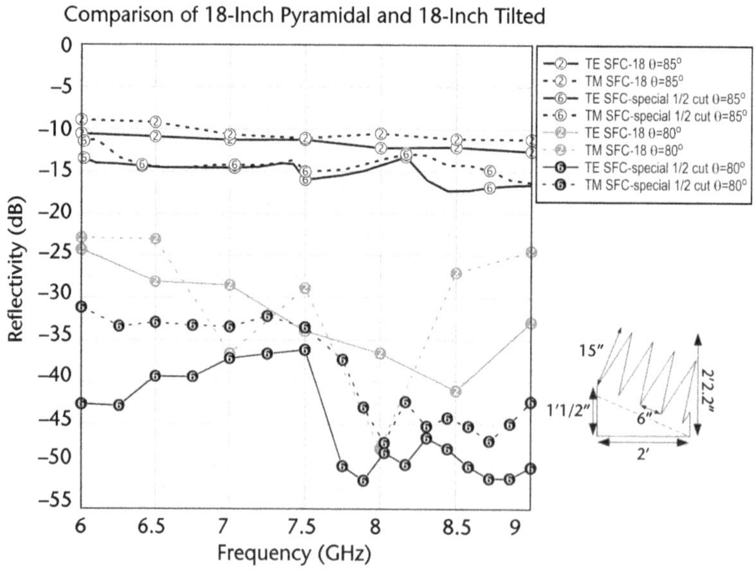

Figure 10.9 Improvement of the tilted absorber at extreme angles of incidence.

The special cuts can potentially increase the reflectivity of the sidewalls a few decibels at these extreme angles of incidence. At some angles of incidence for some of the polarizations, the improvement can be close to 10 dB, (as shown in Figure 10.9) when compared to a standard absorber. This approach can be used to try to improve the reflectivity from the lateral surfaces when the sources move close to the lateral range walls.

10.5.1 Chebyshev Absorber Arrangements

Other approaches, such as the Chebyshev arrangement of the absorber, introduced by Burnside [13], have been used for these highly oblique incidence cases. Chebyshev arrangements are another approach to achieve improved absorption from a standard absorber.

The Chebyshev absorber concept is based on the assumption that the reflection at each step is very small so that the higher order interactions between the steps can be ignored. The theoretical improvement of a third-order Chebyshev can be easily computed. However, this computation assumes that we have an infinite field of absorber and a plane wave incidence. In reality, we have neither so the improvement is always a fraction of the values computed using the approach given here.

If we assume a step d, the propagation phase between the steps is given by [13]

$$\psi = \frac{2\pi}{\lambda} d \cos\theta \tag{10.1}$$

where θ is the angle of incidence onto the absorber field. This gives us a frequency associated with the propagation phase ψ that is

$$f = \frac{c\psi}{2\pi d \cos\theta} \tag{10.2}$$

When $\psi = \pi$, (10.2) gives us the period frequency:

$$f_p = \frac{c}{2d \cos\theta} \tag{10.3}$$

The center of the frequency band is given by $f_p = f_c/2$. Gau et al. [13] give us the third-order Chebyshev normalized reflection coefficient as

$$\tilde{\Gamma}(\psi) = \tilde{\Gamma}_m e^{-jN\psi}\left[\sec^3\psi_m \cos 3\psi + \left(3\sec^3\psi_m - 3\sec\psi_m\right)\cos\psi\right] \tag{10.4}$$

where N is the number of sections or the order levels of the Chebyshev polynomial, $\tilde{\Gamma}_m$ is the pass band ripple, and ψ_m is the lowest passband angular frequency. It is shown in [13] that the normalized reflection coefficient can be written for a third-order filter as

$$\tilde{\Gamma}(\psi) = 2e^{-jN\psi}\left[\tilde{\Gamma}_0 \cos 3\psi + \tilde{\Gamma}_1 \cos\psi + \tilde{\Gamma}_2 \cos(-\psi) + \tilde{\Gamma}_3 \cos(-3\psi)\right] \tag{10.5}$$

So from (10.4) and (10.5), we obtain that

$$\tilde{\Gamma}_0 = \tilde{\Gamma}_3 = \frac{\tilde{\Gamma}_m}{2}\sec^3\psi_m \tag{10.6}$$

and

$$\tilde{\Gamma}_1 = \tilde{\Gamma}_2 = \frac{\tilde{\Gamma}_m}{2}\left(3\sec^3\psi_m - 3\sec\psi_m\right) \tag{10.7}$$

We can then write an equation for the magnitude of the reflection coefficient of a Chebyshev arrangement, which in decibels becomes

$$\left|\tilde{\Gamma}(\psi)\right|(\text{dB}) = 10\log_{10}\left[\left(\left(\tilde{\Gamma}_0 + \tilde{\Gamma}_3\right)\cdot\cos 3\psi + \left(\tilde{\Gamma}_1 + \tilde{\Gamma}_2\right)\cdot\cos\psi\right)^2\right] \tag{10.8}$$

where $\tilde{\Gamma}_0 + \tilde{\Gamma}_1 + \tilde{\Gamma}_2 + \tilde{\Gamma}_3$ and, as shown in (10.6) and (10.7), $\tilde{\Gamma}_0 = \tilde{\Gamma}_3$ and $\tilde{\Gamma}_1 = \tilde{\Gamma}_2$. These coefficients are the percentage of the absorber in the period that is at a given step.

So, let us make the step $d = \lambda/4$ at the center of the X-band (10 GHz). The step is only 0.75 cm. with a maximum extra step of 2.25 cm. It is extremely difficult in reality to maintain these tolerances in absorber manufacturing, especially with polyurethane-based absorber.

If we choose the top step and the no step absorber to be 20% each of the absorber period, and the first and second step to be 30% each (or to put it in terms of the coefficients $\tilde{\Gamma}_0 = \tilde{\Gamma}_3 = 0.2$ and $\tilde{\Gamma}_1 = \tilde{\Gamma}_2 = 0.3$), the absorber arrangement period can be seen in Figure 10.10. If the angle of incidence on the absorber is 80 degrees, then $d = \lambda/(4 \cdot \cos 80°)$ or 4.32 cm or a maximum step of 12.96 cm (5.10 inches).

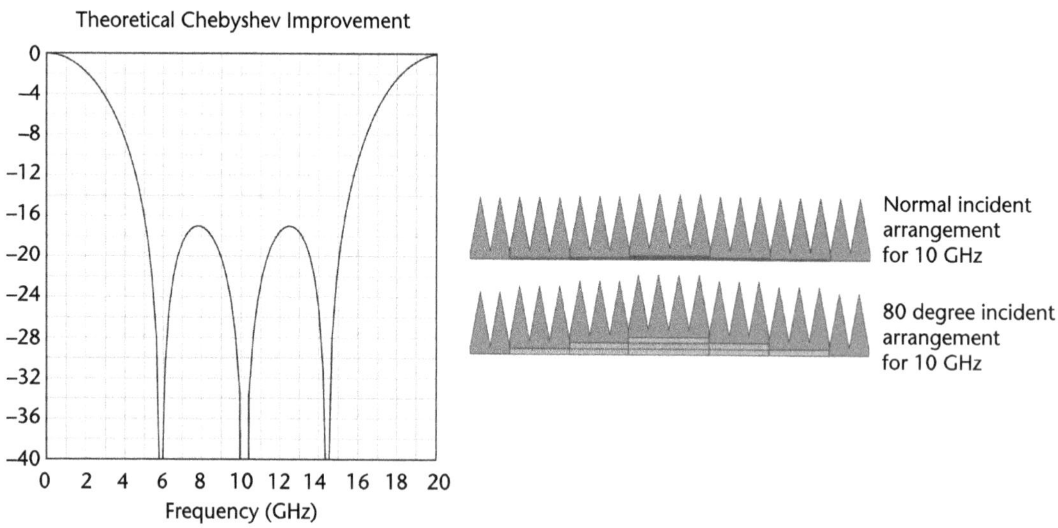

Figure 10.10 Theoretical improvement of a $\tilde{\Gamma}_0 = \tilde{\Gamma}_3 = 0.2$ and $\tilde{\Gamma}_1 = \tilde{\Gamma}_2 = 0.3$ arrangement for normal incidence and oblique incidence. Notice that the step size increases with the angle of incidence.

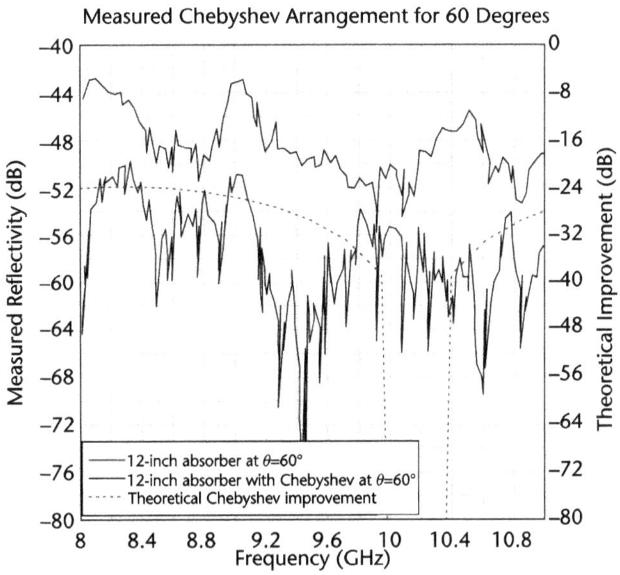

Figure 10.11 Measured Chebyshev arrangement of 12-inch pyramidal absorber.

While according to the data in Figure 10.10, the Chebyshev could have an improvement of –16 dB from about 5 GHz to 16 GHz, we should remember that this is theoretical. Figure 10.11 shows measured results of a 12-inch absorber from 8 to 11 GHz at $\theta = 60°$. In addition to the standard pyramidal absorber, a Chebyshev arrangement of $\tilde{\Gamma}_0 = \tilde{\Gamma}_3 = 0.16667$ and $\tilde{\Gamma}_1 = \tilde{\Gamma}_2 = 0.33333$, was used with a 1.5-cm step or 0.6 inches. The absorber period has one pyramid with no step, two pyramids with one step, two pyramids with two steps, and one pyramid with three steps. Then the arrangement is repeated again through the lateral surface. The theoretical improvement is plotted with the measured data. While the theoretical curve shows a 24-dB improvement across the band with a null at 10 GHz, we can see that the difference in the measured data is about 10 dB on average.

Either tilting the absorber or a Chebyshev arrangement of the absorber on the lateral surfaces may be used to improve the reflectivity of target simulators, especially if there are space constraints that keep the room from being larger. The best approach is to make the room large enough to have adequate incidence onto the absorber.

10.6 Passive Intermodulation

Passive intermodulation (PIM) is a phenomenon that occurs at joints between metals of two conductive materials when two separate different frequencies are injected into the system. The results is that a series of harmonics will appear, especially when the original carrier frequencies are high-power; otherwise the harmonics may be down in the system noise. PIM is important in satellite testing, and unfortunately, sources of PIM include ferrites, nickel and nickel plating, some steels including stainless, and contacts contaminated by dirt.

Typically, PIM rooms required welded shielding to avoid dirt getting on joints between panels. In addition, absorber pieces should be completely encased in paint to prevent the potential dust from handling materials from getting into any gaps.

References

[1] CTIA, "Test Plan for Wireless Device Over-the-Air Performance: Method of Measurement for Radiated RF Power and Receiver Performance V. 307," CTIA Certification Program, Washington, D.C., June 2017.

[2] Rodriguez, V., "Numerical Analysis of Sensitivity of Precision Reference Dipole Performance To Manufacturing Tolerances and Workmanship," in *30th Annual Antenna Measurement Techniques Association Symposium AMTA 2008*, Boston, MA, 2008.

[3] Rosengren, K., P. Bohlin, and P.-S. Kildal, "Multipath Characterizations of Antennas for MIMO Systems in Reverberation Chambers Including Effects of Coupling and Efficiency," in *2004 IEEE Antennas and Propagation Symposium*, Monterey, CA, 2004.

[4] Foegelle, M., "MIMO Device Performance Measurements in a Wireless Environment Simulator," *IEEE Electromagnetic Compatibility Magazine*, Vol. 1, No. 4, 2012, pp. 123–130.

[5] RTCA SC-230, "RTCA/DO-213A Minimum Operational Performance Standards for Nose-Mounted Radomes," RTCA, Washington, D.C., 2016.

[6] Hollis, J. S., T. J. Lyon, and J. L. Clayton, *Microwave Antenna Measurements* (Second Edition), Atlanta, GA: Scientific Atlanta, 1970.

[7] McBride, S. T., et al., "Changes in the DO-213 Standard for Commercial Nose-Radome Testing," in *39th Annual Meeting and Symposium of the Antenna Measurement Techniques Association*, Atlanta, GA, 2017.

[8] Wayne, D., "The 7 Common Habits of Highly Effective Target Simulators," in *39th Annual Meeting and Symposium of the Antenna Measurement Techniques Association (AMTA 2017)*, Atlanta, GA, 2017.

[9] U.S. Department of the Navy, "United States Navy Fact File: AIM-120 ADVANCED MEDIUM-RANGE, AIR-TO-AIR MISSILE (AMRAAM)," 10 March 2017, https://www.navy.mil/navydata/fact_display.asp?cid=2200&tid=100&ct=2. [Accessed 26 March 2019].

[10] Wikipedia, "AIM-120 AMRAAM," 23 March 2019, https://en.wikipedia.org/wiki/AIM-120_AMRAAM. [Accessed 26 March 2019].

[11] Wayne, D., "RF Target and Decoy Simulator," in *Annual Meeting and Symposium of the Antenna Measurement Techniques Association (AMTA 2011)*, Denver, CO, 2011.

[12] Rodriguez, V., B. Walkenhorst, and J. Bruun, "A Method for the Measurement of RF Absorber Using Spectral Domain Transformations," in *2019 IEEE International Symposium on Antennas and Propagation and USNC-URSI Radio Science Meeting—AP-S/URSI 2019*, Atlanta, GA, 2019 (accepted).

[13] Gau, J.-R. J., W. D. Burnside and M. Gilreath, "Chebyshev Multilevel Absorber Design Concept," *IEEE Transactions on Antennas and Propagation*, Vol. 45, No. 8, 1997, pp. 1286–1293.

About the Author

Dr. Vince Rodriguez attended The University of Mississippi (Ole Miss), in Oxford, Mississippi, where he obtained his B.S.E.E. in 1994. Following graduation, Dr. Rodriguez joined the Department of Electrical Engineering at Ole Miss as a research assistant. During his tenure at the department, he earned his M.S. and Ph.D. (both degrees in engineering science with emphasis in electromagnetics) in 1996 and 1999, respectively. After a short period as visiting professor at the Department of Electrical Engineering and Computer Science at Texas A&M University-Kingsville, Dr. Rodriguez joined EMC Test Systems (now ETS-Lindgren) as an RF and electromagnetics engineer in June 2000. During that time he was involved in E-field generator design and the RF design of several anechoic chambers, including rectangular and tapered antenna pattern measurement chambers, some of them operating from 100 MHz to 40 GHz. He was also the principal RF engineer for the anechoic chamber at the Brazilian Institute for Space Research (INPE), the largest anechoic chamber in Latin America, with a fully automotive, EMC, and satellite testing chamber. In September 2004, Dr. Rodriguez took over the position of senior principal antenna design engineer, placing him in charge of the development of new antennas for different applications and on improving the existing antenna line. In the fall of 2010, he served as antenna product manager. In this position, Dr. Rodriguez was in charge of all technical and marketing aspects of the antenna products at ETS-Lindgren. Among the antennas developed by Dr. Rodriguez are broadband double and quad-ridged guide horns, high-field generator horns, stacked LPDAs for automotive and military testing, and printed antennas for wireless testing. While mainly dedicated to antenna design, Dr. Rodriguez continued being involved in anechoic chamber design. In November 2014, Dr. Rodriguez joined MI Technologies (now NSI-MI Technologies) as a senior applications engineer. In this position, Dr. Rodriguez worked on the design of antenna, RCS, and radome measurement systems. During his tenure at NSI-MI, Dr. Rodriguez was involved in designing several antenna and RCS anechoic ranges for near- to far-field compact range and far-field measurements. In 2017, Dr. Rodriguez was promoted to staff engineer, positioning him as the resident expert on RF absorber and indoor antenna ranges at NSI-MI. He continues to be involved in the design of antennas and special RF absorbers to meet the necessary specifications of large antenna, RCS, and HWiL systems. During the period he was involved in the design of over three tapered anechoic chambers and three large RCS ranges with test areas exceeding 6m across. In addition to his duties at NSI-MI Technologies, Dr. Rodriquez has served as Adjunct Research Professor of Electrical Engineering at Ole Miss since the fall of 2017.

Dr. Rodriguez is the author of more than 50 publications, including journal and conference papers as well as book chapters. Dr. Rodriguez holds patents for a hybrid absorber and for a dual-ridge horn antenna. Dr. Rodriguez is a Senior Member of the IEEE and several of its technical societies. Among the IEEE technical societies, he is a member of the EMC Society, where he served as Distinguished Lecturer of the society from 2013 to 2014 and also served on the board of directors of the IEEE-EMC. Dr. Rodriguez also serves on the standard's committee of the IEEE AP-S and as secretary of the working group for IEEE STD 149 and IEEE STD 1128. He is also a member of the Antenna Measurement Techniques Association (AMTA) where he has served on the board of directors of AMTA as meeting coordinator (2010–2011) and vice president in 2012. In 2014, Dr. Rodriguez was named an E. S. Gillepie Fellow of the AMTA. Dr. Rodriguez is also a member of the Applied Computational Electromagnetic Society (ACES), where he served on the board of directors from 2014 to 2017. In 2019, he was elevated to Fellow of ACES. He has served as a reviewer for the *ACES Journal*, the *IEEE Transactions on Antennas and Propagation*, and for the *Journal of Electromagnetic Waves and Applications* (JEWA). He has also served as a reviewer for several IEEE, AMTA, ACES, and EuCAP conferences. He has served as chair of sessions at several conferences of the IEEE, AMTA, and Conference on Precision Electromagnetic Measurements (CPEM). Dr. Rodriguez is a full member of the Sigma Xi Scientific Research Society and of the Eta Kappa Nu Honor Society.

Index

Recent Titles in the Artech House Electromagnetics Series

Tapan K. Sarkar, Series Editor

Advanced FDTD Methods: Parallelization, Acceleration, and Engineering Applications, Wenhua Yu, et al.

Advances in Computational Electrodynamics: The Finite-Difference Time-Domain Method, Allen Taflove, editor

Analysis Methods for Electromagnetic Wave Problems, Volume 2, Eikichi Yamashita, editor

*Analytical and Computational Methods in Electromagnetics,*Ramesh Garg

Analytical Modeling in Applied Electromagnetics, Sergei Tretyakov

Anechoic Range Design for Electromagnetic Measurements, Vince Rodriguez

Applications of Neural Networks in Electromagnetics, Christos Christodoulou and Michael Georgiopoulos

CFDTD: Conformal Finite-Difference Time-Domain Maxwell's Equations Solver, Software and User's Guide, Wenhua Yu and Raj Mittra

The CG-FFT Method: Application of Signal Processing Techniques to Electromagnetics, Manuel F. Cátedra, et al.

Computational Electrodynamics: The Finite-Difference Time-Domain Method, Second Edition, Allen Taflove and Susan C. Hagness

Electromagnetic Waves in Chiral and Bi-Isotropic Media, I. V. Lindell, et al.

Engineering Applications of the Modulated Scatterer Technique, Jean-Charles Bolomey and Fred E. Gardiol

Fast and Efficient Algorithms in Computational Electromagnetics, Weng Cho Chew, et al., editors

Fresnel Zones in Wireless Links, Zone Plate Lenses and Antennas, Hristo D. Hristov

Grid Computing for Electromagnetics, Luciano Tarricone and Alessandra Esposito

High Frequency Electromagnetic Dosimetry, David A. Sánchez-Hernández, editor

Intersystem EMC Analysis, Interference, and Solutions, Uri Vered

Iterative and Self-Adaptive Finite-Elements in Electromagnetic Modeling, Magdalena Salazar-Palma, et al.

Numerical Analysis for Electromagnetic Integral Equations, Karl F. Warnick

Parallel Finite-Difference Time-Domain Method, Wenhua Yu, et al.

Practical Applications of Asymptotic Techniques in Electromagnetics, Francisco Saez de Adana, et al.

A Practical Guide to EMC Engineering, Levent Sevgi

Quick Finite Elements for Electromagnetic Waves, Giuseppe Pelosi, Roberto Coccioli, and Stefano Selleri

Understanding Electromagnetic Scattering Using the Moment Method: A Practical Approach, Randy Bancroft

Wavelet Applications in Engineering Electromagnetics, Tapan K. Sarkar, Magdalena Salazar-Palma, and Michael C. Wicks

For further information on these and other Artech House titles, including previously considered out-of-print books now available through our In-Print-Forever® (IPF®) program, contact:

Artech House Publishers
685 Canton Street
Norwood, MA 02062
Phone: 781-769-9750
Fax: 781-769-6334
e-mail: artech@artechhouse.com

Artech House Books
16 Sussex Street
London SW1V 4RW UK
Phone: +44 (0)20 7596 8750
Fax: +44 (0)20 7630 0166
e-mail: artech-uk@artechhouse.com

Find us on the World Wide Web at: www.artechhouse.com

www.ingramcontent.com/pod-product-compliance
Ingram Content Group UK Ltd.
Pitfield, Milton Keynes, MK11 3LW, UK
UKHW050032170826
474163UK00005B/155

9 781630 81537